– Situated Dialogue Systems –
Agency & Spatial Meaning in Task-Oriented Dialogue

Robert J. Ross

Dissertation

Submitted for the doctoral degree in engineering – Dr. Ing. –
at the Faculty of Mathematics & Computer Science (FB3)
in Universität Bremen

Erste Gutachter: Prof. John Bateman PhD
Zweiter Gutachter: Prof. Dr. Bernd Krieg-Brückner

Kolloquiumsdatum: 11.12.09

Preface

Task-oriented conversation between humans and spatially situated dialogue systems requires a systematic understanding of spatial language, models of spatial representation and reasoning, and theories of intentional action and agency – and that all of these models be made accessible within dialogue processing frameworks that, while modularizing these issues, pull them together within tightly coupled architectures. While such issues pose research questions which are significant, particularly when considered in the light of the many other challenges in language processing and spatial theory, the benefits of competence in situated spatial language to the fields of robotics, geographic information systems, game design, and applied artificial intelligence cannot be underestimated. To progress us towards such overall goals, in this book I develop a modularized agent-oriented language processing framework for spatially situated agents.

In order to determine where boundaries should or should not be drawn between theories of language competence and those of spatial reasoning and agency, the first part of the book reviews the theory and practice behind the modelling and management of natural language dialogue. While existing dialogue models are well developed and highly sophisticated, the particular field of situated spatial language processing requires: (a) greater clarity in the relationship between language representation and domain reasoning; (b) a more systematic approach to situational contextualization to account for the dynamic nature of interpretation in spatial dialogue; and (c) that we look back to issues of agency in dialogue systems to enable a tighter coupling between dialogue processes and the agent's domain-specific capabilities.

In light of this, in the second part of the book I outline a three tiered architecture for dialogue processing that pulls apart layers of language and knowledge representation so as to facilitate reusable and hopefully more scalable communication about space and action. The first tier of the architecture, i.e., the Language Interface, provides the processes and resources which link surface language to the agent's own conceptual models through a spatially rich linguistic semantics that is optimised for the syntax/semantics interface. The second architecture tier, the Agent-Oriented Dialogue Management model, provides a dialogue processing theory which marries a semantics-centric view on dialogue modelling with a practical theory of intentionality, as well as a transparent approach to situational contextualization through functional content resolution and augmentation. The third architectural tier is a concrete situational model against which the Language Interface and Agent Oriented Dialogue Management Model are coupled. I investigate this third tier by developing a model of verbal route interpretation for navigating robots in partially known environments.

This book developed out of research work which I undertook while employed by the *SFB-TR8 Spatial Cognition* research centre in the University of Bremen, Germany. There I worked with a great many people who I am thankful to in preparing this book. First and foremost, I am extremely grateful to my chief supervisor, Prof. John Bateman, for giving me guidance, education, and patience, as well as heaps of diligent proof reading and useful comments on early drafts of this book. Similarly, I am also very much indebted to Prof. Bernd Krieg-Brückner (BKB), who gave me the opportunity to come to Bremen to start working in spatial reasoning, robotics, and linguistics.

Others in the research centre have also been both good company and wonderful collaborators over the years. In particular, I wish to thank Elena Andonova, Joana Hois, Shi Hui, Thora Tenbrink, and Scott Farrar, for stimulating conversations on topics ranging from psycholinguistics to formal methods and ontologies. Also, although a recent addition to the team, I want to thank Kavita Thomas who was a breadth of fresh air in my final months in Bremen. I am also grateful to both Niels Schütte and Nina Dethlefs who contributed sterling work while student assistants. Many other student assistants have also helped me out in grammar development, general system development, or the running of studies over the years. Thank you in particular to David Mautz, David Nakath, Anna Strotseva, and Lea Frermann.

But even more so than colleagues, family and friends are there in person and spirit when you need them most, and that is why they deserve the greatest gratitude of all. Thus, I want to thank my mother and father, Antoine and John Ross, as well as my brother Anthony, for the many years of encouragement that they have given me. All three have supported me through every step along the way, and reminded me always of who I am and why I like to do the things I do. Also, to both Oliver and Chris, I owe you both for the beers and Bremen's never-ending nights. But most of all, I thank Anna, who has been with me in spirit since day one. Anna proof read, offered words of encouragement, listened attentively, and waited patiently during a long-distance relationship. This book is for her.

Finally, I am also thankful to and gratefully acknowledge the support of the Deutsche Forschungsgemeinschaft (DFG) through the Collaborative Research Center SFB/TR 8 Spatial Cognition – Subprojects I3-[SharC] and I5-[DiaSpace]. I also want to thank all those who gave me feedback on drafts along the way. However, mistakes, be them of the conceptual or typographic variety, remain mine and mine alone.

January 2011   *Robert Ross*

Contents

Glossary of Abbreviations

ALPHA - A Language for Programming Hybrid Agents
AODM - Agent Oriented Dialogue Modelling
BDI - Belief, Desire, Intention
CCG - Combinatory Categorial Grammar
Daisie - Diaspace's Adaptive Information State Interaction Executive
DGB - Dialogue Game Board
DL - Description Logic
DM - Dialogue Model
DMM - Dialogue Management Model
DRS - Discourse Representation Structure
DRT - Discourse Representation Theory
DSS - Dialogue State Structure
FOL - First Order Logic
FOS - Frame Object Structure
GUM - Generalised Upper Model
HOL - Higher Order Logic
HLDS - Hybrid Logic Dependency Semantics
IS(S) - Information State Structure
MRS - Minimum Recursion Semantics
NLG - Natural Language Generation
NL(U/P) - Natural Language (Understanding / Processing)
QUD - Questions Under Discussion
SDRT - Segmented Discourse Representation Theory
SDS - Spoken Dialogue System
SFG - Systemic Functional Grammar
SFL - Systemic Functional Linguistics
SPL - Sentence Planning Language
UIO - Upper Interaction Ontologyc

Glossary of Abbreviations

Chapter 1
Introduction

Abstract For 50 years, the natural language interface has tempted and challenged researchers and the public in equal measure. As advanced domains such as robotic systems mature over the next ten years, the need for effective language interfaces will become more significant as the disparity between kinemietic and language behaviours becomes more evident. In this introductory chapter, I argue that the class of situated applications, which encompasses robotics, introduces natural language processing complexities that can only be addressed through a tight, yet modular, incorporation of agency and spatial reasoning theories in dialogue processing.

1.1 The Challenge: Situated Dialogue Systems

In this book we are concerned with the development of language and dialogue processing for the class of situated systems. By *situated* systems, we refer here to those computational applications which are embedded in a real or virtual environment, and which are typically capable of perceiving, reasoning on, and acting within that environment. Examples of such situated systems include virtual agents as used in the gaming or entertainment industries [Gorniak and Roy, 2007], vehicle control and information technologies [Frisch and Stenberg, 2008], and spatially aware assistance applications [Misu and Kawahara, 2007]. But, perhaps the most prototypical examples of situated systems, and the ones which catch the imagination most forcefully, are autonomous and semi-autonomous cognitive robots (see Murphy [2000] for an introduction).

In all of these situated applications, user-system interaction through standard graphical, textual, or tactile modes of communication is either insufficient or simply not feasible for various reasons. First, in the case of entertainment applications, a tactile or visual interface may be available for primary task-oriented activities, however, resorting to the same modes of interaction for communication with other agents often breaks realism and complicates overall interaction structure. Second, for car navigation systems, safety concerns dictate that interaction with the navigation system should not require drivers to direct their visual attention, or hands, away from the primary task of driving. Third, in the case of assistance systems, deficiencies in mental or physical abilities may prevent a user from interacting with a visual or tactile interface. And finally, in the case of robotics, pragmatic concerns mean that a graphical or tactile communication interface can be either too difficult to implement on the robot, or may be too cumbersome to interact with if user and robot are cooperating on a joint task.

These domain-specific motivations strengthen the language interface's longer term appeal to both the research community and the public at large. From a research perspective, natural language dialogue has driven fundamental and applied investigations in computer science, computational linguistics, and artificial intelligence for some 50 years. During the same time, natural language interaction has also caught the imagination of the general public. This is evidenced by the frequency of speaking computers and robots in science fiction works. But, while other fictional technologies such as hand-held communication devices, 3D data visualization, and non-lethal weapons have become realities, naturalistic verbal interaction currently stands alongside faster-than-light travel in the category of true fiction.

Verbal interfaces to both situated and non-situated applications are traditionally modelled in terms of 'Spoken Dialogue Systems'. These systems are architectures of computational components which handle the varied responsibilities of natural language processing. A spoken dialogue system's components are hence diverse, and minimally consist of processing technologies such as language synthesis and perception modules, language analysis and production grammars, as well as high-level dialogue and domain modelling units.

While the general composition of spoken dialogue systems is relatively well understood, any interaction with the majority of prototype dialogue applications quickly reveals a multitude of limitations. At a low-level, speech recognition and speech synthesis quality is still poor. While research continues steadily in these areas, intonation and many of the features of speech that communicate meaning naturally and effectively are rarely produced or recognized accurately. At the level of language analysis and production, grammars which describe written text with considerable accuracy have sizeable difficulties with the disfluencies, self-corrections, and repetitions of natural dialogue. But, perhaps the most telling limitation in current systems is the lack of conversational intelligence at higher level processing – resulting in rigidity in interaction and monotonous scripted contributions from the dialogue systems.

Excluding issues in related fields such as formal semantics, language analysis, and text generation, the particular technical and theoretical problems which warrant such rigidity and monotonous scripted contributions are many fold. First, dialogue theory has yet to definitively answer many questions on the nature of dialogue structure and segmentation. Second, computational dialogue is still coming to terms with non-trivial turn management, dynamic language generation, and the interpretation and production of language's devices of economy such as conversational implicature, elliptical statements, anaphoric reference, and multi-functionality. Third, dialogue engineering continues to face challenges on many fronts. Such challenges include: managing the sheer complexity of dialogue system implementations; assembling and efficiently reasoning on knowledge sources such as application ontologies; and developing integration strategies which balance the organizational principles of modularity against the convenience of tight coupling. Finally, and most significantly, ensuring exchange between these three facets of investigation, i.e., dialogue theory, computational dialogue, and dialogue engineering, can itself be difficult. But since dialogue systems which are ignorant of theoretical fundamentals can never benefit from the breadth of information assembled there, it is vitally important that such exchange is maintained.

While these problems are significant, and require continued intensive research, the nature of situated applications, including robotics, introduces additional complexities to non-trivial dialogue systems development. In particular, the following two issues result in considerable complications:

- **Agency in Application:** Unlike contemporary dialogue applications which revolve around database query and update, situated applications are fundamentally *agentive* in nature. In other words, situated applications often operate in an autonomous manner which entails the adoption of task-oriented and communicative goals and intentions which can be distinct from, or indeed shared with, the user's own goals and desires. Such a factor has implications to dialogue processing at different organizational levels. At the highest level of organization, this requires investigation of models of joint cooperation and shared autonomy in task performance. While at a more pragmatic level, it requires the development of models of mixed-initiative and multi-threading in dialogue competence. And, at the level of content, it entails that the very communicative goals raised in dialogue may be directly related to the situated actions performed by the autonomous system at a given time.
- **Situated Context Sensitivity:** Language processes inherently make use of context to enhance and augment information which is omitted or underspecified in speech. While dialogue theories have paid significant attention to dialogic context, less attention has been given to situated context dependencies. However, in situated domains, such situated context dependencies are frequent as people make reference to objects in their environment, describe spatial relationships with respect to particular frames of reference and perspectives, and describe processes in ways which are in themselves related to the situational context. Understanding and handling these situated dependences then become key to producing and understanding language in natural usage.

These two issues of agency and context dependency are at the heart of developing non-trivial situated dialogue systems. But, it should also be noted that since these two factors in dialogue processing are highly pragmatic in nature, their investigation may well shed light on the very nature of the semantics/pragmatics interface, and are hence useful themes of investigation for dialogue theory as a whole.

Over the next two sections, I examine the two issues above in more detail by investigating the relationships between dialogue and agency, and dialogue and situated language, in more depth.

1.2 Agency in Dialogue Processing

At the heart of any spoken dialogue system is a dialogue management component that is responsible for integrating user contributions into the application state, and deciding upon contributions to be made by the dialogue system at any given time. As we will see in Chapters 2 and 3, there is much variation in approaches to dialogue manager design. Different dialogue management models make varying assumptions about not only the dialogue manager's internal processes, but also about the nature of domain-specific knowledge organization. Given that the majority of deployed dialogue systems are directed at non-situated applications such as phone-based information querying, the resultant dialogue manager designs often assume domain knowledge to be a simple database. However, situated applications require a more complex view of domain modelling which captures the agent-like nature of the situated application.

The application of the intentional agency metaphor to the analysis of both software and human behaviour has a long and varied history. From McCarthy's [1979] description of software as active mental processes, through Minsky's [1985] analysis of human cognitive organization as a society of agents, to Dennett's [1987] analysis of action and behaviour as intentional processes, the abstract notion of agency has proven valuable in the characterization both of human behaviour and system design. More concrete developments such as the creation of programming languages based around the agent metaphor [Shoham, 1993], and the frequent application of the agent metaphor in cognitive robotics [Murphy, 2000] demonstrates the appeal of the abstraction.

Although agent definitions are varied, we can minimally define an agent as an autonomous, situated, pro-active, physical or software entity. While such characteristics already entail other necessary properties such as the existence of sensors and actuators to model and manipulate the environment, we can extend the basic agent notion through additional features ranging from reproductive abilities through compliance with particular communication protocols. But for us the most relevant properties of agency are assumptions of deliberative and intentional action. Putting it succinctly, taking an agent-oriented or intentional perspective on system design entails the existence of models which capture both the actions to be taken by an agent and the reasons for those actions being performed at a transparent symbolic level [Shoham, 1993].

Computational models of agency can be traced back to early work on classical AI. Specifically, formal models of intentionality and agency fused notions of planning and action with models of communicative reason and the more general theories of rationality [Bratman, 1987]. These formal models gave rise to the general class of Belief Desire Intention (BDI) architectures and their realization through more specific theories of knowledge, action, and communication (see e.g., [Georgeff and Lansky, 1987, Wooldridge, 2000, Haddadi, 1996]). Such formal theories developed alongside more pragmatic implementations including various rationalizations of Georgeff and Lansky's [1987] PRS (Procedural Reasoning System) (see e.g., Konolige [1997]), whilst the fields of so-called *Agent-Oriented Programming* also developed from the same principles [Shoham, 1993]. Both agent-oriented programming and more direct BDI implementations can be characterised as allowing the development of an *Intentional Stance* towards the computation and reasoning processes [Dennett, 1987].

The attraction of computational agency models to both software engineering and robotics operates at both micro and macro levels. At the macro-level, computational models of agency can be used to investigate models of communication and cooperation within groups. Whereas, at a micro-level, models of agency provide useful abstractions over very complex system designs. These abstractions thus allow designers to look at individual systems in terms of high-level cognitive concepts such as beliefs, actions and plans. Since situated systems such as those of robotics are inherently oriented towards performing actions with temporal extent, having explicit modelling techniques which can relate these actions to other aspects of computational state becomes extremely beneficial.

Regardless of whether an application is explicitly modelled in such agentive terms, any application which is perceived by a user as having such properties should be modelled from the dialogue perspective in such terms. This in turn makes it possible to: (a) provide minimal models of cooperation on joint tasks; (b) develop appropriate dialogue behaviours given the performance of actions by the system; and (c) allow a user to query and update the agent's planned actions. Fortunately, theories of agency and dialogue processing are far from disjoint.

Rather than having evolved separately, theories of computational agency and dialogue management share a common history and have overlapped frequently in the past 30 years. Specifically, the most formal theories of both rational agency and dialogue modelling originate in work which examined dialogue processes from an agent-oriented perspective, e.g., [Perrault and Allen, 1980, Allen and Perrault, 1980]. These classical *agent-based dialogue management* models apply formal axiomatic theories to express an agent's mental state. This mental state in turn consists of folk-psychological concepts such as beliefs and desires, as well as plan and action definitions which define the mental state pre-conditions and consequences of performing particular speech acts. Even after two decades, such models are still proposed frequently in the research literature. Although, in more recent cases such as Sadek et al. [1997] and Egges et al. [2001], plan-based theories of action are replaced with rational reasoning based models consisting of declarative rules relating beliefs to intentions of the form:

```
BELIEF(request(user,actA)) & other constraints ...
    => INTENTION(Self,Now,SEQUENCE(acknowledge(actA),
                                   do(actA)));
```

which dictate actions to be performed by an agent under certain mental state constraints.

While such a rule is elegant in its simplicity, the problem with these approaches, like the classical agent-based approaches, is that in assuming antecedents of the form BEL(requested(user,actA)) the modelling process presupposes perfect communication. However, issues of grounding and contextualization are what make dialogic communication challenging, and thus by avoiding them, purely agent-oriented models cannot be used directly as stand-alone dialogue managers. Such realizations gave rise to updated theories of agency and dialogue which take into account the non-idealised nature of communication [Traum and Allen, 1994]. While these revised models did much to improve the applicability of agent-based theories of communication, they are considered overly-complex as a general framework for dialogue processing. As a result, agent-based systems have widely been replaced in practical system designs with light-weight dialogue frameworks based on notions such as frames, and are largely devoid of mentalistic and agentive concepts.

Rather than looking at the whole dialogue process as a single agent, a more realistic approach to incorporating agentive properties in dialogue processing is to acknowledge the agent-like features of a domain application when necessary, but otherwise to use more conventional non-agentive models for the core dialogue processing theory. This was, for example, the approach taken by Larsson [2002] and Lemon et al. [2002a]. In his IBiS4 model, Larsson [2002] extended a frame-like account of dialogue processing to incorporate a small number of action-oriented dialogues. The model while simple (no support for system initiated dialogue, and no context sensitivity in action interpretation) did establish the existence of actions and capabilities as distinct entities within the mental state of the dialogue capable agent. Lemon et al subsequently extended the approach and developed a more complex view of action structure in dialogue management. This account, although less formal than Larsson's model, accounts for mixed-initiative dialogue and multithreading when interacting with autonomous systems.

While far from perfect solutions for the integration of agency and dialogue processing, Larsson and Lemon's models clearly demonstrated that it is possible to isolate agent control in a way that it is made accessible to dialogue management, but yet does not become one and the same process as dialogue management. Nevertheless, the views of agency taken by the authors, and the manner in which dialogue processes were linked to the model of agency, were extremely limited. Direct relationships between the agentive state and general dialogue processing were never established, thus meaning that it is impossible to produce or understand language related to the agent's mental state in a scalable way. Most crucially though, the situational context of the application was largely treated as a pseudo-static domain model. However, situated systems employ highly dynamic models of their environment which are continuously updated as the situated application moves around or manipulates its environment. Such dynamic situational context in turn considerably complicates the processes of language understanding and production, as the types of language we most frequently use with such situated systems, i.e., descriptions of space and action, are highly sensitive to such contexts.

1.3 Spatial Language: Modelling & Processing

The nature of situated systems is such that a large proportion of task-oriented dialogue with these systems includes either descriptions of the system's environment, or the actions which the system is to perform in that environment. Since such descriptions of environment or action likely have a high quotient of spatial content, it is vital that we have clear understandings of: (a) the properties of spatial language; and (b) the processes which relate spatial language to the models of situated spatial context. In this section I consider such issues, and briefly explain why the computational interpretation and production of dialogue in general, and situated spatial dialogue in particular, is still in its infancy.

1.3.1 On the Complexities of Language Processing

The weaknesses of computational language interfaces are rooted deeper than some of the technical challenges of spoken dialogue systems considered earlier. Instead, these challenges relate to the fundamental nature of spoken dialogue as a communication medium, and thus span disciplines across theoretical and applied linguistics.

To appreciate the problems faced, consider that the cognitive processes which underpin dialogue as an interactive system allow us to achieve remarkable communicative efficiency across a communication medium which is slow and serialized in comparison to the throughput available in visual imagery or indeed within the communication mechanisms available to artificial agents. This remarkable efficiency in dialogue is achieved through the use of a full and rich potential of language devices, and it is precisely these devices which make the interpretation of language in general, and dialogue in particular, so troublesome for current applications. Such devices, widely studied within linguistic semantics and pragmatics, include *conversational implicature* which allows the incorporation of background knowledge to construct a producer's complete message beyond the shallow meaning of individual messages; *anaphora* which governs referential potential in identifying objects in discourse; and *ellipsis* which allows the non-verbalization of information in individual messages. Moreover, to improve efficiency we also make use of multifunctionality to fuse multiple dimensions of meaning into individual messages so that we can for example express content, our attitude towards that content, and our intentions towards the progress of a conversation in a single message [Allwood, 1997].

The significance of such devices is that in obeying a limited bandwidth restriction, dialogue is strewn with constructions which cannot be understood in isolation since key elements of meaning are either ambiguously stated or completely omitted from the surface form under the assumption that such information can be retrieved from context. Language understanding algorithms must thus apply layers of context application to re-construct information from the surface form, and similarly, language production mechanisms must systematically take such information sources into account when planning and verbalizing information in a naturalistic way. While all language types are subject to such contextual sensitivity, this is particularly true for spatial language.

1.3.2 Spatial Language & Context

Before considering the particular context sensitivity in spatial language, it is useful to first ask what is spatial language, and what are its general semantic features? In this book, we are particularly concerned with so called locative expressions which indicate the position of one or more objects in relation to others, and path and generalized movement expressions which describe the movement of an agent or object within an environment. Another spatial language type of frequent use, but one which will not be considered here, is spatial form descriptions.

Locative expressions are seen frequently in many task types where speakers identify objects with respect to known entities through projective and other spatial relations, e.g.:

(1) the classroom opposite John's office

Object localisation expressions, particularly in the path description domain, also frequently make use of ordering relations with respect to some reference object, e.g.:

(2) the second office

and mixtures thereof, e.g.:

(3) the second office to our left

In addition to locative expressions, motion expressions are frequent in task-oriented dialogue. Such expressions are used to describe movements or turnings in the environment and are typified by prepositional and adverbial constructions which denote motions of movement to be taken. Such expressions necessarily include salient spatial information such as landmarks, directions, and manners of movement:

(4) a. continue along the fence
     b. put the wrench on the table

In addition to handling these overtly spatial constructions, any dialogue system in a situated domain must be able to cope with highly underspecified contributions which can only be interpreted through detailed models of task and spatial relevance; e.g., any of the following might be interpreted in context as either a request to move to a named location, or as a naming of a current location:

(5) a. John's office

    b. John's room

    c. Prof. Smith's room

Given the importance of spatial language content, there unsurprisingly exists a rich literature of models that have attempted to characterize the spatial language types just illustrated. Some of the best regarded work on the relationships between linguistic and spatial structure includes Talmy [1983, 2000], while more recently Levinson [2003b] has worked towards a generalised cognitive grammar of spatial language. Also recently, Tenbrink [2005] has provided a comprehensive recount of the spatial linguistic literature in an analysis of the range of surface, semantic, and pragmatic features of space-centric English and German dialogues. Such works are but a tiny fraction of the breadth and depth of spatial language research, but across these studies we see common themes of spatial semantic schematization. Taking locative expressions as an example, simple expressions such as "the ball is in front of the robot" are characterized by such frameworks as being verbalizations not only of the linguistically marked object to be localized (*locatum:ball*), the object to be localized with respect to (*relatum:robot*), and spatial relationship between the two entities (*relationship:in-front-of*), but are also verbalizations of the often linguistically implicit, but context retrievable, *frames of reference* which identify the underlying coordinate system which the spatial expression is made with respect to (e.g., intrinsic, absolute, relative) [Levinson, 2003b]. The implicit nature of the reference frame is typical of underspecified context-sensitive spatial language.

While spatial language schemas are important resources in constructing models for computational systems, we must also understand the various ways in which language use is sensitive to situational contexts. To illustrate, Figure 1.1(a) schematically depicts an office environment with a robot (a circle with a stubbed arrow to indicate orientation). In such environments, the identification of discourse referents used as landmarks within, for example, movement instructions can depend on a range of spatial factors such as visual saliency, proximity, or accessibility relations. This can be seen from the fact that if the robot was told to "enter the office" then it is highly likely that the office in question is directly ahead rather than behind the agent or to its right.

Moreover, the interpretation process can also involve the application of non-physical context to enrich the surface information provided. To illustrate, Figure 1.1(b) depicts an agent situated at a junction while being given instructions by an operator who has a survey perspective equivalent to the figurative view presented in Figure 1.1(b). In such a case, and where the instructee is aware of the instructor's non-route perspective, an instruction such as "turn to the left" can have alternative meanings as depicted in the diagram. While explicit clarification through dialogue is possible, the more efficient solution is to make use of contextual information such as the frame of reference used historically by the speaker.

Ellipsis in spatial language also provides considerable difficulties in language interpretation. Considering again the situation in Figure 1.1(a), it is not unreasonable for an elliptical statement such as "enter" to be interpreted as an instruction to enter the room directly in front of the agent, while an elliptical statement such as "kitchen" may be interpreted as a statement to go to, or enter, the kitchen. In

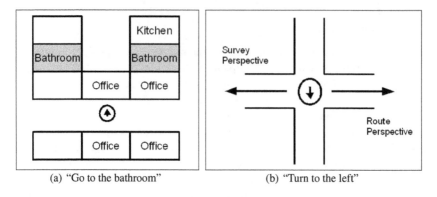

(a) "Go to the bathroom"   (b) "Turn to the left"

Fig. 1.1 Illustration of a spatial situations wherein interpretation of user language is ambiguous because of: (a) multiple salient situational referents; and (b) ambiguous frames of reference.

the spatial domain, elided information can also include other actions, e.g., within the spatial domain the successful interpretation of highly elliptical statements such as "stop", "continue", or "faster" necessarily requires that the action to be stopped, continued, or performed faster, be recoverable from the domain. When the agent can perform more than one action in parallel, this becomes non-trivial.

While complex in itself, the context sensitivity of situated spatial language does not stop at the resolution of individual content. So-called indirect speech acts make use of complex inference systems to allow very simple utterances to be used to invoke quite different behaviours even in a command and control situation. For example: consider the simple utterance "drive forward". Depending on the context of application and the capabilities of the agent which the command is directed at, the desired behaviour can vary widely. For example, for an agent with simple perceptual and planning capabilities a literal interpretation of straight movement along the agent's sagittal plane could be assumed. However, for a more complex agent such an utterance could be interpreted as a command to follow along a winding corridor or road. Even in the case of Figure 1.1(a) earlier, it would not be unusual to interpret "drive forward" as a command to enter a room.

Indeed, context sensitivity not only includes the physical context which surrounds the agent, but can also include the agent's own physical state. To illustrate, consider the simple utterance "turn left", which when used in a static context, i.e., where the agent is stationary, should most likely be interpreted as a request for reorientation on the spot. On the other hand, when used in a dynamic context, e.g., the agent is already moving with a forward velocity, such an utterance should most likely be interpreted as a request to perform a change of direction at a relevant turning place and continue moving straight, i.e., where "turn left" is synonymous with "take the next left".

From all of these examples we see that the relationship between surface spatial language and interpretation is not a simple mapping, but rather a more complex function that is highly dependent on forms of spatial and discourse context. While it is possible to hard-code some choices in language interpretation and planning, in general we should select these parameters from either the surface form as provided by the user, or alternatively from discourse or situational context. This book is concerned with how we organize such layers of spatial information and information processing within practical dialogue systems so that we might develop more principled models of spatial language competence for situated systems. Before considering this point in more detail, let us briefly consider the state of spatial language processing theories which have relevance for this agenda.

1.3.3 Models of Spatial Language Interpretation

Given the difficulty of spatial language interpretation, there have been varied approaches to the computational integration of space and language. We can broadly categorize these approaches into two classes: first, those theories that apply formal spatial models to the interpretation of human spatial concepts; and second, less formal pragmatic accounts which typically assume underlying quantitative models of space.

Within the formal spatial modelling community, well defined mathematical models of spatial configurations, and the spatial relationships which can hold between these entities, are developed as logical theories of space. Such theories, and in particular the so-called spatial calculi (see Cohn and Hazarika [2001] for an introduction) are appealing from a cognitive and computational perspective because of: (a) the relative transparency of these models; (b) the availability of inference algorithms which can be used to compose complex spatial models from elementary models; and (c) the view that these models correspond in some way to the qualitative abstractions of space which human cognition is assumed to be dependent upon. Due in part to these appealing aspects, much research has been conducted on the direct interpretation of natural language expressed in terms of qualitative spatial relations and spatial calculi, e.g., Krieg-Brückner and Shi [2006]. But despite the cognitive appeal, there are several limitations to these formal mapping and interpretation theories. Beyond issues of tractability in models which are built upon first order logics, the most serious issue is that while natural language expressions often use words which seem to map directly onto particular logical relational categories, those logical category definitions frequently embody formal restrictions which simply do not hold for the language term as used in normal conversation.

Rather than adhering to strict formal interpretation, probabilistic interpretations of spatial constructions offer a more flexible view of the possible meaning of language. These approaches, pursued both in the cognitive robotics and computational psychology domains, typically employ: (a) quantitative models of spatial reasoning such as those suited to robotics applications and the visual perception system; and (b) qualitative relation definitions which map linguistic constructions directly onto the underlying quantitative model (e.g., [Kray et al., 2001]). Although appealing in terms of their flexibility in interpreting individual constructions, such models to date are often only concerned with specific cases, or are dependent on a robotics-oriented data processing perspective which assumes high volume input data (e.g., [Mandel et al., 2006]) – a characteristic which is rarely true of natural fragmentary dialogue.

Beyond their individual limitations, it is frequently common in both formalist and probabilistic models of spatial term interpretation to assume as input a representation of language that has already reached a canonical level, and as such, the context sensitivity and underspecified nature of spatial language is rarely acknowledged. In other words, these models are typically developed in isolation from the realities of language processing, and are thus as idealistic as other early language processing theories such as the agent-oriented dialogue management models discussed earlier. Moreover, models such as those above typically focus on spatial relation definition, and are not intended to account for the issues of underspecification and ambiguity highlighted in the last section. Thus, while models of spatial term resolution, either logical or probabilistic in nature, are essential to the overall goal of spatial language interpretation, robust and flexible dialogue management requires that these models be integrated within complete language systems which also provide solutions to reference resolution, reference frame modelling, and process selection.

1.4 Book Objectives & Structure

The spoken language interface is an appealing mode of communication for situated applications. This appeal is due to the interaction constraints imposed by such applications, but also because of the general attraction of dialogic communication. However, while existing dialogue processing is difficult even in simple domains, situated applications introduce significant complications. Namely, situated applications are inherently agentive in nature, and thus require dialogue models which incorporate these agentive properties in a tractable and scalable way. Moreover, the types of language we apply to situated systems, i.e., action and spatial descriptions, are highly sensitive to the situational context of the application. While we can hard code certain assumptions about the processing of situated language, such hard-coding cannot scale to flexible wide-coverage interaction. Thus, for situated dialogue domains such as robotics, we must investigate tractable dialogue processing techniques which integrate models of agency with context-sensitive language processing.

The primary objective of this book is thus to investigate the roles of agency and physical context in the development of verbal interfaces for situated autonomous applications. To achieve this primary objective a number of sub-objectives must be achieved:

1. Investigate dialogue processing models and their relationship to theories of agency and physical context.
2. Develop a dialogue management model that integrates notions of agency and context sensitivity in a tractable and extensible manner.

3. Produce well-defined models of linguistic and situated context representation to support dialogue processing.
4. Develop a methodology for the application of situational context to the interpretation of spatial and action language.
5. Design a dialogue processing architecture which integrates the dialogue management model, context models, and processing algorithms with other units of a complete situated system.
6. Apply the architecture to specific cases of situated language interaction.

While I outline the specific approaches that I have taken to accomplishing these objectives below, I have also employed some more general strategies. In particular, the models presented in this book build upon current trends in knowledge engineering and ontological organization to ensure rich representation of meaning, structure, and context within resultant models. In general, this simply means that the models adhere to the basic principles of ontological engineering, and leverage designs off well-defined ontologies where possible. This will allow us to avoid a more ad hoc approach to the representation of meaning and knowledge, thus helping to achieve the goals of scalable, and hopefully more fine-grained, linguistic control.

The models developed also place a strong emphasis on the modularization of the complete dialogue system, thus ensuring that the resultant designs are tractable in comparison to earlier attempts in the area. This modularization of processes and knowledge is typical of software engineering in general, but is also in keeping with current trends of knowledge segmentation such as that applied within Asher and Lascarides's [2003] Segmented Discourse Representation Theory. The modularization approach not only applies purely to the types of knowledge within the dialogue system, but also extends to how the dialogue management process is in general separated from any notions of rationality and control within the application model itself.

Finally, the development of formal or computational models in isolation from application would lead to systems as academic as the original agent- and plan-based dialogue systems. Thus, while the chapters presented in this book might be linear in nature for the sake of explanatory effect, the approach used for the development of the computational models both in theory and practice was anything but strictly linear. The models presented in this book have come about through trial and error with practical dialogue systems, but as a result are now such that they can be applied to a range of dialogue domains – including robotic agents where timeliness of response is a critical factor not only for reasons of user satisfaction, but also for safety requirements.

Given these objectives, the book is structured as follows. Chapter 2 is the first of two background chapters which together assess existing dialogue modelling and processing techniques. The aim of this review is to provide a clear picture of trends in these areas, and in particular to investigate the incorporation of issues of agency and context in dialogue modelling and management. This first chapter investigates our understanding of dialogue in terms of the semantic and pragmatic underpinnings of dialogue models. Chapter 3 then continues to construct the background picture by reviewing trends in the computational processing or *management* of dialogue.

In addition to reviewing basic principles of spoken dialogue systems and dialogue management, the chapter examines a number of case studies in dialogue system design to ascertain the suitability, or otherwise, of existing dialogue theories to the goals of situated language processing.

On the basis of the analysis provided, Chapter 4 then derives a set of requirements and principles of organization for a theory of situated dialogue modelling and management. The dialogue model takes an 'agent-oriented' perspective on the structure and processing of dialogue, and separates out layers of semantic and context representation. The model is realized through a dialogue processing architecture which employs a three-tier design that addresses issues at: (a) the language interface, (b) intentional action and dialogue processing, and (c) deep contextualization strata. These processing tiers are detailed across subsequent chapters.

Chapter 5 develops the first tier of the architecture, i.e., the language interface. This interface defines the processes and models which link surface language to application-specific semantic meaning. Rather than making a direct mapping between surface language and application meaning, a two-stage semantics is developed and applied. The first stage is a linguistic semantics model that employs a rich context independent account of spatial meaning for the syntax/semantics interface. The second stage is an application-specific model of semantics which is more suited to dialogic and application-specific reasoning. In addition to setting out both these layers, the chapter also develops a context-sensitive transform mechanism which establishes the necessary alignment between these two different layers.

Chapter 6 presents the second architecture tier, i.e., a model of Agent-Oriented Dialogue Management. This model provides the key link between the language interface and spatial and situated reasoning models, but does so in a way which addresses shortcomings in dialogue management for situated systems that were identified in earlier chapters. The first key aspect of the Agent-Oriented Dialogue Management model is a theory of dialogue processing which marries a classical speech act view on dialogue with a practical theory of agency and intentionality. Intentional state and discourse state models are thus developed in detail. The second key aspect of the model developed and presented in this chapter is a general model of situated language resolution and augmentation.

Chapter 7 deals with the third processing tier that addresses the interface between dialogue management and the agent's theories of spatial and domain organization. I explore this processing tier through the development of a model of verbal route interpretation in a partially known environment which builds upon the Agent Oriented Dialogue Management tier. The purpose of this model is to illustrate how layers of linguistic representation relate to models of spatial representation and reasoning. And, as a by-product, it also provides a cognitively plausible solution to an interesting problem in spatial language interpretation.

Chapters 8 and 9 then bring the book to a close. Chapter 8 describes the realization and application of the situated dialogue architecture through two open-source computational frameworks. Finally, Chapter 9 summarises early chapters, links the model outlined in previous chapters to prominent research disciplines, and looks forward to the next class of situated dialogue applications.

Chapter 2
Modelling Dialogue

Abstract In this chapter, we review approaches to dialogue modelling from the linguistics and computational linguistics communities that have had notable consequences for the development of spoken dialogue systems. Specifically, in Sections 2.2 and 2.3 we introduce two topics which underpin the vast majority of dialogue theory and application, i.e., formal linguistic semantics and speech act theory. Then, in Sections 2.4 and 2.5, we move on to review theories of dialogue structure and organization. In particular, Section 2.4 looks at descriptive accounts which typically provide a syntax-like organization of dialogue, while Section 2.5 looks at those accounts which take more of a cognitive dialogue process perspective. We begin however with a very brief introduction to spoken dialogue.

Abstract In this chapter, we review approaches to dialogue modelling from the linguistics and computational linguistics communities that have had notable consequences for the development of spoken dialogue systems. Specifically, in Sections 2.2 and 2.3 we introduce two topics which underpin the vast majority of dialogue theory and application, i.e., formal linguistic semantics and speech act theory. Then, in Sections 2.4 and 2.5, we move on to review theories of dialogue structure and organization. In particular, Section 2.4 looks at descriptive accounts which typically provide a syntax-like organization of dialogue, while Section 2.5 looks at those accounts which take more of a cognitive dialogue process perspective. We begin however with a very brief introduction to spoken dialogue.

2.1 Dialogue: Some Definitions

Definitions of dialogue are many fold, but it is relatively easy to construct a working definition. For example, the Merriam-Webster dictionary defines dialogue as "an exchange of ideas and opinions", or "a conversation between two or more persons", whereas a conversation is in turn defined as "an oral exchange of sentiments, observations, opinions, or ideas". Thus, a reasonable working definition that one might extrapolate from these is that:

A dialogue is an exchange of cognitive concepts between two or more agents

where both the concepts exchanged and the means of exchange are subject to variation.

Other related terms such as *conversation, talk,* and *discourse* are used frequently both in the literature and in our everyday language. Levinson [1983], for example, characterises *conversation* as a type of *talk* which is entered into freely by participants, and is distinct form institutional dialogues which are a type of talk seen in highly socially ordered domains such as staff-customer interaction. *Discourse,* like dialogue, can also be considered an exchange of ideas or concepts, but unlike dialogue, a discourse can also include monologues which are characterised by a receiver or hearer of the talk or text not being able to respond directly to the producer. Some researchers take a stronger view that a discourse is necessarily a coherent activity while neither a dialogue nor monologue need necessarily be so, e.g., Kruijff-Korbayova and Steedman [2003] assert that discourse is "*a coherent multi-utterance dialogue or monologue text*".

Regardless of variety or nomenclature, dialogue achieves the exchange of cognitive concepts through a range of communicative devices which have been studied principally in the field of pragmatics (see Huang [2006], Levinson [1983], Horn and Ward [2004] for useful introductions). These devices, including conversational implicature, anaphora, ellipsis, and multi-functional blending, have already been introduced in Section 1.3.1 as those aspects of language which make it extremely powerful in a slow communication medium via a principle of minimum effort, but which consequentially cause considerable difficulty in language processing applications. Since human users of dialogue systems will inevitably make use of these devices, attention must be paid to the handling of these phenomena in language understanding and production. The successful handling of such complexities in turn relies on the existence of formal or at least pseudo-formal models of language use which can be employed within dialogue systems. In the following sections, we review the modelling techniques that have historically been most commonly applied.

2.2 Formal Linguistic Semantics

In modelling dialogue the symbols we assign to represent utterances and situational models can be quite arbitrary; but formal, or at least quasi-formal, linguistic semantics offers us a means to represent many features of linguistic content in a way that is suited to traditional computational machinery. This notion of formalised linguistic semantics dates back until at least the classical Greek era where philosophers such as Plato, Socrates, and Aristotle developed the basis of what we now know as formal western logic, but it was not until Montague and the birth of modern computational linguistics that systematic approaches to the composition of formal model theoretic interpretations from language utterances were developed. Since such models of semantics permeate all but the most trivial of dialogue models and systems, and since some of the ideas developed later in this book take a semantics-oriented view on the dialogue process, in this section we introduce some of the basic issues in linguistics semantics design.

2.2.1 Linguistic Semantics Requirements

The goal of any linguistics-oriented semantics model is to provide a formally descriptive meaning for words, utterances, and larger chunks of communicative expression in language. For a more concrete definition of minimal requirements, we can turn to Copesteak et al. [2005] who offer four general criteria for computational semantics:

1. **Expressive Adequacy:** the semantics should be sufficiently expressive to capture all necessary features of the surface form, and allow different aspects of semantics to be linked together in a meaningful way.
2. **Grammatical Compatibility:** the semantics should be systematically linked to elements of the grammatical system for analysis or generation.
3. **Computational Tractability:** the semantics must not be so strong so as to prevent practical use such as the determination of equivalence between expressions.
4. **Underspecifiability:** the semantics should allow certain semantic distinctions including scope and category type underspecification when necessary.

These requirements are somewhat axiomatic for a modern linguistic semantics theory, and are hence met at least in part by many semantics accounts. Such compliance is doubtlessly aided by the looseness of the individual requirements. For example, Requirement 1 can be given a weak reading which requires the semantics only to be *capable* of capturing relevant features, but it says nothing itself of the features that are interesting or how they should be organized within the semantics. On the other hand, tractability, as needed by Requirement 3, calls only for an algorithm to be suitable for solving a problem practically, but the definition of practically can vary from one domain to another. A solution which is practical in one domain of use may not be practical in another. Related to this issue is *cognitive compatibility*, or whether the semantics can be related to the non-grammatical elements of a complete cognitive theory. Since such questions are highly relevant to the creation of conversational systems, we restate Requirement 3 and add two additional requirements to Copesteak et al.'s [2005] list:

3. **Practical Tractability:** the semantics should support query and construction operations in practical dialogue systems.
5. **Semantic Feature Coverage:** the semantics should cover a wide range of semantic features necessary, in the limiting case, to capturing linguistic meaning to the point of accurate reproduction of surface form from semantic information. Such features would therefore not only include traditional propositional information, but also details of information structure and interpersonal meaning.

6. **Cognitive Compatibility:** the semantics should be systematically and modularly compatible with non-grammatical elements of a complete cognitive theory such as a rational decision making process or dialogue theory.

2.2.2 First Order Semantics

To give a baseline for semantics models developed in later chapters, here we give a brief introduction to the classical modelling of language through First Order Logic (FOL). Rather than giving a detailed introduction to the syntax and semantics of FOL, we assume the reader has a basic grasp of the topic, and hence we provide only a few illustrative examples to show how particular linguistic constructions are captured in FOL. For more comprehensive introductions to the representations of natural language in formal languages, the reader is directed to Russell and Norvig [1995, Chapter 8], Blackburn and Bos [2005, Chapter 1], Ramsay [1988], or Jurafsky and Martin [2000].

If we say that a **term** is an expression that refers to an object, and that terms can be either simple (constant symbols) or complex (function symbol followed by arguments), then, given simple atomic terms that name a specific table object and a chair object, we may use FOL to express an utterance such as "the table is beside the chair" as follows:

$$beside(Table, Chair) \qquad (2.1)$$

where initially capitalized words are used for constant and function symbols, while initially lower-case words are used for relational symbols.

Through the introduction of FOL's logical connectives, i.e., *not* ($\neg$), *and* ($\wedge$), *or* ($\vee$), and *implication* ($\Rightarrow$), we obtain a more powerful language that captures a far richer underlying model and thus range of natural language utterances. For example, we can represent the utterance "the chair is beside the table and the chair is not blue" with the following logic sentence:

$$beside(Table, Chair) \wedge \neg hasColour(Chair, Blue) \qquad (2.2)$$

Through the introduction of variable terms and FOL's *universal operator* ($\forall$) and *existential operator* ($\exists$) we can also talk about sets of objects and individuals within those sets, and thus represent "all cups are on the table" with the following FOL expression:

$$\forall (?x) Cup(?x) \Rightarrow on(?x, Table) \qquad (2.3)$$

where *question mark* prefixed tokens denotes a variable.

Through quantification we can also model an utterance such as "there is a cup on the table", where we do not actually know which cup is on the table, succinctly as follows:

$$\exists (?x) Cup(?x) \Rightarrow on(?x, Table) \qquad (2.4)$$

In addition to representing the existence of objects and the relationships between these objects, a representation language for natural language must capture the meaning of verbs and the actions that they model. As with all modelling in FOL, there are many different approaches that can be taken to representing a given phenomenon. For example, we might choose to model an action as a relation that existed between objects at a particular time, e.g., we could model the fact that "John drove to the bank" as follows:

$$droveTo(John, Bank) \qquad (2.5)$$

while a more 'ontologically' well formulated approach would be to treat the action as a type of object that was instantiated, and that John and the Bank are related to that *Event* through explicit relations, e.g.:

$$Driving1 \wedge actor(John, Driving1) \wedge destination(Bank, Driving1) \qquad (2.6)$$

where *Driving*1 is a constant symbol that represents the actual driving event in the underlying model.

Such choices available in modelling actions and events points to a general issue in that there are many different choices we can make in a representational *model* of the world. Practically speaking, the question of how our representation should model the world is the topic of *ontological engineering*, which we will return to later, but for the time being it is enough to acknowledge that such choices exist, and that the quality of these choices will naturally have an effect on the representation and reasoning power of formal representation – and hence of the complete dialogue system.

While FOL and other formal languages give us the means to represent particular world states, in language interpretation we require that some process be accountable for composing these semantics for given input utterances and associated semantic chunks. For many years, Alonzo Church's λ-calculus was the tool most commonly used to drive this process. The calculus, consisting of a formal language, and an associated 'reduction' or computation mechanism called λ-reduction, essentially performs variable substitution, thus allowing the controlled insertion of semantic chunks into other statements. While the λ-calculus has historically been the most commonly used means of constructing semantic representations, it is certainly not the only tool available. In recent years it has become more appropriate to apply unification to the semantic construction task. The same basic principles however apply for unification-based approaches. We have individual chunks of semantics, and parts of these semantics chunks are marked as requiring substitution for another semantic entity. The unification process then has to take individual semantic chunks, perform binding, and thus produce large composite semantic chunks. Unification approaches to semantics construction have been particularly appropriate when used within the context of unification-based grammars in general, e.g., CCGs [Steedman, 2000].

With a rich base language and composition mechanisms via λ substitution and unification, First Order Logic (FOL) has had widespread application in linguistic semantics modelling. In such approaches the first order logical forms are typically used to provide literal meanings of utterances against some form of 'world model'. Unfortunately, in practice a number of issues prevent us from mapping an utterance to this model in one move. First, regardless of the grammar applied, a language analysis tool without access to discourse context cannot assign discourse referents to pronouns outside the scope of the individual utterance. In an ideal world, an analysis tool would have access to discourse and situational context, but since many practical implementations of dialogue systems apply a pipelined architecture, this precludes the analysis tool from having direct access to such information sources. Moreover, other ambiguities in language such as role assignment (e.g., *The woman saw the man with binoculars.*), quantifier scope (e.g., *Every man drives a car.*), and lexical ambiguities (e.g., *we* **rented** *the apartment* [Asher and Lascarides, 2003]) are only a handful of the ambiguous features of language that prevent the direct construction of full logical forms directly from the grammar. Although such cases can often be disambiguated through linguistic or extra-linguistic context, some mechanism is required to allow context-less language analysis to produce a representation which is sufficiently expressive without committing too early to one meaning or another. Flat, multi-valued, and underspecified semantics models have become extremely popular means of addressing such problems in recent years, and we consider these briefly in the next section.

2.2.3 Underspecification in Semantics

The most basic approach to handling ambiguity in linguistic semantics is to allow an analysis component to produce multiple parse results for the same input. Thus, the different parses capture all possible variants of interpretation, and these parses can then be analysed in terms of some other heuristics to determine the interpretation which is relevant to the current context. Such an approach is in principle simple, can be related to theories of simultaneous activation in cognitive theory, and shares a common heritage with many other forms of sensory analysis in that multiple hypotheses are composed and then scored. However, as Asher and Lascarides [2003], Copesteak et al. [2005], and others are keen to point out, the mechanics of such an approach are often highly inelegant and do not lend themselves to a formalised treatment.

An alternative to producing multiple explicit hypotheses is to directly encode the underspecification in the semantics model. Historically, the most common approach to this has been through the use of two different levels of semantics; one, un-scoped, and produced by an analyser which represented the surface form of language, and the other, a classically scoped, full semantics model. Using Alshawi's [1992] terminology: in such an architectural approach, the output of the language analyser is said to be a *Quasi-Logical Form (QLF)*, whereas the result following pragmatic analysis is said to be a *Resolved Logical Form* or simply *Logical Form (LF)* .

One problem with QLF treatments is that while they provide a natural treatment of scoping ambiguities, they do not cope well with some forms of lexical ambiguities, plus they can leave considerable un-intended ambiguity in the quasi-logical form. An alternative approach to underspecification as captured by Blackburn and Bos's [2005] Hole semantics, Copesteak et al.'s [2005] Minimum Recursion Semantics (MRS), or Asher and Lascarides's [2003] SDRT, is to make use of a labelling scheme to build a partial description of a clause's logical form. These Underspecified Logical Forms (ULFs) can then be transformed to a true logical form during a discourse semantics construction phase where extra constraints of organization are added such that partial structures are composed into complete trees.

Taking Copesteak et al.'s [2005] MRS as an example of underspecification models, a MRS structure is defined as a 4-tuple including the set of labelled elementary predications, a head label, a set of constraints and a local top for that structure. A fully scope-resolved MRS is syntactically equivalent to a statement of the assumed predicate base language, but an underspecified MRS will instead correspond to a set of statements of the base language – each base language statement corresponding to a reading of an underspecified natural language form. Thus MRS does not define a semantics in terms of what types and roles are defined, but defines a meta-language over a base semantics language. A monotonic composition mechanism can then be provided which allows complete MRS structures to be assembled from an input [Copesteak et al., 2005].

In practice, First Order Logic as introduced earlier is rarely suited to the practicalities of computational semantics such as those just introduced. In addition to the obvious difficulties caused by its undecidability and intractability for certain problems, the very mechanisms just described for underspecification often require operations which are simply not available in FOL (but are available in dynamic logics for example). Also, in many computational dialogue tasks it simply is not necessary to invoke a full logical approach. In such cases, frame languages have been used to capture un-scoped surface form semantics either in language generation or analysis. The Core Language Engine's Quasi-Logical Form is such a frame language, as are Kasper and Whitney's [1989] Sentence Planning Language (SPL) or O'Donnell's [1994] Macro Planning Language. The advantage of such frame languages is that they provide a simple specification language for asserting certain semantic features of language analysed, or to be produced, but they are in themselves not suited to the composition of complete discourse models.

In Section 2.5.1 we will return to models of semantics and underspecification in the context of complete theories of discourse and dialogue. But, before we can move on to these *discourse semantics* theories, or any theories of dialogue organization, we must first address the second building block of meaning-centric dialogue systems, i.e., speech act theory.

2.3 Speech Act Theory

While linguistic semantics captures the objects and events that are the content of language, speech acts allow us to explicitly capture an interaction in terms of the subtle actions we perform in communication; thus providing a route to understanding the intentions behind a given utterance, and equally importantly allowing us to update a hearer's mental state in a rational way based on the interpretation of an utterance. While speech act theory stands as its own theoretical framework, it is so widely accepted as a principle of communication that, as we will see later, the notion of speech acts are pervasive in many other theories of dialogue structure and management – and consequently highly relevant to our purposes here.

Given this relative importance of speech acts, in this section we provide an introduction to speech act theory in terms of its history, theoretical concerns and the various approaches to speech act classification and definition which have been pursued. However, since the literature contains extensive coverage of speech act theory, here we aim to capture salient points only. Readers looking for more depth are directed toward Huang's [2006] approachable introduction to speech act theory which was itself based on a number of chapters within the more thorough *Handbook of Pragmatics* [Horn, 2004, ch. 3,21,26]. From the computational perspective, readers might also consider Jurafsky and Martin's [2000] chapter on dialogue structure, which presents communicative acts in the context of complete language systems.

2.3.1 Historical Development

The Theory of Speech Acts originated principally in the work of the philosopher J.L. Austin, who, in his book *"How to do things with words"* [Austin, 1962], set out a theoretical framework for the study of language as action rather than purely truth conditional statements. Austin's view developed from his observations that there were a group of utterances, which he termed *performatives*, that while looking initially like simple declarative statements or *constatives*, have very different properties. For example, unlike a declarative utterance, the performative utterance "I declare war" is: (a) better characterized as changing the state of the world rather than describing it; and (b) better characterised as living on a spectrum between felicitous and infelicitous rather than true or false.

However, Austin's attempts to concretely characterize the performative and establish clear boundaries to constatives were only a means by which he set-up a more general theory of language where any utterances, not only performatives, when given in context have the effect of actions. Thus, all utterances performed sincerely in dialogue are *speech acts* which can describe, query, and change the world. One cornerstone of this proposal was that a speech act naturally consists of a number of different aspects or acts, i.e., the *locutionary* act, or the act **of** speaking; *the illocutionary* act, or the act **in** speaking; and the *perlocutionary* acts, or the act **by** speaking. The locutionary act is the production of the speech act, e.g. saying 'I declare war', and thus encompassing a large number of actions performed in the production of language, these ranging from the semantic and syntactic construction of messages to the uttering of certain sounds or the writing of certain symbols. On the other hand, the illocutionary act describes the interactive goal of the act, e.g. the declaration of war. Finally the perlocutionary act is the consequential effect of the illocutionary act – either in the mind of the hearer or in the environment. To look at it from the perspective of an agent who is about to speak, he or she will be motivated by a perlocutionary effect which they wish to achieve, e.g., to make their partner adopt some new belief about the world, and to achieve this perlocutionary effect they choose some illocutionary act, e.g., persuasion, which they realize through some locutionary act of speaking a particular utterance. Moreover, while a speaker often has the direct ability to perform both the locutionary and illocutionary acts (given the correct rights and privileges), she may not directly achieve the perlocutionary act, but must instead hope that the illocutionary act leads to the desired perlocutionary effect.

While Austin set out the basis of the theory, it was one of his students, Searle [1969, 1975], who defended and refined Austin's work, thus producing the theory of speech acts as we now know it. Searle's [1969] chief observation was a very distinct bipartite nature of utterances, where certain features realize the propositional content of the speech act, while other features realize the illocutionary force itself. Moreover, this distinction between propositional and illocutionary features of speech acts was central to a rule-based characterization of individual speech acts. In this approach, Searle extended Austin's earlier notion of *felicity conditions* which determined the *happiness* of a speech act by elevating the general conditions to specific *constitutive rules*, which in themselves define a given speech act. Thus, individual speech acts were defined in terms of their particular (a) propositional content conditions; (b) preparatory conditions; (c) sincerity conditions; and (d) essential conditions. This approach to the definition of speech acts later gave rise to Searle's [1975] highly influential characterization of speech acts to which we will return in Section 2.3.3.

Before moving on to considering some of the properties of speech acts in more depth, it should be noted that Searle and Austin originally used the term *speech acts* for the illocutionary acts we are concerned with here, and this continues to be the favoured term in the pragmatics and philosophy of language communities [Huang, 2006, Sadock, 2004, Bach, 2004, Vanderveken and Kubo, 2002]. However, the computational linguistics community introduced more general terms such as *dialogue act* [Bunt, 1994] or *conversational move* [Power, 1979] to model "more kinds of conversational function than an utterance can perform" [Jurafsky, 2004, p. 588]. Thus, terms such as dialogue act are intended to capture a wider range of communicative actions than can be achieved through more than natural language alone. Throughout the rest of this book we will use the generic term dialogue act unless we wish to explicitly distinguish between either speech or conversational acts.

2.3.2 Indirect and Surface Level Speech Acts

One aspect of speech act modelling which is particularly relevant to natural language pragmatics is the notion of indirect and surface speech acts. The basic idea of indirect speech acts, like conversational implicature, is that a given utterance can be used to convey intent that goes far beyond the simple surface form of an utterance. For example, the utterance "it's too hot in here" can simply be interpreted as a direct statement regarding the temperature in a room given a surface interpretation, or it can be interpreted as an indirect request to have the air conditioning turned up or a window opened. Computational approaches to speech act modelling have often focused on the distinction between the intended indirect speech act and the so-called surface speech act. Here, a *Literal Force Hypothesis* was assumed which argued that while the principle illocutionary act, i.e., the indirect act, is only visible in the context of a complete dialogue (e.g., the indirect request for a window to be opened), a surface speech act can be read directly from an utterance in isolation (e.g., the statement regarding room temperature).

This notion of a surface speech act with a subsequent deep interpretation has considerable appeal for a number of reasons. First, it was directly related to much of the work of Grice on logical conversational models, and in particular on his model of *conversational implicature* [Grice, 1975]. Second, from a computational perspective, it allowed a traditional semantic analysis of a sentence to yield not only the propositional content of the sentence, but also a surface speech act which is identified through lexical and grammatical features of the isolated utterance. Indeed, in the absence of propositional content, e.g., "yeah", such an analysis would still be able to produce a surface speech act specification, e.g., "ACCEPT".

However, there are a number of problems associated with the Literal Force Hypothesis – particularly with respect to the interpretation of speech acts. For one thing, as Appelt [1985] and others note, the relationship between surface speech acts and illocutionary acts is complex for two reasons: first, the same surface speech act can be recognized as the realization of any number of illocutionary speech acts; and second, an illocutionary speech act may be realized directly with a surface speech act or embedded within another surface speech act through action subsumption (i.e., fusion of illocutionary acts into single surface speech acts). Moreover, one surface speech act may realize one illocutionary act in one context and another illocutionary act in a different context. For example, the utterance "Yeah" can be conceivably interpreted as a check question, an acceptance of a fact, or as a simple answer to a yes-no question. In this case a traditional grammar could only commit to an underspecified feature. The response to such criticisms of the inference-based approach to speech act interpretation gave rise to a rejection of the Literal Force Hypothesis within the dialogue act interpretation community, and the development of probabilistic *cue-based* approaches to speech act interpretation. Unlike classical models, these approaches make use of several information sources, i.e., grammatical features, prosody, and dialogue structure to directly derive the principle speech act without assigning an intermediate surface speech act [Jurafsky, 2004, MacWhinney et al., 1984],

2.3.3 Classifications & Definitions

Austin [1962] proposed an initial classification of utterances into a small number of illocutionary force groups: *verdictives*, the giving of a verdict; *exercitives*, the exercising of powers; *commissives*, commitment to action; *behabitives*, miscellaneous expressions of attitudes and social behaviour; and finally *expositives*, acts of exposition (of views). Regarding his illocutionary force classes, Austin was, in his own words, "far from equally happy about all of them" [Austin, 1962, p. 151]. Indeed, the preliminary classification was intuitive rather than well-structured, left great overlap between the categories, and failed to address issues of multiple illocutionary forces being applicable to a single utterance. In the decades since Austin's work there have been a multitude of revised proposals on the classification and definition of speech acts. We consider some of these proposals over the following sections.

2.3.3.1 Speech Act Formalisation

Searle [1975] extended Austin's work with a highly influential classification of speech act types based on a multi-dimensional rule-based analysis. That analysis resulted in five speech act classes: (a) *representives* (or assertives), which are truth-value carrying statements which assert a fact about the world; (b) *directives* which are used by a speaker to influence the future actions of a hearer; (c) *commissives* which are used by a speaker to commit the speaker to some future action; (d) *expressives* which express a psychological or social attitude; and finally (e) *declaratives* which are explicit performatives which have some tangible world changing effect. Although Searle's classification was an improvement on Austin's, it was not without criticism. For example, Ballmer and Brennenstuhl [1981] note that Searle's use of a *declarative* category was ultimately self-defeating in that the *declaratives* became at best a waste-basket of awkward classifications, and at worst corresponded directly to the *performatives* which Austin had initially set out to abolish through his speech act classification.

In subsequent years, there were numerous attempts to augment and systematically re-structure Searle's original speech act classification. Bach and Harnish [1979] for example produced a well-regarded Searlean style work with formal definitions for types based on an attitudinal system. Coming from a more lexical perspective, Ballmer and Brennenstuhl [1981] developed a lexical analysis methodology to produce a Searlean style categorization of German speech acts. These two early variants on Searle's work characterise what became a data-driven versus rule-driven distinction in how speech act categorizations were described and used.

Formal definitions of speech acts grew out of two bodies of work. On one hand, the linguistics community developed formal definitions of speech acts to iron out classification problems seen in Searle's early proposals, e.g., Bach and Harnish [1979]. On the other hand, the computer science community developed formal models of speech acts so that notions of speech acts could be included in computational models of language understanding and production [Cohen and Perrault, 1979]. In both cases, the specification of speech acts built on a formalised notion of action and attitudinal frameworks involving cognitive concepts such as *beliefs* and *intentions*. In turn, through these formalizations, speech acts could be planned like any other physical actions, and thus used in means-end and deliberative reasoning systems.

The most widely cited formal specifications of speech acts are due to the work of Cohen, Pernault and Allen. In particular, Cohen and Perrault [1979] applied a formalization of speech acts to the goal-directed planning of natural language utterances. Later Perrault and Allen [1980], Allen and Perrault [1980] applied the same style of formal speech act specification to the interpretation of indirect speech acts in language. Both these approaches made use of Fikes and Nilsson's [1971] STRIPS planning system by defining the illocutionary acts in terms of pre-conditions and effects. A number of inform acts were for example also defined. For example, the vanilla inform had a precondition that the speaker Know(P), while the effect stated that the hearer would then Know(P), and furthermore the hearer would *Know* that the speaker knows P. It should be noted that while these early works and their close derivatives, e.g., [Appelt, 1985], looked at issues in natural language generation and interpretation, these same formalizations were later applied and extended by Cohen and Levesque [1991] in their seminal works on agent communication and cooperation.

Despite their appeal from both a philosophical and computational perspective, formal speech act definitions have however been subject to several criticisms. One early and frequent criticism centres on the many assumptions which are necessarily made by the definitions, i.e., communication is assumed to be perfect, and it is always assumed that the hearer is well-behaved on hearing the speech act. However, these assumptions in fact reflect the general constraints placed by Searle on his speech act definitions, i.e., it is not valid to talk about a speech act having been committed unless it was completed successfully. Furthermore, following subsequent work on grounding and social theories of speech acts (described later in this chapter), this class of criticisms has been addressed.

But the more problematic issues with speech act theory have always centred on the theories of attitudinal state required for their definition. Such arguments typically assert that standard attitudinal state models are too simple to capture anything but veridical and literal statements, and that defining mental state models to handle more natural situations will require considerable research effort – and since such models are not yet available, they cannot be used to provide definitions of speech acts. This argument has, for example, recently been put forward again by Mann and Kreutel [2004], who criticize Searle for relying on a strict relationship between belief in the propositional content of a statement and the sincerity of the statement. The authors argue that other language phenomena, such as irony and exaggeration, can be characterized as the speaker not believing the truth of the statement, but yet these would not count as insincere statements in any reasonable treatment. While Mann does not propose any specific solution to the problem, he suggests that a considerably more structured semantic network description of mental state will be essential to any meaningful dialogue interpretation theory.

2.3.3.2 Dialogue Act Taxonomies

Due to such criticisms, attitudinal state based formalisations of speech acts have been rejected over the last 15 years in favour of data-driven classifications which we will refer to as the *Dialogue Act* centric methodologies. These dialogue act approaches on one hand make use of feature-based descriptions of communicative acts based on surface form and adjacency to other acts, and on the other hand rely on act classifications which, while being motivated by assumed attitudinal state, are not expressed formally in terms of such attitudinal state. These classifications, the so-called *Dialogue Act Taxonomies*, grew out of a body of research which rejected the inference model of speech act interpretation, and also argued that communicative acts which were broader in definition than speech acts, e.g. by including conversational analysis concepts (see e.g., Allwood [1997]), were more appropriate for computational models of conversational agents.

Dialogue act taxonomies offer definitions of speech acts in terms of a taxonomy rather than a formal structure, and thus are limited to capturing hierarchical relationships between act types. As a result, there has been a proliferation of dialogue act taxonomies which have offered various interpretations on how the hierarchy should be organized. The most noteworthy of these offerings are DAMSL [Allen and Core, 1997] and Bunt's [2006] DIT++. Other notable dialogue act taxonomies include Verbmobil 2 [Alexandersson et al., 1998], HCRC [Carletta et al., 1996], LinLin [Larsson, 1998, Jönsson, 1995], and TRAINS [Traum and Hinkelman, 1992]. As an example of a dialogue act taxonomy, Figure 2.1 depicts the DAMSL (Dialog Act Markup in Several Layers) multidimensional dialogue act tag set which was created as part of an annotation schema for the markup of task-oriented dialogues [Core and Allen, 1997, Allen and Core, 1997]. One of the motivating factors in DAMSL's inception was to overcome a limitation of assigning only single dialogue or speech act labels to a given dialogue contribution. While Searle had foreseen the need to assign both surface and primary illocutionary speech acts to given utterances, DAMSL was intended to address issues of multiple primary illocutionary force ascription. DAMSL thus allowed a given contribution to be annotated simultaneously across a number of different layers. For each individual utterance, a successful annotation is then to assign a number of utterance-tags, where each tag is taken from one of several quasi-orthogonal dimensions of communicative function. While DAMSL was devised with an eye toward task-oriented dialogue [Allen and Core, 1997, p. 2], it was intended that the coding schema be extensible to other dialogue types and application-specific domains (see Jurafsky [2004] for details of specific extensions).

Among DAMSL's features, the most pioneering was its explicit provision of a notion of layering and dimensionality in dialogue act tagging. However, while DAMSL is usually presented in terms of its four organizational levels or categories, in effect it is a 16-dimensional tag set, thus providing sixteen distinct feature sets for annotators. Some of these feature sets contain explicit sub-categories, while other sets are simple binary choices between the aspect holding or not holding, e.g., *Info-Request*. Moreover, since DAMSL was developed primarily as a taxonomy of features to be used by human annotators on texts that are available in advance, DAMSL's annota-

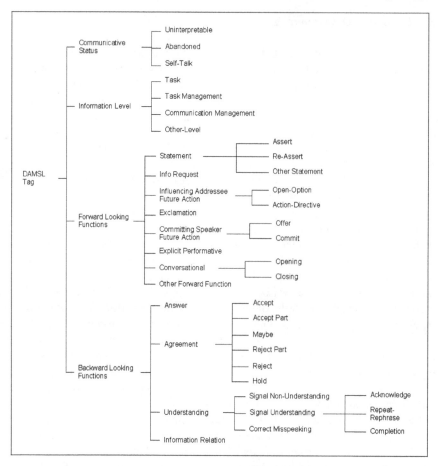

Fig. 2.1 The DAMSL Speech Function Tag Taxonomy (Adapted from Allen and Core [1997])

tions are highly dependent on deep intuitions about meaning asserted by annotators rather than surface features of the dialogue. This immediately has the effect of rendering DAMSL categories unsuited for direct use within context-less grammatical formalisms.

As structured collections of speech act like labels, dialogue act taxonomies have found considerable success over the past ten years through their application in statistical models of dialogue modelling and management (e.g., [Keizer, 2003]). However the proliferation of dialogue act taxonomy proposals suggests that the basic organizational principles applied may not be optimal.

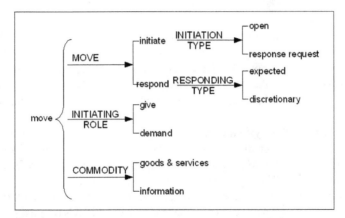

Fig. 2.2 Halliday's Speech Function Network. Individual move types can be characterised as a cross-combination of a number of features selected in parallel over the network. (Adapted from Halliday and Matthiessen [2004])

2.3.3.3 A Systemic Functional Perspective

Coming from a different philosophical perspective, Halliday's Systemic Functional Linguistics school rejects the treatment of speech acts as a distinct pragmatic entity stratified away from linguistic semantics. Instead, this school assumes an explicit representation of interpersonal meaning within linguistic semantics, and that this interpersonal meaning is principally captured by units termed *moves*. The move itself, a central unit in a more complex theory of exchanges and negotiation, is in turn said to have an associated *speech function* which is broadly equivalent to the more orthodox notion of illocutionary force [Halliday and Matthiessen, 2004].

While this treatment of moves rather than speech acts relates to a very comprehensive theory of interpersonal communication (see Martin [1992, ch. 2] for a detailed introduction), for us here the most significant differentiating factor between systemic moves and the speech act is the manner in which the different speech functions that a move can have are defined. The approach taken to this issue dates back to an early proposal by Halliday who, adopting grammatical arguments, organized move types not on the basis of a traditional taxonomy, but rather on a set of three paradigmatic distinctions which can be used to cross-classify a given move. The first opposition is based on whether a move in an exchange is involved in the **giving** of a commodity versus **demanding** as exemplified by the difference between *offering* versus *demanding* an action. The second paradigmatic opposition reflects the choice of commodity which can be negotiated in an exchange, i.e., **information** versus **goods & services**. And the final opposition set up **initiating** versus **responding** moves within the grammar. These paradigmatic distinction are in turn formalised in terms of a Systemic Network[1] which defines the available choices in defining the

---

[1] A type of decision tree where at each decision point some well defined external operation is

properties of a move.

Figure 2.2 depicts such a systemic network for an updated version of Halliday's model. A network is composed of a set of systems which are traversed from left to right. Squared braces denote *systems* where a concrete choice must be made in moving from the input of the system (the left) to the system's output (the right). Curled braces on the other hand denote a system where all systems immediately to the right must be traversed if the system was entered. Thus, referring for example to the *move* system, if this system is entered during the course of a complete system traversal, then the traversal must continue on to the three systems to its right; whereas having entered the *MOVE* network, only one of either the *INITIATION TYPE* or *RESPONDING TYPE* network is subsequently entered.

Many alternative characterizations of dialogue moves based around paradigmatic choice and systemic networks have been proposed since Halliday's original suggestion. Among these, one notable contribution is due to Martin, who in his development of a text (rather than clause) semantics, suggests an extension of Halliday's proposal through the inclusion of conversational move types by introducing *calling*, *greeting*, and *reacting* choices [Martin, 1992]. The speech functions proposed by Martin, organized as a systemic network, can be cross classified out to produce seven different adjacency pairs, totalling fourteen different move types.

While certainly not as wide ranging as current dialogue act taxonomy proposals, nor as formal as the true intentional logics dating back to Searle's proposals, the systemic view on dialogue moves is significant for the broader topic of speech acts in two ways. First, the organization proposed by Halliday, and later extended by Martin and others, focused on the use of grammatical evidence as an organizational principle rather than subjective cognitive concepts, and thus arguably suggests a sharper razor which can be applied in developing classifications of speech act like entities. Second, the systemic network approach effectively proposes a form of detailed cross-classification in the description of individual speech act like categories, and thus avoids the proliferation of dialogue act taxonomies where taxonomy alternatives were often proposed on the basis of disagreement over primary organizational factors.

2.3.3.4 Discussion

The great appeal of speech act theory is that actions provide a natural analysis of language and communication which can be linked to notions of mental state and general theories of deliberative and intentional action. But, just as the examination of the structural qualities of utterances reveals considerably more about grammar than a more theoretical consideration of parts of speech, looking to dialogue as a whole in terms of its structural and usage qualities can provide considerable insight into the nature of dialogue. To that end, we move on in the next section to consider the structural properties of dialogue.

---

performed as a traversal step is made.

2.4 Structural Discourse & Dialogue Models

Having reviewed the individual characteristics of speech acts, in this section we move on to models of dialogue as a whole. We can characterise such models as being classed into either *structural* or *cognitive* accounts. Structural accounts of dialogue are concerned with the characterization of surface form dialogue as an external phenomenon from a third party perspective. Such characterizations are expressed in terms of rhetorical relations, speech acts, and exchange structures, but make few or no reference to either domain-specific or internal mental state dynamics. On the other hand, cognitive accounts capture complete models in terms of the semantics, internal attitudinal states, and the processes involved in dialogue. Naturally such cognitive accounts often assume the existence of structural models, but these are often implicit and difficult to distinguish within the greater cognitive theory. In this section we consider the set of structural models before moving on to cognitive models in Section 2.5.

Structural models of dialogue can essentially be traced back to the notion of *discourse analysis*, a term first suggested by Harris [1952]. Since its inception, discourse analysis has evolved into a collection of methodologies for the analysis of texts in order to investigate how language is used to communicate for a purpose in a given context [Brown and Yule, 1983]. Common characteristics of the discourse analysis methods, regardless of which of the many academic disciplines they are applied in, include an emphasis on the study of language beyond the clause or sentence, and the importance of studying natural real-world language rather than artificially contrived segments. Of the spectrum of Discourse Analysis sub-disciplines, the branches which are of most relevance here, and which will be considered below, are those of coherence and cohesive structure, exchange structure, and dialogue network descriptions.

2.4.1 Rhetorical & Cohesive Structure

Rather than looking at language as simply a set of disconnected speech act like entities, Halliday and Hasan [1976] argued that well formed language obeys rules of coherence and cohesion that bind individual contributions into a whole which has a concrete meaning that is shared by producer and interpreter. Within the framework offered, a coherent discourse must make sense, i.e., an observer or reader of a discourse, with the requisite background knowledge, should be able to interpret the discourse's meaning, and thus view the text or dialogue to be a coherent unit of communication rather than simply a collection of sentences. A concept closely related to coherence is the notion of cohesion. A cohesive discourse is one which is well formed in that it makes proper use of the many mechanisms of discourse structuring available in language – those mechanisms resulting in a discourse body within which any given unit is connected to other units directly or indirectly through one or more *discourse* or *rhetorical* relations. Discourse relations which couple the elements of the text together are typically semantic in nature and may or may not be marked explicitly in the surface form.

While certainly not the oldest, Mann and Thompson's [1988] Rhetorical Structure Theory (RST) is probably the best known account of discourse organization through a systematic assignment of relations that hold between discourse units. Although developed in the context of systemic language generation systems, RST is linguistically agnostic in inception, thus making it widely applicable across both computational and discourse analysis fields [Taboada and Mann, 2006].

A successful RST analysis decomposes a text into a hierarchically connected structure where individual parts or *units*, which broadly correspond to clauses, are connected via discourse or rhetorical relations to other units or composite units called *spans*. Twenty-four relations were originally defined for RST, and these can be organized under different dimensions into particular categories. One dimension against which relations are organized is as either list-based or nucleus-satellite, where in the latter case one text span has a specific role relative to the other. A second dimension against which relations are categorized is as either *informational*, where the rhetorical relation reflects a true logical relationship between two text units, or *presentational*, where the relation is used as a form of presentation style or argumentation.

While RST, like other theories of coherence and cohesion, has been most widely applied to monologic discourse, the same principles have been assumed to hold for spoken and written dialogues – albeit with the addition of extra dimensions of structure. For example, in an early approach Fawcett and Davies [1992] applied RST in the analysis of individual turns in dialogue, thus treating each turn as a monologue for conventional RST analysis. Daradoumis [1995] however proposed Dialogic RST as an extension of classic RST with relations across turns.

As a single instance of a coherence relation framework, RST itself has a number of shortcomings. For example, Martin [1992] argues that RST fails to give unambiguous accounts of many features where a satellite-nucleus structure is not naturally present, e.g., sandwich structures where an initial message is used to set up a culmination which together sandwich argumentation points. Worse still from a computational perspective, RST is a top down analysis approach which relies on human analysers making highly cognitive – and sometimes arbitrary – decisions in the process of coding a text. While this may be reasonable for text analysis, and even as a tool for planning texts to be generated computationally, it can pose notable difficulties for automatic computational construction.

An alternative, more fine-grained analysis of cohesive semantics, which is arguably more suited to an automatic linguistic analysis, was proposed by Martin [1992, ch. 4]. Specifically Martin's analysis looked at the *Conjunctive Relations* which can hold internally between clauses in a clause complex, or across the bounds of clause complexes, but based his classification of such relations to the most part on available surface cues such as types of conjunction. Martin proposed a conjunction relation classification which placed conjunction relations along two dimensions of analysis: the first dimension reflecting the underlying logical nature of the relationship, and the other categorizing the relation as either *external* or *internal* to the text – a notion very similar to RST's informational/presentational distinction. Like an RST analysis, a Conjunctive Relation analysis makes use of a dependency-based model of the logical structures created in text. In such an analysis, which is typically conducted diagrammatically, the distinction between internal and external relations is focused through two orthogonal structural decompositions. In addition to the internal/external distinction being made prominent, some of the other modelling features of significance are that messages can relate to more than one other, and that relations may be either forward or backward looking (in the case of hypotaxis), but need not be contiguous.

RST and Martin's conjunctive relations are just two of a raft of coherence relation classifications and analysis methodologies. Others of note include Halliday and Matthiessen's [1999] model of *Ideational Sequences*, Hovy's [1990] taxonomic collection of discourse relations, and the semantics centric organizations such as SDRT [Asher and Lascarides, 2003] which we will return to in a later section. Although these coherence-based theories seem quite *academic* when considered in isolation, works such as SDRT have shown that having an understanding of the rhetorical relationships between units in dialogue is important not only to demonstrating that language is understood, but is key to facilitating the language understanding process itself.

However, while coherence and rhetorical theories are an important aspect of language structure, such theories have tended to focus on logical relations between elements in text, and have thus veered away from interpersonal considerations. As a result, these theories say little about dialogue as action or cognitive process, and offer no explanations as to why we find it difficult to assign speech acts to dialogue in a consistent way. Although structural considerations do appear to have importance for language understanding, we require a more dialogue centric approach to structural analysis. In the next section we turn to the first of two classes of analysis that have been developed to address such needs.

2.4.2 Exchange Structure

Exchange Structure theory is a class of dialogue-focused discourse analysis methods that provides a constituency-based decomposition of a dialogue. This theory draws on, and is broadly in keeping with, the functional view of language organization as advocated by Halliday [Halliday, 1985]. Although originating in sociolinguistics, exchange structure theory has had significant impact on computational approaches to dialogue, and still plays an important organizational construct in dialogue coding proposals such as DAMSL and the HCRC schema. Given this impact, in this section we will take time to introduce exchange structure as originally proposed by Sinclair and Coulthard [1975], and examine some of the many refinements and modifications proposed in the literature. For comprehensive introductions to this field of analysis the reader is directed to Eggins [2004], Thompson's [1996] introduction to functional linguistics, or Martin's [1992] comprehensive recount of exchange structure modelling.

2.4.2.1 The Birmingham School

Starting out from an analysis of student-teacher dialogues in a classroom setting, Sinclair and Coulthard [1975], and later Coulthard and Brazil [1979], proposed a discourse analysis methodology based around a decomposition of dialogue into constituents taken from a *rank scale* of organization including *acts, moves, exchanges, transactions*, and *interactions*. The central unit of this organization is the *exchange* within which some piece of information or service is negotiated by dialogue participants. Exchanges are in turn constructed co-operatively by participants through a sequence of contributions termed *moves*. These moves, which may be realized through clauses or clause complexes with specific illocutionary force-like speech functions, are in turn constituently composed of *acts*. The act is the smallest unit of interpersonal meaning potential, and may be realized through the clause. Moving back up the rank scale from the exchange, *transactions* and *interactions* structure exchanges into higher order socio-organizational entities.

The basic Exchange Structure approach, which is sometimes referred to as the *Birmingham School*, not only proposes a constituency-based decomposition of dialogue, but also defines a structural potential which licences a number of different configurations of elements in exchanges. Two classes of exchange structure are suggested, corresponding to Halliday's distinction between the negotiation of action and the negotiation of propositions in language. These exchanges may maximally comprise a structure consisting of three moves: *Initiate, Response*, and *Feedback*. We illustrate this potential, and the exchange structure methodology in general, with the following example:

	Exchange Initiate	where is the bed?
(6)	Response	on the right
	Feedback	ok

Thus, within a single exchange which negotiated the propositional fact that "the bed is on the right", there are three moves involved, with each of the moves playing a specific role, or function, within the exchange.

The success of Sinclair, Coulthard, and Brazil's exchange model, as measured by its influence on the many works derived from it, is that it provides a simple but linguistically motivated model of interaction structure that fits well within an established linguistic system, i.e., Halliday's Systemic Functional Linguistics. But given its simplicity, a number of limitations in the model were soon pointed out. Among these issues, Berry [1981] criticises Sinclair and Coulthard's model for not adhering strictly enough to Halliday's linguistic framework, and thus subsequently provides an analysis which breaks moves down in a manner more in keeping with the Systemic Functional Linguistics methodology, i.e., in terms of their interpersonal, textual, and ideational meta-functional contributions to meaning (see Section

5.3.2.1 for more information on meta-functions.).

With respect to the interpersonal meta-function, Berry for example proposes some six distinct interpersonal micro-functions for knowledge negotiation[2]:

- **k1 - Primary Knower's Role:** Marks a move played by the primary knower of the exchange proposition.
- **k2 - Secondary Knower's Role:** Marks a move played by the secondary knower in the exchanged unit.
- **dk1 - Delayed Knower's Role:** Marks a move where the primary knower of the proposition delays their assertion of the proposition within the exchange.
- **k2f - Secondary Knower's Final Knowledge:** Marks a move which indicates the final state of the secondary knower's knowledge state.
- **qk1 - Question Primary Knower's Information:** Marks a move within which the primary knower's information is questioned.
- **qk2 - Question Secondary Knower's Information:** Marks a move within which the secondary knower's information is questioned.

These micro-functions can be used to characterize the interpersonal contributions of moves within a typical exchange. To illustrate, consider again the simple exchange used above recast in Berry's terminology:

Interpersonal Surface

(7)
k2	where is the bed?
k1	on the right
k2f	ok

Here, the first utterance is marked as *k2* since it is a move by the secondary knower (i.e., he or she that does not actually know the state of the exchange's target proposition at the beginning of the exchange). The second utterance is marked k1 since it s a move by the primary knower of the exchange's proposition, and k2f is used to mark the third utterance as it indicates the final knowledge of the secondary knower at the end of the exchange.

Such a structural description seems quite unusual with respect to the typical speech act like descriptions which we have come to expect in dialogue characterizations, but the clarity of the analysis, and the relationship to speech act type models becomes more clear if we introduce the textual and propositional micro-functions of analysis. Three ideational micro-functions were introduced by Berry: *pb* (proposition base), which is used to mark sentences in which a propositional skeleton is introduced that should become filled through the course of the exchange; *pc* (propositional completion), which marks moves in the exchange which complete the proposition under negotiation in the exchange; and *ps* (propositional support), which marks moves within which the negotiated proposition is supported or confirmed. The textual micro-functions proposed simply denote the sequential contributions by each participant. Taking again the same example from above, we can see the complete meta-functional analysis by Berry is:

---

[2] Although not discussed here, Berry's account was equally applicable to the negotiation of goods & services

Ideational	Interpersonal	Textual	Surface
pb	k2	ai	where is the bed?
pc	k1	bi	on the right
ps	k2f	aii	ok

(8)

Thus, within this exchange which is said to concern the proposition which we might paraphrase as "the bed is on the right", the first exchange move is said to set up the propositional base (pb), and is a move by the secondary knower of information, i.e., the person who is not aware of the complete proposition at the start of the exchange. Similar readings can be given to each of the other two propositions in accordance with the particular micro-functions introduced above. We can thus see that the resultant analysis is multi-functional in that each utterance is analysed in terms of its contribution to different dimensions of meaning. Such a multi-dimensional analysis shares motivational justification with the newer dialogue act taxonomies, even though the particular dimensions of analysis differ.

Another style of modification made to Sinclair & Coulthard's original proposal introduced explicit roles between units in exchange analyses. To illustrate, consider Example 9 below from Butler [1985], which depicts one complete exchange with interpersonal meaning captured simultaneously in terms of acts and moves, and the roles that these acts and moves play in the units ranked directly above them:

Exchange Role	Move	Move Role	Act	Surface
Initiation	opening	Signal	summons	John,
		Pre-Head	preface	about that drink you mentioned,
		Head	directive	could you get me a beer, please?
		Post-Head	comment	It's getting late.
Response	supporting	Pre-Head	accept	– Sure.
		Head	react	(hands over drink)
		Post-Head	comment	Nice and Cold.

(9)

Within the exchange structure and functional linguistics communities, the structures in this analysis are referred to as *multivariate*. To put it simply, a multivariate structure is based on each lower order rank object playing a well-defined role within the higher order rank of which it is a constituent. Although this multivariate view is appealing in its consistent organizational principles across ranks of units in exchange structures, it faces significant scaling limitations in its application to more natural conversational situations. Essentially the argument against the multivariate view as pushed within the Conversational Analysis school by Levinson [1983] and also within the more traditional Discourse Analysis school by Martin [1992] and Fawcett and Young [1988], was that dialogue structure is a far more free and open-ended phenomena than can be accounted for by the multivariate approaches. To paraphrase Martin, interlocutors can at any time 'dig in' – continuing an exchange with clarifications, elaborations, or corrections, to a length and complexity which goes well beyond the structural potential afforded by multivariate approaches, yet remaining undoubtedly *grammatical*.

Exchange Structure	Speech Function	Mood	Utterance
K1	statement	declarative	That's pretty
K2	command	interrogative	Can you tell me how many cups that would hold?
K1	response offer to command	elliptical declarative	- Em.. about five or six, I think
cl			- five or six?
rcl			- yeah
K1	statement	declarative	Doesn't Say
K1	statement	declarative	Looks like it'd hold about five
K2f	acknowledge statement	paralinguistic	Mmm

Fig. 2.3 Complete example from Martin [1992].

2.4.2.2 From Multivariate to Univariate Structure

Following on from the arguments of the Conversational Analysis school, as well as from some of the earlier proposals from Ventola [1988], Fawcett et al. [1988] and others who proposed models which place emphasis on univariate and sister-dependency relations between units of the same rank, Martin [1992] developed a hybrid dialogue modelling account that looks to multivariate structure for overall exchange organization, but falls back to univariate structure to account for support statements, clarifications, and situations where an interlocutor breaks off an ongoing exchange.

The first aspect of Martin's model concerns the variety of supporting statements which a speaker can make in an exchange. Martin's argument was that, rather than connecting all meaning-carrying moves directly into the exchange structure, some moves can be linked into the structure through ideational, or propositional content, relationships that can hold between moves in the exchange. Thus, when viewed as a move within a single exchange, a *request* such as the following:

(10) I'm getting cold, can you open the window?

should not be captured in terms of two moves playing well-defined constituent relationships within the exchange. Rather, only one move, the primary "can you open the window" plays an explicit role within the exchange, while the supporting move is linked to the primary move via an ideational relationship.

The second aspect of Martin's model concerned classes of moves which he argued were better integrated into the exchange through dependency rather than constituency relationships, despite the fact that these relationships were capturing truly interpersonal meaning, unlike in the case of the supporting statement above. Specifically, Martin proposes two classes of moves which fall into this category: *tracking moves*, which are used to clarify or confirm the ideational meaning that is being negotiated within an exchange; and *challenging moves*, which denote negotiation of interpersonal commitment to the exchanges themselves. Tracking moves are typically realised through back-channelling, while challenging moves are either when

an interlocutor breaks off communication or declines to perform an action or produce an evaluation of the proposition within an exchange.

To illustrate the hybrid-model, Figure 2.3 replicates a small segment of an example presented by Martin. This example analyses a short passage in terms of three
aspects of interpersonal analysis, i.e., mood, speech function and exchange function. The exchange structure description (first column) depicts the exchange functional contribution of each utterance, i.e., K1, K2, cl, rcl, with the multivariate constituency structural relations on the left of the moves, and dependency (univariate)
relations to the right of the move descriptions. Thus, the first utterance has a mood
contribution *declarative*, speech function *statement*, and sits alone in a one element
exchange structure as a K1 function. The next four utterances comprise a single exchange structure with K1 and K2 moves present in the structure, and the K1 move
connected via dependency relations to a clarification side-sequence.

While Martin's general approach is appealing in that it draws together a number
of modelling perspectives, it begins to break
down at certain points – particularly concerning
the reliance on univariate structures to address
clarification sub-dialogues. While examples such
as that shown in Figure 2.3 are well behaved,
other clarification sequences which diverge from
the propositional content of the dialogue can result in analyses with large proportions of content
outside of a constituency analysis (see Figure 2.4
for an example).

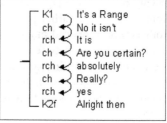

Moreover, the *challenging* moves were proposed as a means to allow interlocutors to get
out of exchanges, either by explicitly removing
themselves or refusing to provide judgements or

Fig. 2.4 Problematic analysis from
Martin [1992].

actions. But if we look to a simple example where the second speaker refuses a
request such as:

(11) a. K2 get me a drink will you?
      b. no

then the second move is not actually interpreted within Martin's analysis as a constituency role within the exchange but is rather modelled as a challenging move,
which is itself only related to the K2 move by means of a dependency relation.
Similarly, in Example 12 below, the two consecutive turns between speakers cannot be modelled as a simple exchange consisting of a question with a reasonably
good answer provided. Instead, it must be treated as either an exchange with one
constituency move which happens to have an attached dependency move, or as two
complete exchanges which have negotiated two independent pieces of information:

(12) a. K2 Is John coming?
      b. I have no idea

2.4.2.3 Discussion

The constituency-oriented analysis presented in this section tries to posit the kinds of structures allowed in dialogue. While such a constituency-based analysis has a natural appeal, the problems seen with Martin's model point to general issues with exchange structure modelling. Although social units such as exchanges and speech-act like moves seem to have a natural organizational role in dialogue, hand-crafted categorizations of the classes of exchanges and moves allowed have proven difficult to determine. Indeed, given cases such as that shown in Figure 2.4, it is questionable whether attempting to classify exchanges concretely in terms of the moves that are played within them will prove to be a useful exercise. Moreover, the problem with any sort of static description is that it only outlines the dialogue structure post-production, i.e., as a product viewed from an external perspective, and says little about the specifics of choices available to a speaker at a particular point in an exchange. Such issues are however fundamental to language production and interpretation, and this has led to the observation that dialogue is in fact mostly a locally organized phenomena rather than a more structured top-down entity. As we will see in Section 2.5, this local organization view is prominent in cognitive theories of dialogue. But before we examine these theories we first consider non-cognitive organizations of local dialogue potential.

2.4.3 Dialogue Networks & Grammars

Dynamic potential based accounts of dialogue view an exchange as a process within which an interlocutor is participating, and thus set out in situation specific terms what production or interpretation choices are available to the interlocutor. Using Halliday's [1984] terminology, dynamic potential models can be classed into two groups: *Single Point Potentials* that organise dialogue on a global basis into a set of predetermined states; and *Generalised Potentials* that view dialogue as a strictly locally organized phenomenon wherein an interlocutor chooses their next dialogue move based on a more global and non-predetermined state description. We will address what Halliday refers to as the generalised dialogue potentials within the context of cognitive dialogue theories in the next section. Here we consider the static point potentials as manifest in state diagrams and flowcharts.

State-based dialogue models are highly intuitive graph-based accounts of dialogue that typically organize dialogue into a set of states, linked together by arcs denoting speech acts. Thus, the arcs from a given state enumerate all possible actions available to an agent at that given dialogue state. These state-based models, sometimes referred to as dialogue games and grammars, attempt to capture recurrent interaction patterns or regularities in dialogue at the illocutionary force level of speech acts. State diagram based models range from extremely concrete instances, such as those applied in state-based dialogue management systems [McTear, 2002], to far more abstract *illocutionary structures*, or *Generalized Dialogue Models* [Sitter and Stein, 1996] which capture specific regularities in dialogue without reference to target domain or mode of communication.

As an example of a more abstract state-based model, Figure 2.5 depicts Sitter and Stein's [1996] 'Conversational Roles' (COR) Model. This model is an example of a Generalized Dialogue Model which has been applied successfully in dialogue interaction [Dilley et al., 1992, Stein et al., 1997]. Sitter and Stein formulated the COR model as a Recursive Transition Network (RTN) composed of nested instances of a primary dialogue network along with three smaller networks – each of which captures individual dialogue moves within the complete RTN. Transitions can be due to either atomic dialogue acts, jump or empty moves, or dialogue moves including possibly complete instances of the initial *Dialogue(A,B)* network (thus showing the notion of recursive networks). One noteworthy feature of the COR model is that, following the principles of exchange structure, moves can be realised as either aggregate or atomic acts.

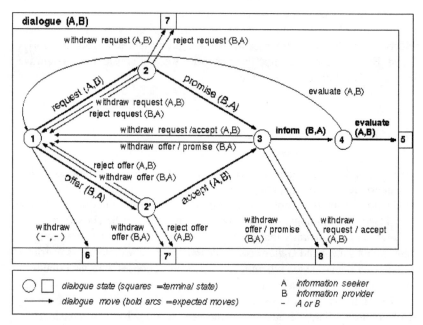

Fig. 2.5 Primary COR model network (Adapted from Sitter and Stein [1996])

The COR model's three smaller schemata for the description of dialogue moves potentials are shown in Figure 2.6. The first of these, *promise(A,B)* is the simplest of the three and can be used to capture either *promise* or *accept* moves through either an explicit act or a jump, where the filled circle stands for the terminal state. The second network schema, *inform(A,B)*, is used to model both *inform* and *assert* moves, and may consist of an atomic inform act which may or may not be followed by a traversal of the main COR network if a clarification sub-dialogue is entered into. Finally, the *request(A,B)* schema applies to all COR moves apart from the previously mentioned *promise*, *accept*, *assert*, and *inform* moves.

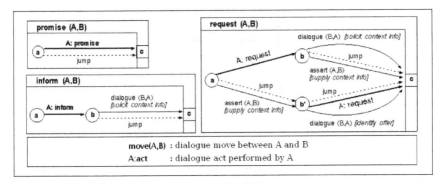

Fig. 2.6 COR model move schema (Adapted from [Sitter and Stein, 1996])

Due to their visual appeal and simplicity of design, state diagrams have been applied extensively in discourse analysis and computational linguistic work. However, in their pure form, state-based specifications such as the COR model are non-deterministic, saying nothing of why a particular move should be selected by an interlocutor in a given state. Thus, operationalisation of these models in practical applications requires the addition of determinism to the basic model. This is most commonly handled through the use of standard computational mechanisms that enhance the state machine semantics to allow conditioning of moves based on specific aspects of context, and to describe the particular changes of context warranted by the choice of a particular arc (see Woods [1970] for an example of one of the earlier concrete extensions of state diagrams in this way). An alternative approach, used for example in application of the COR model by Stein et al.'s [1999] MIRACLE or Teich et al.'s [1997] SPEAK! system, is to supplement the declarative dialogue model representation with scripts or other directly interpretable plans that are based loosely around the state models [Belkin et al., 1995].

Rather than extending the state transition formalism to account for context conditioning and the effects of a move – or worse still, using scripts which have vague relationships to the posited state description – an alternative is to base the modelling of the dynamic potential on a formalism which was explicitly developed to capture the necessary semantics. Flowcharts are one such modelling metaphor which became highly popular as a computational tool in the 1980s, and which were ideally suited to capturing abstract dialogue behaviour as they allow the labelling of the decision process in a straight-forward way.

Due to this descriptive appeal, flowcharts were for example used by Ventola [1987] to concisely explain the allowed behaviours in a dialogue through the conditioning of dialogue states with internal interlocutor context and the latest interlocutor moves. In this approach the flowchart effectively becomes a form of dialogue plan specification that makes the complex decision processes behind moves

far more transparent, yet allow a state machine specification to be abstracted from
the flowchart if necessary. However, in their classical form flowcharts are limited
to rather informal descriptions and are not directly executable. That said, in recent
years low-level variants on the flowchart metaphor have become extremely popular
in the development of commercial dialogue systems, but such systems are far removed from any notion of generalized dialogue models which might be applicable
across different dialogue domains – or even within the same dialogue.

Whereas Ventola and modern flowchart-based systems conflate the turn taking
decision and the content planning decision into single decision points within the
flowchart, Fawcett and Young [1988] developed a flowchart variant which separated out these decision processes. Specifically, Fawcett employed a hybrid model
which makes use of classical flowchart mechanisms to describe the overall flow of
an exchange between interlocutors, while falling back on systemic choice networks
to capture the choices made by a speaker once they have acquired the turn. Figure
2.7 illustrates one such systemic flowchart. Lines with arrows, i.e., traditional flow
chart lines, are used to indicate a changes of speaker, while lines without arrows
move between networks to outline the production choices available to a speaker at
that time.

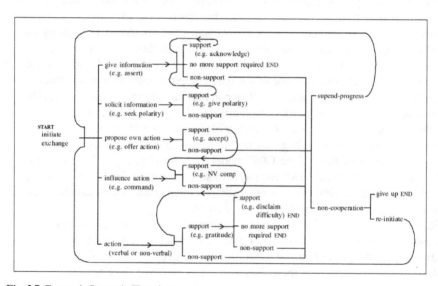

Fig. 2.7 Fawcett's Systemic Flowcharts.

Fawcett's model is appealing in that it teases apart two aspects of language planning by fusing the notion of systemic networks with the flowchart in a straightforward way. The particular systemic flowchart approach was however less true to
the original advantage of flowcharts in that no context modifications were captured
by Fawcett's flowchart, nor was there any dependence on general context in move
choices. Another limitation of Fawcett's modelling – although perhaps not of sys-

temic flowchart potential in general – is that the flowchart focused on the rank of move, without developing anything substantive in terms of a relationship to the rank of act. Nevertheless these disadvantages were specific to the particular account provided by Fawcett and it is easy to see how the general approach offers considerable descriptive and computational appeal.

Single point potential based dialogue models are extremely transparent and, when cast appropriately, are highly amenable to computational modelling. However, as we scale up dialogue sizes and complexities we encounter an explosive proliferation of states. Similar states with identical possible subsequent moves tend to be replicated across the modelling medium. The application of a notion of sub-network does help to reduce the number of individual states, but when we consider that natural dialogue allows a great range of interaction possibilities – even if we do find that there are preferred paths through interaction – then we must conclude that more natural dialogue would eventually lead to an unmanageable number of individual states.

Generalization of structures as typified by the COR model can help to reduce the number of states, but for natural dialogue styles any minimal collection of generalized states would necessarily have a vast array of conditionalised moves between these minimal states. Successful generalization thus requires a systematic approach to the conditionalization of moves based on context. But, for such an approach to be applicable to anything but trivial dialogue systems, this conditionalization in turn depends on well informed models of both mental state and deliberative processes. In the following we move on to theories of dialogue anchored in this perspective.

2.5 Semantic & Cognitive Theories

Whereas the structural and state-based accounts discussed in the last two sections focus on dialogue without reference to internal mental state dynamics, in this section we turn to the class of 'mental state'-oriented cognitive theories. Within the analysis presented here, this class of theory includes models ranging from theories of discourse semantics, through the agent-based theories of dialogue competence, to some accounts originating in the discourse analysis community that focus on generalized dynamic potentials rather than exchange structure descriptions. These models all place significant focus on descriptions of internal state, context, and processes, and thus are of considerable importance to situated dialogue modelling.

2.5.1 Discourse Semantics Frameworks

A criticism of some of the early agent-oriented theories of dialogue by Cohen and Perrault [1979] and Perrault and Allen [1980], Allen and Perrault [1980] were that they focused on the cognitive process of dialogue act planning and recognition to the detriment of features of surface language. Conversely discourse analysis approaches and classical compositional semantics as introduced earlier in this chapter cared little of why or how surface form was produced or interpreted in dialogue – and certainly cared little for how surface form related to internal conceptual structures. A middle-ground approach to the modelling of surface language was required instead. In this section we turn to models of *discourse semantics* which are designed to be this middle-ground.

A discourse semantics is a representation of a complete surface discourse in terms of a formal semantics structure. Such approaches, in contrast to full first order models, attempt only to capture the discourse itself rather than the world, and usually do so in a logic weaker than first order. Such models, unlike pure discourse analysis, also aim to be dynamic in the sense that they provide, or at least facilitate, operations necessary for the model to capture dialogue as a dynamic ongoing process rather than simply a product. Such devices include update mechanisms for the introduction of new statements into the model, presupposition introduction, anaphoric reference, and elliptic construction. Facilitation of such devices in turn requires that the models draw on theories of discourse analysis such as those introduced in the last section.

Discourse Representation Theory (DRT), originating in the work of Kamp [1981] and Kamp and Reyle [1993], is a theory of discourse modelling which has been proposed to overcome many of the failures of first order logic in the representation of a wide variety of language phenomena. In vanilla DRT, a given discourse is represented in terms of a Discourse Representation Structure (DRS) which contains two fields: one for representing discourse referents; and another for representing conditions or predications over those referents. Conventionally these DRSs are depicted as box diagrams, with referents noted along the top of the box, and the conditions and predications on those referents in the bottom field. For illustration, Figure 2.8 illustrates a DRS with box notation for a single simple utterance.

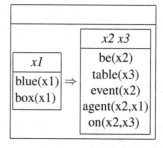

Fig. 2.8 Discourse Representation Structures for "Every blue box is on a table".

DRT is however far more than a representation model. Rather, the representation structures only provide the state of the discourse model at any given time. The key aspect of the theory is a well formalised account of how the discourse model is updated during the course of a dialogue. The update procedures not only account for very basic idealised utterances, but also handle a wide variety of discourse phenomena. This dynamic account thus goes far beyond what can be handled by a first order logic theory, but yet can be linked to a first order theory at a later stage as demonstrated by Bos and Oka [2002].

Basic DRT is not however a perfect model of discourse representation for dialogue processing. DRT emphasises the representation of context semantics and abstracts away from pragmatic information such as speech act theory and coherence relations. On a related note, DRT's update dynamics fail to take into account the complexities of the language grounding and contextualization processes in dialogue. DRT also provides a limited handling of underspecification issues, and the original DRS construction algorithm was based on the direct incorporation of syntactic transformations of complete sentences, whilst most dialogue segments are fragments smaller than complete sentences. Such problems have motivated a large body of research within the dialogue and pragmatics communities to extend vanilla DRT with features relevant for its use in more complex situations. One influential extensions of DRT in this regard was Poesio and Traum's [1998] axiomatization of a DAMSL subset in a DRT style dynamic logic. However, others argued that DRT in its original form is simply incapable of overcoming some of the limitations just identified. In the next section we consider an alternative discourse semantics theory which grew out of attempts to address DRT's limitations.

Asher and Lascarides's [2003] Segmented Discourse Representation Theory (SDRT) is a formal theory of discourse interpretation which emerged initially as an extension of DRT, but is now independent of the DRT formalism. SDRT provides the representation and compositional logics to build complete models of discourse from the shallow semantics of individual clauses. A key motivating factor in this development was the view that the meaning of a given discourse depends upon and interacts with its rhetorical structure. This use of rhetorical relations as a structuring mechanism marks a great difference with DRT; Asher & Lascarides argue that the disciplined use of these discourse relations is key to achieving many pragmatic issues such as presupposition, indirect speech act interpretation, and anaphoric reference resolution. Thus, SDRT not only presents a purely semantic representation of meaning, but also constitutes a semantic/pragmatics interface. As such the axioms of the various logics which together constitute the complete SDRT model include notions of pragmatics and contextual effects that go far beyond the surface representation of the dialogue.

Moreover, while SDRT recognizes that domain knowledge and cognitive state influence the construction of discourse logical forms, SDRT does not attempt to capture all elements of composition, representation, and pragmatic consequence within a single logic. Instead, the theory is spread across a number of different logical models, which, while being related to each other, possess a number of very different qualities. In summary these logics are:

- **Information Content Logic:** The principle logic for representing complete discourse interpretations defined in terms of dynamic semantics.
- **Underspecified Logical Form Logic:** A labelled language for representing underspecified semantics for individual clauses and discourses. The labelled language is assumed to be the output of an analysis grammar.
- **Lexicon:** A lexical semantics model used to cover a number of underspecification effects.
- **Cognitive Modelling:** A logic used to represent the attitudinal states of participants and typical of a BDI style modelling of speaker intent. SDRT does not however attempt to directly construct and affect such models; rather SDRT draws upon an assumed model when necessary to aid the construction of the shallow form SDRS structures.

- **World Knowledge:** An account of general knowledge used in the construction of SDRS structures through limited access via the glue logic.
- **Glue Logic:** A processing logic which binds all other logics together in the construction of complete SDRT models.

As indicated the glue logic provides the mechanisms necessary for constructing and augmenting discourse structures as new clauses are entered into the discourse structure. Specifically, this logic has access to each of the other logics to allow the construction of SDRSs to leverage off information from a number of sources. However, unlike the earlier cognitive discourse model by Hobbs et al. [1993], the glue logic's access to other logics is limited. Specifically, while the glue logic can see many of the predicates in the other logics, it cannot make all of the same inferences that are possible in the more powerful logics that it accesses, e.g., the Cognitive Modelling logic.

The discourse structures themselves, expressed in the Information Content Logic, are logic representation structures that are somewhat similar to DRSs. In SDRT however relations do not hold directly between propositions. Instead a label is used to tag contents of a clause as well as bigger linguistic units, e.g., $Relation(\pi_1, \pi_2)$ holds between π_1 and π_2 where they are labels of logical forms. Thus, labels capture occurrences of propositions, and two labels can serve as arguments to more than one discourse relation. It should be noted that labels in this case are part of the logical form of discourse itself, not just a part of a labelling language as is the case for approaches to underspecified semantics. Formally then, a discourse structure is a triple (A,F,LAST) where:

- A: is the set of labels
- LAST: is a label in A which marked the last clause
- F: is a function which assigns each member of A a well formed SDRS formula.

For example, a well formed SDRS might be as follows:

- $A = (\pi_0, \pi_1, \ldots\ldots \pi_7)$
- $F(\pi_1) = K_{\pi 1}$
  $F(\pi_5) = K_{\pi 5}$
  $F(\pi_0) = Elaboration(\pi_1, \pi_6)$
  $F(\pi_6) = Narration(\pi_2, \pi_5) \wedge Elaboration(\pi_2, \pi_7)$
  $F(\pi_7) = Narration(\pi_3, \pi_4)$
- $LAST = \pi_5$

where $K_{\pi i}$ are DRSs. While this is formally the discourse model, the SDRS can also be presented in a DRS box style as shown in Figure 2.9.

One interesting feature of SDRT is its treatment of speech acts and dialogue structure. Namely, SDRT attempts to provide a bridge between the traditional illocutionary analysis of dialogue and a rhetorical analysis. Specifically, the discourse relations included aim to capture not only relations between statements as are typical of rhetorical analysis, but also the relations between different speech act types such as questions and commands. While the authors claim that this leads to a more expressive set of illocutionary forces than is traditionally made use of in speech act theory, in fact only three speech acts are used in the analysis, and instead 'speech act'-like rhetorical relations make up the majority of the illocutionary force representation. Thus, as in classical speech act theory, an utterance can be thought of as having a surface speech act and a contextual speech act. However in SDRT this contextual speech act is captured in terms of the rhetorical relation which holds between it and other utterances.

In moving SDRT from its monologic text basis towards the interpretation of natural conversation, there has been some focus in recent years on the interpretation of non-sentential utterances in SDRT. Among these works, Schlangen's [2003] exploration of "content conveying fragmentary responses" to propositions and questions within an SDRT framework is probably one of the best known efforts. The main contrast between this work and previous fragment interpretation approaches was the application of coherence as a principle source of information in resolving underspecification, and the avoidance of using contextual information directly within the grammar. Importantly from the perspective of SDRT application, Schlangen also developed RUDI, one of the first concrete computational SDRT frameworks. More recently, Luecking et al. [2006] present an extension of SDRT aimed towards handling situated communication including gestural reference and the processing of non-sentential input. But, instead of following in the direction of Schlangen, Luecking et al extend recent proposals by Asher which aim to treat non-sentential input through the direct use of contextual norms.

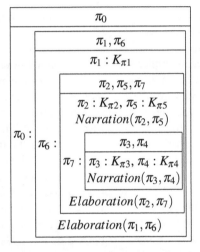

Fig. 2.9 Example of Segmented Discourse Representation Structures

Whereas SDRT is an extremely powerful mechanism for the construction of discourse semantics, it is not without limitation. Namely, many argue that the computational power required for such a deep analysis is still unattainable, and that other syntactic methods might alternatively present a means to produce discourse structure without the need to resort to full interpretation [Egg and Redeker, 2007]. More generally though, one of the most bleak aspects of SDRT application is that the selection of rhetorical relations depends on a particularly deep understanding at a cognitive level which the SDRT mechanisms themselves do not develop nearly completely enough. To overcome such problems a tighter integration of SDRT with more cognitive theories of communication and practical reasoning will be necessary.

Models such as DRT and SDRT provide well formulated semantics-oriented accounts of dialogue structure and the mechanisms which compose these structures

over the course of a dialogue. While these construction mechanisms are appealing
in that they support discourse model based handling of linguistic phenomena such
as reference resolution and elliptical expansion, these discourse semantics models
only account for the surface form of language and do not in themselves offer a decision theory which accounts for why an agent might make a particular move in
dialogue at any given time. To have such a decision process, which is vital to building computational models of dialogue competence, we also need to move beyond
the surface form of language and consider models of the interlocutors's extra linguistic knowledge and how that knowledge affects the language interpretation and
production process. In the following we consider such theories.

2.5.2 Agent-Based Theories

In this section we turn to the class of *agent-based* dialogue theories which, for our
purposes here, we define as those models which are mental state oriented in nature
but which are principally concerned with cognitive models and rational decision
making rather than focusing on low-level linguistic phenomena such as those handled by a discourse semantics. Such competence-oriented theories of dialogue are
crucial to the development of computational models of dialogue processing, and
as such have manifested themselves in a variety of formalisms which vary widely
in both function and formulation. While many dialogue system implementations
necessarily include such an agent-like model of competency, we will hold off on
concrete dialogue system based implementations until the next chapter.

Here we consider the more basic agent-based dialogue processing models that
have had most influence on our understanding of computational models of dialogue processing. Although many of these agent-based models can be traced back
to the seminal works of Cohen and Perrault [1979] and Perrault and Allen [1980] on
speech act based language production and understanding, we will not consider these
models directly here since they have already been introduced in Section 2.3.3, and
moreover, because these models abstracted too far from the difficulties of language
processing. Instead, we skip past these accounts and begin with the first notable
models of discourse which aimed to integrate a model of agency with the realities
of language processing.

2.5.2.1 Grosz & Sidner's Tripartite Model

One of the earliest and still best known approaches to agent-based theories of dialogue competence is the Tripartite Model of Grosz and Sidner [1986]. This model
grew out of criticisms of the early agent-based models of Pernault and Allen and
argued that rather than capturing an interlocutor's mental state purely in terms of an
intentional logic, a diverse set of structures were required to capture the various aspects of our communicative state. Specifically three separate structural components

were proposed, i.e., intentional, attentional, and linguistic elements. In summary, the intentional component records the tasks and subtasks to be performed by an agent in terms of beliefs and intentional states. The attentional component, on the other hand, captures the focus of attention in a dialogue thorough the use of a stack of focus spaces that are pushed and popped as discourse progresses. Finally, the linguistic component decomposes a given discourse into a number of discourse segments, which can broadly be thought of as the surface semantics of the interaction.

While the early work on the Tripartite model focused on the structural representation rather than the interpretation or production processes, subsequent developments such as Grosz & Sidner's SharedPlans formalisation, Sidner's Artificial Discourse Language (ADL), and Lochbaum's Discourse Interpretation Algorithm, pushed the theory towards a more complete communication account. Grosz and Sidner's [1990] SharedPlans provided a collaborative planning model which allowed two cooperating agents to coordinate their activities without the need for a notion of irreducible joint intention. Sidner's [1994] ADL, like other *agent communication languages*, defined a number of communicative acts expressed in terms of the effects of individual messages on the interlocutors' mental states – specifically in terms of *Beliefs*, *Mutual Beliefs*, and *Intentions*. Lochbaum's [1998] algorithm, which modelled the effects of discourse events on the tripartite dialogue model e.g., a discourse event completes the current purpose and, thus, eventually pops the focus stack, attempted to predict how interlocutors follow conversation flow based on their understanding of each other's intentions and beliefs.

The Tripartite model gave rise to one of the best known pronoun resolution algorithms, i.e., Grosz et al.'s [1995] Centering Algorithm, and was applied somewhat successfully within the domain of collaborative discourse agents (i.e., Collagen [Rich et al., 2001]). However, the approach's applicability outside of domains involving discourse regarding complex plans is unclear. For one thing, in practice plan recognition is simply not often needed in practical dialogue tasks, since the plan is signalled explicitly in many cases. Furthermore, while the Centering algorithm did grow out of the Tripartate model, Grosz & Sidner's work paid little attention to the realities of spoken communication which requires a significant focus on grounding and contextualization.

2.5.2.2 Discourse Obligations & Conversational Acts

Following trends in the artificial intelligence community towards theories of cooperative action, e.g., Levesque et al. [1990], the early 1990s also saw a re-development of agent-like models by Poesio & Traum that moved beyond basic goal inference models to explain instead, from a joint process perspective, how the decisions we make in dialogue are related on the one hand to our social obligations in dialogue, and on the other hand to the realities of imperfect communication. With respect to social obligations the core argument made by Traum for example was that the hearer of a message should acknowledge their partner's utterances explicitly in order to make clear that the grounding and contextualization process is successful, but on the other hand the language grounding process itself must acknowledge that any contribution in a real dialogue may exist in a number of different states before being accepted by both parties [Traum and Allen, 1994, Traum and Hinkelman, 1992]. Such social obligations and the realities of imperfect communication therefore give rise to processes of communication management which supplement the core task-oriented influences on our dialogue processes.

Traum subsequently presented a computational theory of grounding that reflects the fact that illocutionary acts such as the assertion of a fact are not often established immediately in the common ground, and need instead to be explicitly discussed before becoming mutually agreed upon [Traum, 1994]. Thus, instead of an agent-based model operating solely in terms of ground beliefs and task-oriented speech acts, Traum argued that the dialogue model must also operate in terms of: (a) units which capture an utterance before it has been fully understood and agreed upon (*Discourse Units* in his terminology); and (b) dialogue moves which serve not to directly progress the core task, but instead serve to facilitate the grounding of individual dialogue contributions.

While an improvement on classical agent theories, Traum's models remained at a significant distance from the realities of the surface of spoken language and the models of linguistic semantics which have been designed to capture it. Such limitations were later overcome in part through an extension of a conventional discourse semantics theory towards a conversational scoreboard which provides means to model locutionary, illocutionary, and perlocutionary acts in a grounding process [Poesio and Traum, 1997]. Specifically, Poesio and Traum applied DRT as a basis for their discourse theory but augmented it to support pragmatic information such as the assertion of speech acts (of a number of different varieties and granularities). This model was further extended to axiomatize many of the DAMSL dialogue acts [Poesio and Traum, 1998]. It should be noted though, that while such an axiomatization of DAMSL acts seems similar to the earlier non-linguistic work of Cohen and Levesque [1990], the inclusion of social obligation concepts, the use of the DRT framework, and the explicit inclusion of the non-instantaneous grounding model clearly sets Poesio & Traum's work apart from the earlier idealised proposals.

2.5.2.3 Ginzburg's Dialogue Game Board & Derived Models

Coming from a linguistic rather than computational linguistic perspective, Ginzburg [1996] essentially arrived at a similar result to Poesio and Traum [1998], but with the objective of interpreting sub-sentential utterances in context rather than providing specific accounts of the grounding process. Ginzburg's argument, motivated by the communicative power of sub-sentential utterances in context, was simply that utterances rather than complete sentences are a principle token of context structuring, and therefore "The notion of context required for explicating dialogue must be intrinsically richer than that required for written text" [Ginzburg, 1996].

Ginzburg's concrete model, which assumed a Stalnakerian view of mental state organization [Stalnaker, 1978], proposed a structured organization of context in dialogue. Rather than assuming a global view of context, whereby all information is of equal strength in determining the preconditions for the next dialogue move, this structured model keeps track of the dialogue state in a terms of salient information types. For example, rather than assuming the mental state to be a *bag of predicates*, Ginzburg proposes the following concrete structure as a discourse model for information-query style dialogue:

- FACTS: The Set of Currently Accepted Facts
- LATEST-MOVE: The representation of the syntax and semantics of the latest move made in dialogue – thus key to capturing a notion of locality in dialogue organization.
- QUD (Questions Under Discussion): A partially ordered stack that specifies issues which are currently subject to discussion.

Such a structure, which Ginzburg refers to as a Dialogue Game Board (DGB), provides a locally oriented mental state view on the dialogue at any given time. Ginzburg's model however went beyond suggesting a structured mental state, and was instead more concerned with developing an algorithmic style *Utterance Incorporation Protocol* that captures the semantics of how a given utterance should be processed with respect to the DGB at any given time. Ginzburg's complete theory, termed the KOS model, thus placed emphasis on the local organization of dialogue, and argued that in interpreting and planning dialogue, what is ultimately most important is not the sequence of moves performed up to a given point, but rather the information that has been collected on the DGB.

Notably, in developing the KOS model, Ginzburg questioned the integrity of a strict semantics/pragmatics distinction, and pointed out that we often cannot claim to understand the meaning of an utterance without having identified the objects referred to in the utterance – and that this process in turn requires full pragmatic instantiation. Thus, while the DGB contains conceptual slots which are traditionally viewed as highly pragmatic units, the contents of the slots can include directly encoded meaning from the language surface, as well as content inferred through other extra-semantic means. Such an approach contrasts in part with the view taken by the Discourse Semantics community where models such as DRT were often treated as a semantics only stratum [Fernndez and Purver, 2004]. However, recent developments do point to a possible unification of the discourse semantics and dialogue game board theories [Maudet et al., 2004, Muller et al., 2006].

Implementations of the KOS model (described in the next chapter), as well as the theory of Poesio & Traum, focus on a structured mental state and express dialogue progress potential not in terms of arcs between nodes on a state-diagram or flowchart, but in terms of axioms which licence particular moves under various mental state conditions. While such rule-based accounts meet the goals of moving away from overly simplistic state-based dialogue models, axiomatic descriptions of move potential can be cumbersome. Before bringing this section to a close, in the following we briefly consider a planning approach which provides a notably clearer view on dialogue move licensing.

2.5.2.4 Systemic Move Planning

As indicated earlier in this chapter, within the Systemic Functional Linguistics school, Halliday [1984] argued against organizing dialogue in terms of a set of pre-determined states, and instead favoured viewing dialogue as a locally organized phenomenon wherein an interlocutor chooses their next dialogue move on the basis of cognitive state (which may include dialogue structure aspects) rather than some pre-determined set of non-cognitive dialogue states. Such a model, which he termed a *Generalised Dialogue Potential*, therefore has much in common with the mental-state oriented views of dialogue competence. However, unlike mainstream dialogue theories such as Ginzburg's KOS model or Poesio & Traum's theory, Halliday saw the licensing of moves at any point in the dialogue in terms of a single systemic network decision process network which is traversed by the agent at every step in the dialogue to determine the next move to be made. Concretely, Halliday set out a relationship between social context and functions which a dialogue move was to achieve at any given point in the dialogue. Halliday's proposal was however embryonic rather than complete, and suffered in that only a subset of context to speech function relationships were established. Moreover, no context update effects were specified for individual speech functions – thus preventing the model from being applied in a true dynamic sense.

With computational goals in mind, O'Donnell considerably extended and developed Halliday's view by proposing concrete models of dynamic discourse context and by significantly enhancing the systemic networks offered by Halliday. The principles of this model were re-worked extensively by O'Donnell over almost a decade (cf. O'Donnell [1999, 1994, 1992, 1990]). However, for the purposes of discussion here we take the networks and approach presented in O'Donnell [1992] since that work arguably amounts to the most complete treatment provided. The dynamic context model that was queried and potentially updated at each point in the dialogue is thus similar in inception to Ginzburg's Dialogue Game Board in that it included *stacks* and *queues* of context state which could be *pushed* and *popped* to handle exchange embedding. However the key feature of the model was that at each point in the dialogue, a traversal through a speech function network was made to decide upon the speech function features of a move to be made at that time. Speech function feature selection was determined by the context model at that time, and subsequently resulted in updates of the context model – thus making the approach a truly dynamic account.

In the context of complete dialogue theories, O'Donnell's speech function network does however suffer one major limitation in that O'Donnell was primarily concerned with the language production process, and thus paid little attention to the realities of dialogue move grounding or contextualization. While some work has been conducted by O'Donnell on the use of systemic network based grammar models for language analysis, it is not yet clear if the systemic network based move planning approach can be *reversed* to account for move interpretation in a tractable way. Nevertheless, since move planning and interpretation do not necessarily make use of the same processing mechanisms even in Ginzburg and Traum's models,

O'Donnell's structuring methodology may yet offer considerable advantage in organizing the planning of dialogue moves in cognitive dialogue theories.

2.6 Discussion

Over the course of this chapter we have seen a wide variety of modelling techniques which can be applied to dialogue. These techniques analyse dialogue from the perspectives of referenced content, actions, structure, and the processes we apply in interpreting and producing language. But despite the breadth of analysis possible with these theories, the models presented here remain only a fraction of those available. A more comprehensive analysis would necessarily include all theoretical perspective on the pragmatics of language (see Huang [2006] for an approachable overview), and numerous other concrete dialogue theories such as Carberry's [1988] Plan Recognition Model, Chu-Carroll and Carberry's [1998] models of collaborative negotiation, or Mann's [2002] Dialogue Macrogame Theory, all of which can offer helpful insights into dialogue structure and processing. Rather than attempting to cover dialogue from all these perspectives, this chapter has aimed to introduce those theories which have had the most significant impact on dialogue processing.

Nevertheless, it can be difficult at first to see the relevance of sometimes abstract dialogue theories to situated language processing. However, such theories of dialogue processing are the cornerstones of practical computational dialogue. For example, as we will see in the next chapter, state diagram and structure-based dialogue models have been extensively applied to patterning interaction in both simplistic and complex dialogue systems; while, on the other hand, semantic-oriented theories of content and context are finding increasing application where non-trivial interactions are desired.

While more advanced theories such as discourse semantics and agentive dialogue models have aimed to address the role of linguistic context in language production and interpretation, the incorporation of extra-linguistic context remains in its infancy. The most promising developments in this direction undoubtedly come from SDRT which has, since its inception, acknowledged the importance of a clear interface to extra-linguistic context in the resolution of coherence relations. However, the context models assumed by SDRT are on one hand too powerful for practical dialogue application, and are on the other hand far too simplistic for the realities of context modelling for situated language applications. Moreover, there are no direct explanations in such models as to how extra-linguistic context should be organized or applied to surface language in a systematic yet practical way. And since, as we saw in the last chapter, the processing of spatial language in the situated domain is highly subject to extra-linguistic context, it is clear that we cannot hope to directly apply existing dialogue theories to the task of spatial language interpretation and production without significant extensions in this regard.

Although not perfect, the dialogue models which we have considered here do offer a basis within which purely linguistic principles can be applied to both describe dialogue as a structure and as a process which we engage in. Arguably, the theories considered are not mutually exclusive, but should fit within a complete theory of dialogue competence which any practical dialogue processing must build upon. However, many of the models which we have considered are either too abstract in nature, or based on too strong a formalism, to be applied directly in practical systems. Rather, these theories must be rationalised for the realities of computational implementations. Such rationalizations come under the scope of *dialogue management theory* to which we turn in the next chapter.

2.7 Summary

In this chapter we have reviewed the state of dialogue structure analysis within the linguistics and computational linguistics communities. The aim of this analysis has been to provide a firm foundation for dialogue system design which meets the particular requirements of interaction set out earlier. We saw that theories of dialogue modelling make use of two core modelling frameworks: first, formal semantics theories that model the states, events, objects, and qualities which we describe in language; and second, speech act theory which characterizes dialogue as sequences of actions performed by interlocutors. We also saw that dialogue analysis theories can vary widely in the stance they take on how to model dialogue. On one hand structural accounts such as rhetorical analysis, exchange structure, and network-like descriptions of dialogue provide a third-party perspective on dialogue patterns and regularities. On the other hand more cognitive approaches such as discourse semantics, agent-based, and information state accounts, postulate models of dialogue competence within an interlocutor. This difference in focus is based not only on perspective, but also leads to contrasting views as to whether dialogue itself can be viewed as a locally or more globally organized phenomenon.

Chapter 3
Dialogue Management

Abstract Having established the theoretical basis of dialogue modelling, this chapter analyses the state of art in spoken dialogue systems to determine how well existing systems handle the context-dependent features of situated language. We begin in Section 3.1 with a brief review of the prototypical design of spoken dialogue systems. Section 3.2 then provides a more concrete introduction to dialogue management by establishing a general set of requirements for a *dialogue manager*. In Section 3.3, we review the most common types of *dialogue management model* and relate these back to the theories of dialogue organization discussed in the last chapter. Section 3.4 then moves beyond these idealized models by presenting a number of dialogue system case studies. We bring the chapter to a close by introducing a number of requirements for situated dialogue systems.

3.1 Spoken Dialogue Systems

McTear [2002, p. 2] defines Spoken Dialogue Systems as "computer systems that use spoken language to interact with users to accomplish a task". Such systems typically make use of a wide range of resources and processes of heterogeneous types to move fluidly from spoken word to string, symbol, concept and back again. A characteristic architectural feature of these systems is in turn the separation of the process which manages the dialogue interaction, i.e., the dialogue manager, from, on one hand, the processes which map between surface language and semantics representations, and on the other hand, the domain knowledge.

The specific processing components which comprise a non-trivial dialogue system can vary considerably depending on many factors. While most variations are due to variation in processing component capabilities, one factor which is of particular consequence to overall system design is the choice of communication mode. While the spoken interface has been the most prominent mode of communication, in the last decade there has been considerable interest in the development of multimodal dialogue systems which augment or replace the speech interface with graph-

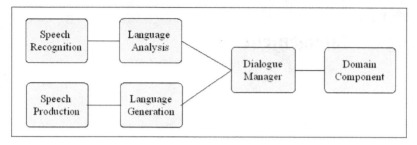

Fig. 3.1 Prototypical Spoken Dialogue System architecture.

ical and tactile modalities. While multiple modes of interaction will of course be of interest to situated systems, the investigation of multi-modal interaction is a considerable research topic in of itself, and hence, with the exception of a discussion in Chapter 9, will not be addressed here.

With respect to variation of components in *spoken* dialogue systems, we can however abstract over low-level choices and define a prototypical *spoken* dialogue system design, thus allowing a discussion of general ideas without committing to one specific architectural composition. Figure 3.1 presents such a prototypical dialogue system architecture consisting of the following processing components:

- **Speech Recognition:** Translates audible speech into raw textual surface form or n-best lists of speech hypotheses using statistical methods such as Hidden Markov Models [Rabiner, 1989].
- **Speech Analysis:** Constructs logical forms, i.e., models of surface form information including grammatical and semantic features, from surface language through the use of statistical methods, and/or formal grammars such as Head-Driven Phrase Structural Grammars (HPSG) [Pollard and Sag, 1994], Tree Adjoining Grammars (TAG) [Joshi, 1987], or Combinatorial Categorial Grammars (CCG) [Steedman, 2000].
- **Domain Component:** The information or functionality over which a dialogue is to be conducted. This can range in nature from a database application to a robot control agent.
- **Dialogue Management:** Updates and maintains dialogue state following user utterances, and select dialogue actions to be performed by the system.
- **Language Generation:** Determines *how* to express particular dialogue goals which have been provided by the Dialogue Manager as logical forms. Systemic Functional Grammar [Halliday and Matthiessen, 2004] has been applied widely in strategic language generation. Other formalisms including TAG and CCG are also widely applied.
- **Speech Synthesis:** Produce audible speech based on the output of Language Generation.

Within specific dialogue systems, some of the components above are frequently merged, split, or even omitted. But for our interests here, it is the nature of Dialogue

Management that is of most relevance to us, and it is this component which has the most variation in assumptions of design and responsibilities. Unlike language analysis or generation components, dialogue managers are given numerous varied responsibilities including language contextualization, system dialogue move selection, and state maintenance. Each of these responsibilities in turn unfold into sets of other duties; for example, state maintenance includes the upkeep of discourse semantics, pragmatic dialogue state, and/or literal dialogue history depending on the design choices taken. For this reason there is considerable variation in how dialogue management duties are embodied within dialogue systems. Dialogue managers are often replaced by, or operated in conjunction with, explicit tactical generation, discourse maintenance, and/or contextualization components. Moreover, there is a growing trend towards the use of discourse and extra-linguistic context information as a solution to problems in speech recognition and language analysis; thus, even at the lower level, the boundaries between dialogue system components are increasingly blurred.

But, regardless of the blurring of architectural lines, at the heart of the Spoken Dialogue System is a Dialogue Manager, charged with the maintenance of dialogue state and the control of the dialogue processing model. The complexity and effectiveness of these responsibilities varies principally in accordance with the *dialogue management model* adopted. The simplest dialogue management models, i.e., *State Based*, provide an effective but minimalist approach by encoding discourse data and control within state transition networks. *Frame Based* models improve on this by abstracting control away from frame-held data, but are yet constrained to very simple domains. The most complex models, i.e., *Information State Based* or more generally *Agent Based* models, have the potential to adequately meet the requirements of dialogue management, but the necessary complexity of these models, along with their reliance on other research activities such as formal semantics and discourse structure has made their application difficult. Before detailing these different models and their realizations, let us first establish an understanding of the meaning and responsibilities of dialogue management.

3.2 Defining Dialogue Management

For a definition of dialogue management, we turn to the concrete proposal made by Traum and Larsson [2003] who define dialogue management as the following functions within a dialogue system:

1. Updating the dialogue context on the basis of interpreted communication.
2. Interfacing with task/domain processing.
3. Providing context-dependent expectations.
4. Deciding on what context to express, and when to express it.

While these provide a good first step towards a definition of dialogue manager responsibilities, this list of qualities is quite dependent on the particular configuration of components that Traum and Larsson [2003] used in their dialogue systems. For example, the third responsibility of providing context-dependent expectations is highly dependent on the particular configuration of analysis and dialogue manager components used. Furthermore, arguably it is the dialogue context itself which provides the expectations for the analysis and contextualisation system; thus this provision of expectations is implicitly a result of updating dialogue context.

Dialogue management is difficult to define since most definitions are highly dependent on one hand on the particular configuration of components which make up a dialogue system, while on the other hand on the granularity to which we explain the dialogue manager's responsibilities. Rather than restating many other definitions of dialogue management, we instead propose the following list of dialogue manager responsibilities at this point:

- **Contextualizing User Actions:** While semantic analysis provides a linguistic semantic representation of users' utterances, further contextualization is necessary to account for the fact that people do not generally give well-specified direct single clause based accounts of their intent. Explicit contextualization is a process which is frequently omitted in many dialogue system implementations, but yet is one that is essential to the construction of dialogue systems with a level of conversational intelligence necessary for acceptable use in wide-coverage domains. Furthermore, as stated above, it is a general desire that application components be disjoint from linguistic semantic representations. Therefore, a process of contextualisation or enrichment is required to: (a) resolve meaning against discourse and application context; and (b) transform linguistic semantics to application-dependent representations. Part of the contextualization process is also the identification of speech acts.
- **Updating Dialogue State after User Acts:** Updating and maintaining a consistent dialogue state which represents the actual discourse to date is arguably the single most critical responsibility of a dialogue manager, and the one which should at all times be considered in dialogue system design and research. As we saw in the last chapter, many different approaches to dialogue state description exist and these various approaches often manifested themselves directly in dialogue manager designs. For example, while some dialogue managers can assume a discourse semantics based model of contributed state, others may assume a more pragmatics-oriented view of dialogue state such as provided by the dialogue game board style models, while yet more dialogue management designs include only a *control state* which is used by the dialogue manager to decide what action may be performed next without reference to semantic representations of specific utterances.
- **Application Synchronization:** With the exception of trivial state-based dialogue systems, the dialogue manager will act as a broker between the user and an application model. Thus the configuration of the interface between the dialogue manager and that domain component can have far reaching consequences for the scalability of a dialogue application and the level of natural interaction which can be provided. For example, if a database query application can only answer questions without providing clues as to what parameters might be var-

ied so as to provide either more specific or alternatively more general results, then the dialogue manager cannot hope to provide such alternatives to a user, and thus the relative intelligence of the system will be limited. A dialogue manager is thus responsible for providing the correct form of interfaces to domain applications and for maintaining a connection with those applications.

- **Planning System Acts:** Dialogue planning is the processing of choosing appropriate contributions to be made by the system at any point in the dialogue. In a state-based dialogue manager, dialogue planning is essentially an implicit process that is strictly dictated by dialogue model transitions. In more sophisticated dialogue systems, dialogue planning can become a far more complex task, with dialogue management taking responsibility for deciding when dialogue acts are to be made, and application-specific logic deciding on the relevant content. Examples of such systems include dialogue-based information systems where a dedicated content manager selects relevant information to present to a user dependent on a number of contextual factors.
- **Updating Dialogue State after Application Acts:** Just as dialogue state must be updated with a newly recognized user dialogue act, so too must the dialogue state be updated on the basis of application acts.

Whether these responsibilities are undertaken by the dialogue manager component as such, or are instead undertaken by alternative explicit components will vary considerably from implementation to implementation. In fact, as we will see later, as dialogue systems become more complex and sophisticated, the responsibilities of a dedicated dialogue manager actually decrease to the point that the dialogue manager ceases to exist as a distinct component. This happens as different aspects of dialogue management functionality are instead undertaken by dedicated processing units.

With the basic requirements of dialogue management in hand, in the next section we briefly review the most common dialogue management model types, and in so doing explore where the various requirements of dialogue management are spread across different processing components according to the management models.

3.3 Dialogue Management Models

We define a *Dialogue Management Model* is an architectural recipe or design pattern used to construct dialogue managers. Broadly speaking, there are five accepted dialogue manager models: *Finite-State Based*, *Frame Based*, *Agent Based*, *Information State Based*, and *Probabilistic* models.

3.3.1 Finite-State Based Dialogue Management

Finite-state based dialogue management models are the most widely deployed models in commercial dialogue systems. Most finite-state based systems guide a user through an interaction that has been scripted in advance as a graph of dialogue states, with arcs between states describing acceptable input phrases. While an acceptable input could be specified in terms of a semantic representation, most finite-state based implementations only specify input in terms of single words or numbers, thus improving the accuracy of speech recognition. Recursive Transition Networks (RTNs) provide a more flexible dialogue control strategy, where allowed dialogues and sub-dialogue may be recursively embedded. However, regardless of the range of input allowed in dialogue arcs, or whether recursive arcs are used rather than finite models, state-based dialogue management models follow a well scripted structure that makes it difficult to provide flexibility in interaction. Another characteristic feature of state-based dialogue managers is that the dialogue planning, or tactical generation, process is encapsulated directly within the dialogue manager through the state transition descriptions. Some level of information selection may be provided through additional macro expansion, but in general such an additional process is more akin to tactical rather than strategic generation. For practical dialogue systems, it is perhaps more interesting to separate out the information that is required for a particular task from the exact pattern of interaction, and this is the strategy that is pursued in frame-based dialogue management models.

3.3.2 Frame-Based Dialogue Management

Frame-based, task-based, and template-based dialogue management models are widely deployed within commercial and research systems, and are particularly suited to well defined task domains where a user must supply a number of pieces of information to a system. The frame-based approach decomposes an interaction specification into two orthogonal components: (a) a specification of the information required by the system from the user; and (b) a specification of the generic dialogue strategies that control the overall dialogic interaction. The application-dependent specification of required information is typically defined in a frame-like structure which details a number of slots that indicate the attributes and perhaps acceptable values required by the system. Frames can be either a list of required values, or may be specified in a hierarchical manner to provide more structuring and coherence to an interaction. The application-independent dialogue strategy that operates over the frame specification is often encoded as a simple conversational game or pattern-based dialogue model, along with the language analysis technology to extract relevant information from user's utterances.

To illustrate the structure and use of such frames, let us consider Ward and Pellom's [1999] CU Communicator system. The Communicator's dialogue manager operates over a set of frame structures, each of which captures in a single struc-

```
Form: car
   Name: company_name
   Parse: [Rental_Company]
   SQL: ``car_comp_name like '%!%' ''
   English prompt: ``What car company would you like?''

   Name: car_type
   Parse: [Car_Class]
   SQL: ``car_type_code in (select rental_code from
           car_rental_types where rental_type like '%!%') ''
   English prompt: ``What type of car would you like?''
```

Fig. 3.2 CU Communicator frame structure.

ture all information which must be ascertained from the user in order to complete a meaningful database query or update. Each slot in turn is divided into a series of "pointers" which specify different types of information necessary to complete the filling out of the slot. Figure 3.2 depicts two such slots for a single frame. As can be seen, pointers include those for the name of the field, the value extracted for the field, parser extracted token strings that map to the field, and pointers to natural or artificial language strings that can be used to express the field or make queries to a database.

While the operation of frame-based dialogue managers can vary somewhat, typical properties include: (a) the dialogue strategy is principally system-initiative driven in that the system poses a series of questions to the user to elicit information required in a frame; (b) the ordering of issue handling is dependent on the form's organization; and (c) a limited amount of user initiative is often allowed in that the system will attempt to fill out multiple slots from a single utterance if a user over-answers in response to a question.

While a clear improvement over state-based dialogue systems in that there is a separation between application data and dialogue processing logic, the frame-based dialogue management models are not well suited to all dialogue processing scenarios. Namely, frame-based dialogue management models have been developed in response to the needs of interfacing with database applications, and as such certain processing assumptions are biased towards such scenarios at the expense of any application which can behave either more pro-actively, or which has dynamic context models which affect valid interpretations of user contributions.

3.3.3 Agent-Based Dialogue Management

Unlike frame-based models, agent-based dialogue management models have been specifically formulated from the perspective of applications which are assumed to take a more pro-active view on interaction. Also, unlike frame-based models which form a relatively small but well defined set of approaches, agent-based dialogue management models are a broad collection of dialogue management approaches, all of which are based upon more complete theories of discourse competence that are characterized by explicit mental state representations which include notions such as *beliefs*, *plans*, *intentions*, and *goals*. More specifically, agent-based dialogue management models aim to provide cognitively inspired dialogue processes through the following techniques:

- Modelling and reasoning about the mental state of dialogue participants in terms of folk psychological concepts
- Providing mixed-initiative in dialogue control
- Providing non-trivial natural language analysis and understanding techniques
- Outputting un-canned natural language in conversation

Agent-based dialogue management models are directly related to the agent-based theories of dialogue organization discussed in the last chapter. Thus, early agent-based models owe much to the language understanding and generation work of Perrault and Allen [1980] in their development of relatively complete cognitive models to investigate the interpretation of indirect speech acts in language. Moreover, one class of agent-based models which has received close attention are the task- and plan-based models which not only rely on notions of folk psychology and attitudinal state, but also on the assumption of explicit plan structures that on one hand influence the actions of an agent, and on the other hand must be *recognized* by an interlocutor as part of the dialogue process. Carberry's [1988] Plan Recognition Model is probably one of the best known plan-based accounts.

In comparison to frame-based and state-based accounts, the early agent-based models aimed to provide cognitively plausible views on the dialogue process which integrate well with notions of situated reasoning such as planning and deliberation. But this cognitively plausible perspective could not be realised to the point of useful dialogue systems because, on one hand, far too many assumptions are made about the mechanisms of language understanding, while on the other hand the processes which are proposed are either intractable or make unrealistic assumptions about the existence of third party processing capabilities. However, in hindsight, probably the biggest limitation of the early agent- and plan-based dialogue management approaches was that they attempted to achieve too much within a single computational metaphor. The dialogue agent was in fact the only deliberative processing entity present, and was responsible for both dialogue processing tasks and application modelling functionality.

3.3.4 Information State Update-Based Dialogue Management

Derived from the more theory-oriented work on Discourse Semantics and Utterance Interpretation from Poesio and Traum [1997] and Ginzburg [1996] respectively, Information State Update (ISU) based dialogue management models [Larsson and Traum, 2000, Traum and Larsson, 2003] advocate dialogue manager construction based around discourse objects (e.g., questions, beliefs) and rules which encode update-relationships between these objects. In practice, these semantic objects, their inter-relationships, and update rules are in turn computationally modelled using declarative programming techniques which provide a middle-ground between tractability and the expressivity of early agent-based dialogue management models. Thus, ISU-based systems may be viewed as practical instantiations of agent-based models, where, on one hand, the broad notions of beliefs, actions, and plans, are replaced with more precise semantic types and inter-relationships, and, on the other hand, the complexities of language grounding are factored into the processing theory.

With respect to the ISU theory, Larsson & Traum define the Information State of a dialogue as "the information necessary to distinguish it from other dialogues, representing the cumulative additions from previous actions in the dialogue and motivating future actions" [Larsson and Traum, 2000, p. 1]. An information state-based theory of dialogue is then said to consist of:

- A description of the informational components
- A formal representation of those components
- A set of dialogue moves
- A set up update rules
- An update strategy for applying those update rules

A formal representation of informational components is typically specified by a record-like set of value-argument pairs. Dialogue moves are similar to Searlean speech acts, but no direct connection is assumed. Update rules are *a coherent bundle of change* that operate on the information state under certain conditions. These rules are in turn specified by an update rule application strategy.

As indicated, there is a strong relationship between information state-based dialogue management models and those *cognitive* dialogue modelling theories we considered in Section 2.5. In particular, ISU techniques owe much to models such as Poesio and Traum's [1997] discourse semantics model, and Ginzburg's [1996] Dialogue Game Board (DGB) – both of which draw much from earlier dialogue accounts such as the *Conversational Scoreboard* [Lewis, 1979, Stalnaker, 1978]. It should be emphasised however, that within the ISU approach the *Information State* is assumed to only consist of the dialogue state as such, and not include application state, which is assumed to be contained within a separate component, or as distinct representational structures from the dialogue state. It is this difference along with the explicit handling of grounding issues which sets the ISU approach apart from the early agent-based models.

In terms of the relationship between the Information State approach and dialogue structure state approach (i.e., the classical state-based approaches), Larsson and Traum [2000] make the point that it would be very difficult to formulate an information state approach as a dialogue state model; while, conversely, it would be relatively easy to encode either finite state or recursive transition networks as an

Information State model. Specifically, the dialogue state can be encoded as a register for finite state networks or a stack of values for an RTN. The dialogue moves would then be encoded as update rules. This ability to provide implementations of many different dialogue models is a useful point of the ISU paradigm, and is similar to Green's [1986] argument that, with respect to User Interface Construction, transition networks and context free grammars are subsumed by event models; thus meaning it should only be necessarily to provide an event system interpretation given that transition networks and context free grammars are transformable to event-based specifications.

While declarative specifications of dialogue systems are generally considered to be more advantageous than procedural implementations, experience with information state implementations – and declarative systems in general – has shown that rule-based system are often opaque and scale poorly in design complexity. Specifically, the use of a potentially large number of rules to define information state transitions can lead to systems that are difficult to design and debug, with unforeseen logic errors which are tedious to trace, and which may lead to potentially serious side-effects. Moreover, it should be noted that within his formalization of the DGB and subsequent KOS model, Ginzburg [1996] did not call for the application of an update rule-based account for the rules of his dialogue game; he instead used a clear procedural description of the interpretation process.

In the last ten years, a number of dialogue systems based more or less around the information state-based dialogue management model have been developed. The original systems which were used to develop the management model include GoDiS [Bohlin et al., 1999, Traum and Larsson, 2003], EDIS [Matheson et al., 2000], and IBiS [Larsson, 2002], each of which were developed using the TrinkdiKit Information State development toolkit [Traum and Larsson, 2003]. While these were idealized models, more pragmatic developments based roughly around the ISU approach were also pursued. Lewin [2000] for example developed an ISU-based Conversational Game Theory which both described and made use of explicit structured dialogue models, and maintained parallel game parses to facilitate backtracking of dialogue models depending on the outcome of alternative utterance interpretations. Further rationalising of the ISU concept towards practical dialogue systems came through the WITAS [Lemon and Gruenstein, 2004, Lemon et al., 2002a] and SCoT-DC dialogue systems [Schultz et al., 2003, Pon-Barry et al., 2004a,c,b]. Other extensions for example include Kreutel and Matheson's [2003] distinction between context-dependent and context-independent operations on the information state structure.

3.3.5 Probabilistic Dialogue Management

In applied dialogue systems, language interpretation errors can be introduced anywhere from the speech perception to speech contextualization processes; and these errors, sometimes compounding each other, can cause serious difficulties to dialogue manager performance. Arguing that conventional dialogue management models such as frame-based and ISU-based theories are simply too logical and brittle to handle the realities of errors in language interpretation, there has been considerable research aimed at augmenting existing dialogue management theories with probabilistic machinery. These methods generally fall into one of two categories: (a) the application of probabilistic methods in concept interpretation; and (b) the application of probabilistic methods to dialogue planning.

Rejecting deep understanding and inferential dialogue act interpretation, the application of probabilistic methods, and Bayesian networks in particular, to the recognition of dialogue contributions has received considerable attention. Keizer et al. [2002] for example applied Bayesian networks to dialogue act classification based on a number of surface and contextual features. Lemon et al. [2002b] on the other hand take the Bayesian approach further by applying it to anaphora resolution in dialogue. While Porzel and Gurevych [2002] have proposed the application of Bayesian networks to the deep contextualization of user utterances. Across all of these approaches probabilistic methods are used to improve the reliability of selecting a valid interpretation of a contribution, but yet, once selected, a given interpretation can be factored directly into a classical frame-based or ISU-based dialogue approach.

The second body of work in probabilistic dialogue management takes a very different perspective on the application of statistical methods. In this second group the state of the dialogue is assumed to be inherently a probabilistic structure; thus requiring the dialogue planning process itself move from a logical production system to a stochastic system. One of the earliest such approaches comes from Pulman [1996], who developed an extension of classic *conversational game theory* to enhance its game-like qualities in terms of probabilities and decision theory between individual moves. More recent research in probabilistic dialogue management has stemmed from work on speech recognition and action recognition in robotics. Here, Partially Observable Markov Decision Processes (POMDPs) have been used to provide a universal means to describe probable dialogue states, which are in turn used to parameterise a dialogue planning process which aims to select at any given time the dialogue move to perform which maximises pay-off given the spread of possible dialogue states [Williams and Young, 2007].

Probabilistic dialogue management models essentially enhance the robustness of existing dialogue management models, but this enhancement comes at a cost and is not yet a perfect solution to dialogue management issues. While probabilistic contextualization undoubtedly enhances interpretation reliability, these processes still rely on highly structured content and context theories which in turn have proven to be equally difficult to formalise in either probabilistic or non-probabilistic frameworks. With respect to probabilistic dialogue planning, the basic POMDP approach has been found to scale badly in that larger systems suffer a state space explosion which makes it prohibitively difficult to chose the next action to perform in dialogue [Bui et al., 2007]. Moreover, while traditional dialogue management approaches may require considerable efforts in developing logical models, the cost of annotating and building the data sets necessary for statistical methods are expensive and not generally re-usable unlike their counterparts in speech recognition or speech synthesis work. Nevertheless, probabilistic methods remain a promising enhancement of existing dialogue management theories.

3.3.6 Discussion

Regardless of the particular approach taken, dialogue management models serve to perform the same role within a spoken dialogue system, and thus have a set of shared core responsibilities. These responsibilities include: the maintenance of a dialogue state representation, the updating of that state based on what user or system moves are performed, the interpretation of user dialogue moves against that dialogue state, and the decision process which chooses the most appropriate system move to make at any given time. As we have seen however, the way in which these responsibilities are handled varies widely across the dialogue management model approaches. This variation results in stark contrasts between the classic state-based accounts within which all responsibilities reduce to an explicit state diagram with allowed inter-state transitions, and the probabilistic and ISU-based accounts which offer rich dialogue state models and sophisticated dialogue interpretation and planning methodologies.

Despite the contrasts between these idealised dialogue management models, the lines of differentiation begin to break down as these dialogue models evolve and are put to work in practical spoken dialogue system. In particular, frame-based and the most common ISU-based dialogue management models serve to perform a common dialogue type where information must be collected from the user to perform an update or query of the domain application. In both cases the distinction between, on one hand, the specification of what information is required from the user, and on the other hand the general dialogue processes which govern how this information is to be obtained, are paramount modelling concerns that are handled similarly in both management types. Even probabilistic accounts have at their core the same modelling constraints, but simply extend the underlying model with statistical elements to improve robustness. And yet, while agent-based dialogue management models would appear to be rather different to the frame-based, ISU-based, and probabilistic dialogue management models, the agent-based models include types of functionality, such as indirect speech act interpretation, which will have to be incorporated by the other three accounts before they are applicable beyond simple knowledge-base query domains.

Moreover, in practice, many differences in how spoken dialogue system components operate, and are interacted with, mean that the idealised dialogue management models must augment their behaviour for the particular constraints imposed by either the features of the domain application, or the particular choices made in language input and output technologies. Thus, dialogue managers deployed in practical spoken dialogue systems can depart from the ideals discussed above. Therefore, in the next section we consider a number of dialogue system case studies so that we have a clearer picture of dialogue manager use in practice.

3.4 Dialogue System Case Studies

Since dialogue management responsibilities are wide, and are achieved in ways which are highly dependent on the overall component selection in a dialogue system, in this section we survey a number of state-of-the-art dialogue systems to evaluate key issues in dialogue management. We analyse dialogue systems as a whole rather than only the dialogue management component since it can be very difficult to isolate management aspects in real-world dialogue systems. The survey is however focused on those aspects of the dialogue system most relevant to the topic of dialogue management.

Of course, given the long history of dialogue systems research, a number of surveys of dialogue management and dialogue systems design already exist. Among these surveys, one of the best known is McTear's [2002] analysis of spoken dialogue systems. This analysis looks at spoken dialogue systems as a whole, considering both architectural issues such as integration methodologies, and also the particular technologies and techniques applied in typical dialogue system components. With respect to dialogue managers, this survey looks in detail at state-based, frame-based, and agent-based methods, considering in particular the control methodology and representation of dialogue state employed. While McTear's survey is extremely comprehensive, the actual case studies considered tended to reflect the most idealised instances of the state-based, frame-based, and agent-based models, and hence did not reflect the more common realities of dialogue system design. Moreover, since McTear's survey aimed to be broad coverage in that no particular domain application was focused upon, the models considered in that survey placed little emphasis on issues related to situated dialogue system design such as we require here.

Therefore, unlike McTear's broad coverage analysis, the survey presented here places emphasis on research prototypes which have either considered the issue of integration with physically situated systems, or addressed the problem of contextualization more generally – but in ways which were aimed at being computationally tractable. A result of these constraints is that we will not consider any instance of pure state-based or frame-based dialogue systems, but look instead at a collection of implementations which vary in complexity from ISU-based to agent-based dialogue management.

3.4.1 The Rochester Interactive Planning Scheduler

The Rochester Interactive Planning Scheduler (TRIPS), like its predecessors including the TRAINS system [Allen et al., 1996], is a dialogue processing architecture which places considerable emphasis on deep contextualization and intention recognition [Allen et al., 2001a,b, 2000, Ferguson and Allen, 1998]. Developed for interactive cooperative planning tasks, both TRIPS and its predecessors have had considerable impact on the practical dialogue systems community as instances of agent-based dialogue architectures which retain an element of tractability in processing.

In the context of modern practical dialogue systems, the most unique aspect of the TRIPS approach comes through the explicit analysis of the properties of a collaborative problem solving model, and the use of those properties to constrain the language interpretation and production processes. Specifically, TRIPS provides an abstract problem solving model which can be customised for a given domain by dialogue system developers. This abstract problem solving model includes key concepts such as: (a) objectives – goals, sub-goals, constraints; (b) solutions – courses of action; (c) resources – objects and abstractions; and (d) situations – world states. Utterances in a practical dialogue are interpreted as manipulations of these different concepts, e.g. creating, modifying, deleting, and evaluating problem solving objects. The abstract model can be specialised to a particular domain through an instantiation of a *Task Manager* which defines the details of the objects in the domain, as well as the operations which may be performed on the objects.

Of course an abstract problem solving model which pays no attention to the difficulties of language processing and grounding would not be far different from the classical agent-based dialogue management models. Thus, the second major feature of TRIPS is a multi-tiered discourse representation model which is built upon by language interpretation and generation mechanisms. This Discourse Context model captures a discourse in terms of: (a) turn status; (b) outstanding discourse obligations; (c) salient entities in the discourse; (d) structure and interpretation of the immediately preceding utterance; and (e) the discourse history consisting of speech act interpretations and a history of ground utterances. Thus TRIPS's Discourse Context is typical of an agentive dialogue model such as that introduced in Section 2.5.2.

Assuming true speech act based utterance interpretation, and a fine-grained mental state, TRIPS and TRAINS's discourse model provides a highly multi-tiered representation of individual dialogue contributions. For example, in the context of the TRAINS project, Traum and Andersen [1999] describe a six layer utterance semantics consisting of the following levels:

- **L-req:** The locutionary or literal level
- **I-req:** The illocutionary or interpreted level
- **D-req:** The domain or application level
- **P-act:** The plan called
- **E-act:** The executable effect
- **O-act:** The observable effect

Within this model, the first three levels correspond rather neatly to the notion of a surface form, shallow semantics representation, and domain-specific semantic level as we will apply later in this book. The latter three levels on the other hand are more concerned with action representation within the system, and can arguably be considered part of the internals of a highly agent-oriented application. This level of plan detail is unsurprising given the importance of action and plan modelling within the TRAINS and TRIPS projects, but need not necessarily be useful to domains where explicit user plan modelling is not so vital.

However, while considerable attention is given to the description of each of the

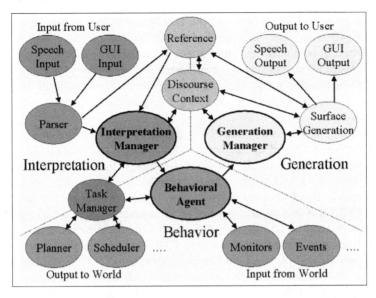

Fig. 3.3 The Rochester Interactive Planning Schedule – system architecture. (Taken from [Allen et al., 2001b])

six representation levels by Traum and Andersen, considerably less detail is available on the transitions between, for example, the *I-req* and *D-req*, or the *D-req* and *P-act* levels. With respect to the first of these transitions, Traum states that the process involves "reinterpreting the useful information of the I-req in the ontology of whatever domain reasoner is used" and that this may involve "disambiguation techniques for underspecified information". However, it fell to Dizikovska et al. [2007] some eight years later to give a concrete view on the mapping processes which exist between interpretation level and application level utterance representations in the TRIPS architecture.

Both the collaborative problem solving model and the multi-tiered discourse representation are employed within the complete TRIPS architecture. Figure 3.3 depicts the TRIPS's architecture as presented by Allen et al. [2001b]. Whereas earlier versions of the TRIPS and TRAINS systems located all application and dialogue management functionality within a central dialogue management component [Ferguson and Allen, 1998], this architecture takes a considerably more distributed view, spreading dialogue management functionality across dedicated *interpretation*, *generation*, and *behavioural* components.

To illustrate the application of the architecture, let us consider in particular the language input process. Here, the Interpretation Manager interprets incoming parsed utterances – namely a sequence of literal speech acts – and directly updates the discourse context. In doing so, the Interpretation Manager must identify the intended speech act, a collaborative problem solving act that it furthers, and the system's obligations arising from the interaction. The Interpretation Manager therefore performs interpretation in context, interacting with the Task Manager, to determine the most probable interpretation. Once interpreted, the Interpretation Manager notifies the discourse context of both interpreted user utterances and discourse obligations. The Behavioural Agent is then notified of the collaborated problem solving act initiated by the user. Conceptually, the language production process is similar to the inverse of the input process, but in practice processing responsibilities are not necessarily simply reversed.

The TRIPS architecture is a powerful dialogue processing system and arguably remains at the forefront of deep language understanding capabilities in dialogue system development. However, TRIPS's power comes at the cost of complexity, and it is not generally a straightforward task to apply either the TRIPS code or the underlying models to new domains of application. Indeed, one of the most constant criticisms of TRIPS is that the mechanisms of an abstract problem solving model are simply not needed in the majority of practical dialogue system domains, and that the existence of this model at the core of TRIPS only serves to obfuscate the language grounding and planning models which have a more widespread appeal. In the next section we consider a dialogue system, which while being born in part out of TRIPS and TRAINS language grounding work, foregoes a detailed problem solving model in favour of a more frame-oriented view on the dialogue domain task.

3.4.2 IBiS

The TRAINS and early TRIPS systems were the forerunners of the broad class of Information State Update (ISU) based dialogue management models which recognised the importance of the mental state modelling perspective taken by TRAINS and TRIPS, but yet sought to have this same mental state organized view of dialogue implemented in a more light-weight and extensible framework. Among these ISU-based systems, Larsson's [2002] IBiS series of systems are arguably the most completely detailed of the dialogue systems which adhered most closely to the ISU methodology.

IBiS, based on the earlier GoDiS system [Bohlin et al., 1999] and Ginzburg's [1996] KOS model, was developed to study how issues, modelled semantically as questions, can be used as a general basis for dialogue management. Building on the TrindiKit implementation, IBiS includes features such as multi-level language grounding, accommodation of over-answered questions, and information state revision in information-seeking and limited action-oriented dialogue. While Larsson described IBiS in terms of four systems with progressively increased functionality, the description here will treat the four systems as a complete unit.

Figure 3.4 depicts the IBiS system architecture which is a specialisation of the general TrindiKit architecture. The system includes standard language input and output modules, an information state container, a dialogue move engine which includes update and selection modules, a number of domain-specific resources, and an algorithm which controls the overall flow of processing throughout the system. In

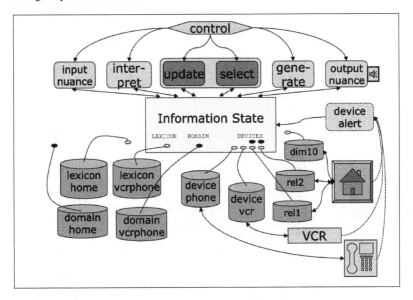

Fig. 3.4 IBiS system architecture. Taken from Larsson [2002].

keeping with the ISU-based methodology, IBiS's key features are its use of a well-structured declarative information state representation, and a rule-based language update and planning process.

Like the other classical ISU-based approaches, IBiS makes use of a declarative dialogue state model which is dependent on a small but well defined set of semantic types which can be used to form a record-style structured representation of the dialogue state at any given time. A range of basic data types are defined to allow the composition of the information state container and the various rules which manipulate it. These data types minimally include lightly sorted semantic types used to define the content of moves and attitudinal state. IBiS also makes use of a number of *move types* which denote the dialogue act types which can be handled by the system, as well as a set of special communication management moves which are used to implement the range of possibilities allowed by a multi-tiered theory of grounding.

Semantic types also include constructs for the description of dialogue plans which are processed by the IBiS implementation to handle the complexity of individual domains. In addition to a number of basic plan constructs for sequencing and conditioning actions, plan operator types includes both dialogue moves, and more abstract actions which manage system dialogue moves in a more controlled way. For example, the dialogue plan construct *findout(q)* is a strong-minded behaviour which will pose a question q to the user that will not be removed from the dialogue manager's agenda until the question is resolved; whereas *raise(q)* on the other hand poses the question q to the user but will not persist in asking the question if not immediately answered. Each dialogue plan can also include a number of domain

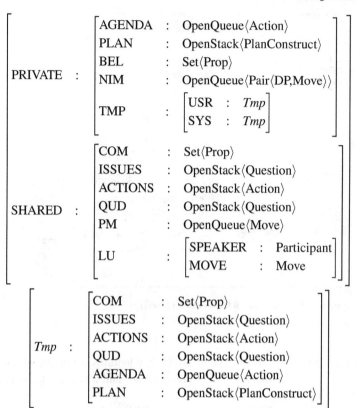

Fig. 3.5 IBiS Information State. Taken from Larsson [2002].

actions such as *consultDB(q)* where q is a question which triggers a search of an attached domain-specific database. Dialogue plans are part of domain knowledge, can be quite complex through nesting, and in some way correspond to an imperative view of frame information as used in frame-based dialogue managers. Larsson himself states that similarities do exist there, but claims that ISU-based dialogue management is more scalable to dialogue types such as negotiation and tutorial than are frame-based systems.

The dialogue state representation structure, or *Total Information State*, includes the Information State proper (depicted in Figure 3.5), and a range of interface variables to domain resources and language processing components. As stated earlier, IBiS's information state model is a derivative of Ginzburg's Dialogue Game Board with respect to the partitioning of the agent's attitudinal state, but is significantly larger and more detailed than the original theory. Following Ginzburg, the information state is partitioned into public and private information. Private information includes *AGENDA*, a set of dialogue plan actions which are to be taken by the agent in the immediate future; *PLAN*, the contents of a domain-specific dialogue plan which has been loaded to deal with a particular user question or request; *BEL*, a set of private beliefs; *NIM*, the set of user moves which have not yet been processed, or *integrated* into the information state; and *TMP*, a temporary store where copies of parts of the agent's and user's information state models can be put aside for roll-back in the event of miscommunication on the part of either interlocutor. Shared information on the other hand includes: *COM*, the set of shared common beliefs; *ISSUES*, the set of questions which are open; *ACTIONS*, the set of one or more actions which have been requested by a user and which are to be executed by the system; *QUD*, the set of questions which are open and to be processed in the immediate future; *PM* a queue of previous moves; and *LU*, a record of the last move made by user or system.

In addition to the information state definition, the core of IBiS's dialogue management approach is defined in terms of two sets of functionality which are responsible for integrating performed dialogue moves and planning system dialogue moves respectively. These functionality sets, i.e., the *update* and *select* modules, are implemented via a set of declarative rules, and a control algorithm which orders the application of these rules to the information state. Moreover, the rule definitions make use of a diverse set of functions which manipulate information state, interface with domain components, and perform semantic operations such as the combination of questions and answers. For example, in addition to a *combines(A,Q)* function which combines a question and an answer to form a proposition, the IBiS rules are highly dependent on domain-specific functions such as *findPlan(A)* and *relevant(A,Q)* which are responsible for selecting a dialogue plan for a given proposition and determining if an answer is possibly relevant to a given question respectively.

To illustrate the structure of rules and their interaction with information state elements and domain-specific functions, consider Figure 3.6 which correspond to IBiS's Rule 5.1 as presented in Larsson's [2002] account. The rule, an update rule of type integrate, is intended to process a user's request for a particular action if the action is available to the system. More specifically, the pre-conditions of the rule first check if a request move is currently outstanding on the queue of non-integrated moves (*NIM*) and assigns it to the variable *request(A)* if such a move exists. The pre-condition also checks that the last move was made by the user, and that the interpretation accuracy is at least 0.7. Finally, the pre-conditions check that there exists a plan in the domain resources which can be used to process the specific request. These pre-conditions are processed from top to bottom, and if any of these checks fail, then no subsequent conditions will be checked and the rule fails. The *effects* of the rule, if triggered, first pop the move from *NIM*, before setting the latest utterance variable, and checking the levels of communication management feedback to be given. Finally, the shared actions and the agenda are updated to note the specific action which has been requested.

Although difficult to capture here without an overly long recital of the many rules in IBiS's update and selection processes, IBiS's usefulness comes in terms of a relatively clear model of dialogue state and update which acknowledges the difficulties of language grounding as highlighted by Traum and Ginzburg. This clarity

RULE: **integrateUsrRequest**
CLASS: integrate

PRE:
$$\begin{cases} \$/PRIVATE/NIM/PST/SND = request(A) \\ \$/SHARED/LU/SPEAKER == usr \\ \$SCORE = Score \\ Score > 0.7 \\ \$DOMAIN :: plan(A, Plan) \end{cases}$$

EFF:
$$\begin{cases} pop(/PRIVATE/NIM) \\ add(/SHARED/LU/MOVES, request(A)) \\ push(/PRIVATE/AGENDA, icm.acc * pos) \\ if\_do(Score < 0.9, \\ \quad push(/PRIVATE/AGENDA, icm.und * pos : usr * action(a))) \\ push(/SHARED/ACTIONS, A) \\ push(/PRIVATE/AGENDA, A) \end{cases}$$

Fig. 3.6 IBiS Update Rule 5.1. Taken from Larsson [2002].

and power comes through the distinction between issues of dialogue state representation, move integration, and dialogue planning (selection), and results in a dialogue system that, at least within information-query domains such as ticket purchase and time-table checking, is reusable across a number of applications.

While the IBiS dialogue system is laudable in that it provides a concise implementation of dialogue management which takes grounding issues seriously, the IBiS dialogue systems, like all classic Information State Update dialogue manager implementations, suffer a number of weaknesses which have prevented their designs being replicated across a range of dialogue domains. One readily apparent weakness is that the *update* and *selection* modules rely on a large number of declarative rules. The problem here being both the quantity of rules – some seventy-three rules are present in the complete IBiS implementation – and the fact that these rules can be quite involved, thus making it difficult to interpret the overall operation of the system. A more serious weakness is the dependence on a range of dialogue functions which, while being explained in broad terms, are given little in the way of an operational semantics. These functions such as *findPlan*, *combine*, *resolves*, and others, are defined in a relatively static domain-dependent way, but for the IBiS style systems to become robust and extensible to a wider variety of dialogue types, these operations must be treated in a considerably more systematic manner.

3.4.3 SmartKom

The SmartKom project [Wahlster, 2003, Buehler et al., 2002] was a large scale multi-modal dialogue systems project which developed a series of dialogue platforms and implementations which combined a semantically rich discourse modelling approach with multiple interaction modalities and robust interpretation. While the dialogue management techniques developed do not fit neatly into any of the dialogue management model categories introduced in Section 3.3, the techniques deployed represent a modern pragmatic view on how dialogue resources must be organized so as to provide a degree of robustness, contextual sensitivity, and reusability.

SmartKom's processing backbone, integrated through a custom blackboard-based messaging system, included a large number of components that are responsible for managing various input and output modalities, the fusion and fission of information from those modalities, the management of discourse state, and interaction with a range of domain application back ends. Figure 3.7 presents a highly simplified view of this processing backbone to highlight the overall dialogue system structure. Within the backbone, dialogue management is broadly split between discourse management and action planning modules. The discourse management module is responsible for semantic enrichment, intention identification, and the update and maintenance of the system's dialogue state. The action planner, described by Lockelt et al. [2002], updates the overall task state on the basis of the acts identified by the discourse manager, and subsequently plans dialogue and presentational acts to be performed by the dialogue system. While this distinction between discourse integration and action planning broadly follows the integration and selection distinction applied by IBiS, SmartKom's novel features included the use of a rich multi-modal semantics model, and well defined computational machinery which allowed the contextual augmentation of semantic inputs. In the following we analyse SmartKom in terms of these features.

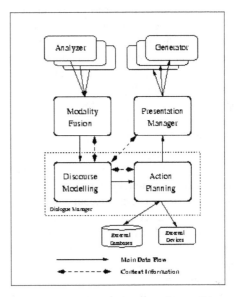

Fig. 3.7 SmartKom's language backbone. Taken from [Pfleger et al., 2003].

At SmartKom's heart is a three-tiered discourse model that has been tailored for multimodal interactions [Pfleger et al., 2003, Lockelt et al., 2002]. This model, illustrated in Figure 3.8, consists of a domain layer, a discourse layer, and a modality

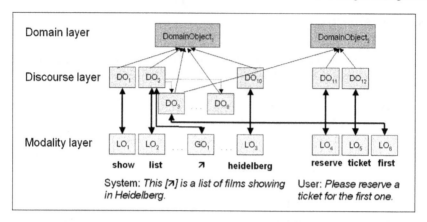

Fig. 3.8 SmartKom's 3-tiered representation of multimodal discourse.

layer. Objects in the modality layer include concrete linguistic, visual, and gestural objects that span a number of input channels. Each concrete modality object is associated with one or more discourse objects which abstract over concrete modality objects. The discourse objects, stored in the discourse layer, are in turn linked to conceptual domain objects in the domain layer. Thus, the discourse layer is equivalent to a standard discourse semantics in that it provides a semantic abstraction over concrete modality expressions. The domain layer on the other hand is an application-specific view on the discourse structure in terms of well defined application query and update structures which correspond broadly to traditional frames. Objects within the discourse model are expressed as typed feature structures backed up by a well specified domain ontology.

Following in the style of Grosz & Sidner's Tripartite Model, SmartKom's discourse manager maintains a set of global and local focus spaces which provide a view over elements in the discourse model. A global focus space is specific to a single topic under discussion, and is used by the discourse manager to guide reference resolution and other contextualization processes. Each global focus stack is in turn associated with a local focus space which acts as a salience ordered list of discourse objects which have been referenced recently within utterances associated with the given global focus space.

During operation, SmartKom's multiple recognition modules e.g. prosody and semantic analysis, produce individual recognition hypotheses with associated scores. The most important of these is a production-rule based language analysis system which composes frame-like representations of surface language [Engel, 2002]. The modality fusion modules bring together a number of these hypotheses to form an intention hypotheses lattice, which is then passed onto the discourse manager, which subsequently applies a two stage *validation* and *enrichment* process to select the most probable hypothesis before that hypothesis is integrated into the dialogue state. These validation, enrichment, and update functions made use of well defined semantic operations which are described in the next section.

During validation, the fitness of each hypotheses sequence with respect to the current discourse state is determined by the discourse manager. This process, like enrichment, makes use of an *overlay* operator [Alexandersson and Becker., 2003, Alexandersson and Becker, 2001, Romanell et al., 2005], which is a binary operation over typed feature structures. The general principles behind the use of overlay are that we can consider the discourse context to be a set of constraints on a frame which have been established in the course of a dialogue; these constraints can then be combined with the semantics of a new user contribution to create a new domain object and enrich the incoming semantics object such as to perform ellipsis and anaphora resolution. The overlay operation is thus somewhat similar to a unification function, but unlike unification, overlay is non-commutative, never fails, and makes use of a scoring function for its application in discourse modelling [Pfleger et al., 2002].

Overlay can however only be applied in the case that both arguments contain a complete application object. If, however, a semantic input is partial in that it consists of a set of *subobjects*, e.g., names, numbers, etc., then a second operation, *bridging* must first be used to compose a complete application object from the available semantic inputs. Bridging makes use of expectation values posited by SmartKom's *action planner* to build a probable application object. This newly created application object can then be examined with the overlay operator in the normal way.

The overlay and bridging operations make extensive use of SmartKom's concrete application ontology, which was originally based on the upper ontology proposed by Russell and Norvig [2003]. Some interesting features of the SmartKom ontology, outlined by [Gurevych et al., 2003], include its extensive Process Hierarchy, which is its main extension of the original work by Russell & Norwig. Moreover, spurred on by the ideals of simplicity and re-usability, the SmartKom ontology developers argued strongly in favour of a single all encompassing ontology rather than different ontologies for the representation of discourse layer and domain layer objects. This approach puts them in stark contrast with the approach taken in the TRAINS project which advocated unique ontological definitions for different levels of representation.

The SmartKom project was a large-scale research endeavour which allowed a very systematic approach to the investigation of issues related to the development of multi-modal dialogues that have a rich semantics-oriented organization. This systematic investigation allowed a number of interesting directions to be pursued, not least: SmartKom's rather straightforward yet powerful view on discourse semantics organization; the in-depth analysis of the structure and role of domain ontologies in the design of the dialogue system; and the development of non-trivial semantic comparison and composition operators which allowed a constraint-based view on dialogue structure composition to be investigated.

The level of detail which was possible in component development and integration reflected the scale of the SmartKom project, but this project scale also had the effect that the approaches taken to system development could afford to be insular in that there was no need to leverage off existing technologies. While this certainly raises interesting questions about the way in which research-oriented dialogue system construction should move if we are to make more serious inroads into dialogue systems research problems, the side effect of this perspective is that it is not generally possible to take the SmartKom system and link it to dialogue types which do not map well to SmartKom's research agenda – or indeed, to make use of language technologies other than those directly used within SmartKom. This is particularly troublesome for us here since SmartKom's application domain, while concerned with multi-modal interaction, was not situated in the sense we outlined earlier, and hence did not have to consider the types of spatial context sensitivity outlined in the introductory chapter.

Moreover, SmartKom, like the TRIPS and IBiS systems, was concerned with non-situated dialogue domains, and hence placed minimal research focus on issues related to interaction with potentially autonomous systems. With this in mind, in the next section we turn to a dialogue project which provides some insight into situated systems dialogue.

3.4.4 WITAS & The Conversational Intelligence Architecture

Building upon earlier work on information state update based dialogue systems, Lemon et al. [2002a] and Gruenstein [2002] developed a dialogue enabled control system for unmanned aerial vehicles as part of the larger WITAS project [Doherty et al., 2000]. Since the targeted unmanned-vehicle was semi-autonomous in nature, and should initiate dialogue on the basis of changes in its environment, the dialogue management aspects of WITAS focused in particular on the issues which arise in adapting ISU style dialogue systems to the realities of mixed initiative dialogue in the autonomous vehicle domain. As such, WITAS is particularly relevant to our interests here.

Figure 3.9 illustrates the WITAS system architecture in broad terms. To support integration between dialogue components and embedded system code such as used in the WITAS autonomous vehicle, WITAS was implemented as a distributed agent-based model built around the Open Agent Architecture (OAA) [Martin et al., 1999]. Agents used include a GEMINI [Dowding et al., 1993] based system for language analysis and generation; a speech synthesis agent based on the Festival Speech Synthesis engine [Taylor et al., 1998]; a speech recognizer based on the Nuance speech recognition; a graphical user interface which supports limited multi-modal interaction; an *Activity Model Interface* which defines the interface to device specific capabilities; and a Dialogue Manager which incorporated dialogue state representation, language understanding, and language planning functionality.

As indicated, WITAS's dialogue manager, the Conversational Intelligence Architecture (CIA) [Lemon and Gruenstein, 2004, Lemon et al., 2002a], is based broadly on the principles of information state update based dialogue management. Figure 3.10 depicts the information state elements and the relationships that hold between them. Solid arrows represent possible update functions, while the dashed arrows represent query functions. Specifically, the CIA's information state is composed of:

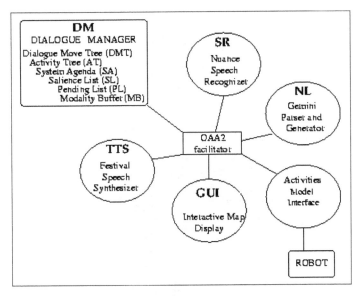

Fig. 3.9 The WITAS system architecture. (Adapted from [Lemon et al., 2002a].)

- **The Dialogue Move Tree:** Used as a message board to keep a record of the utterances made by both the system and the user. The Dialogue Move Tree is a rooted tree, in which each sub-tree of the root node represents a thread in the conversation, and where each node in a sub-tree represents an utterance either made by the system or the user.
- **The Active Node List:** Keeps track of which nodes are active, i.e. a conversational contribution that is still in some sense relevant to the current dialogue. A node that is higher on the list is said to be more active than another; thus is more likely to contribute to the current conversation.
- **The Activity Tree:** Tracks current, past, or planned activities. Each node describes an action in terms of a set of slots, including: type, parameters, and state. The device and the dialogue system share this same model of the device's activity state; thus, they are always properly coordinated.
- **The System Agenda:** A structure containing all utterances that the system intends to produce.
- **The Pending List:** Tracks questions the system has asked, but the user has not yet answered.
- **The Salience List:** Keeps track of noun phrases used in conversation.
- **The Modality Buffer:** Used to store graphical display gestures given by the user in multi-modal interactions.

Although in many ways, the CIA was typical of ISU-based systems, the particular treatment of the interface between dialogue management itself and the agent-like domain application is worthy of further elaboration. Noting the task-oriented rather

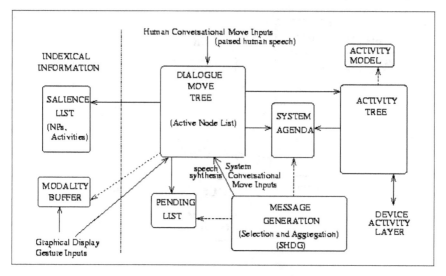

Fig. 3.10 The Conversational Intelligence Architecture (Adapted from [Lemon et al., 2002a].)

than information-seeking nature of vehicle control dialogues, Gruenstein's [2002] contributions to the CIA established a tighter linking between the ISU dialogue management model and rational action theories than had been achieved to that date. Specifically, Gruenstein's contributions included the models of plan recipes and the *Activity Tree* already mentioned, and also a formalized conversational constraints model that allowed globally defined rules to be used in filling unspecified parameters in user commands.

At the core of Gruenstein's approach was the assumption that domain actions with temporal extent, and structures of those actions, were core to task-oriented dialogue. Under this basic assumption, he argued that detailed models of actions, or *activities*, were required to bridge traditional agency models and dialogue management processes. The purpose of this bridging is not only to allow activities to be instantiated through dialogue, but that once instantiated, the dialogue manager would be able to keep track of these activities along with their resultant sub-goals and sub-actions. This information on the agent's intentional state would then facilitate both the answering of questions on the state of the agent, but would also enable the regular updating of the user on the status of requested tasks so as to avoid potential mode-contusions. In light of these goals, Gruenstein developed an interface between the domain application and the dialogue system in terms of a *task-recipe*. This recipe is essentially a frame-like structure which included features such as the activity type, mandatory and optional slot information, natural language semantics corresponding to these slots, pre-conditions on activities, as well as possible default values. As typical of frame-like dialogue management models, the dialogue management process then becomes responsible for filling these recipe slots. Unlike a standard frame however, the recipe also contains a body which decomposes a given activity into sub-activities in a manner very similar to that applied in agent-oriented programming methodologies.

The inclusion of the plan body in recipe specification was coupled to the notion of the Activity Tree as an explicit construct in the dialogue manager's information state. The basic idea was that once a recipe had successfully been filled and initialised to an activity through dialogue management, the state of the activity – including the state of any sub-activities if the activity was not a primitive – would be tracked within the information state. It would thus be this explicit intention-like structure that would be used to answer explicit questions on what the system was doing. However, rather than being the actual intentional structure of the domain agent, the Activity Tree is intended as a pseudo-mirror of the agent's actual intentional state. The effectiveness and nature of this mirror was however left relatively unclear by Gruenstein as he himself states that the activity body and resultant structures in the Activity Tree are intended to reflect how a user would conceptualize the performance of an activity – as distinct from how the domain agent might in practice pursue the activity. It remains to be seen how these different conceptualizations of the same task-goals would be aligned in practice.

In addition to developing the basic notion of activity recipes and the Activity Tree, Gruenstein's second major contribution was a model of formally defined conversational constraints in the parameterization of actions. Here the basic idea was that in addition to primary task dialogue moves, there is a second class of dialogue moves in task-oriented dialogues which are used to impose constraints on the parameterization of activities. Such constraints may be as simple as the semantic correlate of "fly quickly" or "fly at a high altitude". Within the CIA, such constraints are essentially modelled as global rules which are incorporated into the filling of unspecified parameter types in new activities. Importantly, it should be noted that the organization of the Activity Tree allowed such constraints to not only update the top-most node in an activity, but also any dependent secondary activities. The CIA's dialogue manager not only directly incorporated such constraints into the parameter filling process, but also detected conflicts between constraints and initiated dialogues to overcome such conflicts.

WITAS, and in particular the Conversational Intelligence Architecture, represents a notable departure from classical ISU-based dialogue management in a number of ways. The principal differences lie not on the focus on mixed-initiative multi-threaded dialogue management – as Larsson's IBiS model was also capable of handling the same dialogue type – but rather on the inclusion of an explicit model of rational action in the information state, and a departure from a strict record style structures modelling based on TrindiKit.

Given our goals of developing situated dialogue management, the importance of Gruenstein's incorporation of some notion of domain action and intentional structure into the dialogue management model cannot be overstated. Although Gruenstein's work did not ultimately meet his desired goal of naturally answering questions on the state of intentions and the reasons behind those intentions, the CIA's model did illustrate how dialogue management for situated applications may make use of agency-like notions in the information state. Gruenstein's model was how-

ever limited in a significant and important number of ways. First, while the recipe
construct did allow static pre-defined default values to be used in the automatic specification of activity parameters, e.g., to specify that the default speed of an action is
a "medium speed", the CIA incorporated no way of using truly situated contextual
knowledge to fill unspecified content that was more dynamic. Thus, the CIA had no
way to determine what the destination might be for a highly elided command such
as "enter". Related to this, while the CIA did incorporate a basic model of discourse
referent resolution, this model was seemingly independent of situated context and
thus was incapable of handling truly exophoric references as are used frequently
in situated dialogue. Also significantly, the notion of an Activity Tree as a distinct
entity from the domain agent's true intentional state introduces a number of difficulties in terms of aligning these obviously similar constructs. Although it is clear
that many aspects of a task's implementation would exist below a so-called intentional level, the tighter coupling of intentional and dialogue states through a single
construct may make the relationship between the agent's actual intentions and the
information provided to the user more transparent.

3.4.5 Discussion

With a number of dialogue system case studies in hand, it is worth taking a moment
to examine some of the more common properties of these dialogue systems, and
question where these systems are potentially lacking with respect to the overall goals
of situated dialogue systems which we introduced in Chapter 1.

Beyond the more obvious features such as resource re-use, the abstraction of
dialogue processes for specialization across domains, and the increased interest in
integrating multiple modes of interaction in spoken dialogue systems, there are a
number of trends in the dialogue systems discussed in the last section which are
worthy of note:

- **Separation of Language Integration and Planning:** Rejecting state-based dialogue management, there exists an unequivocal separation of concerns in dialogue management, and in particular a distinction between the integration of
  user moves in dialogue, and the planning of the next move to be taken by the
  system. Within the case studies this separation is very evident, e.g., IBiS's use
  of separate *update* and *selection* modules, SmartKom's distinction between a
  Discourse Modelling module and an *Action Planner* module, and TRIPS use of
  distinct *Interpretation* and *Generation* managers. The reasons for this distinction seem to the most part to be to allow a separation of concerns in dialogue
  manager design which would otherwise become too large and unwieldy when
  scaled to complex dialogue domains.
- **Local Dialogue Organization:** Connected to the separation of interpretation
  and move selection, applied dialogue management theory increasingly assumes
  a locally motivated rather than globally structured view of dialogue. The mo-

tivations for this design feature come from the cognitive dialogue modelling community (see Section 2.5) as well as the Conversational Analysis community [Sacks et al., 1974], and hinges on the point that what constitutes a 'correct' dialogue is not some global structure which is similar in nature to a syntactic parse, but is instead based on what *made sense* in that particular context. Such local organization can in turn give rise to global organization as emergent in simple single threaded dialogue such as those studied within the discourse analysis community. This local organization feature is an explicit assumption of the information state update approach to dialogue, and is hence manifest in IBiS and WITAS; it is also evident in SmartKom's dialogue planning component which principally assumes an initiate-response based dialogue organization.

- **Multi-Layered Discourse Modelling:** Rather than assuming that a single layer of semantic representation is suited for all aspects of linguistic and domain-specific reasoning, there is a growing tendency towards the use of multiple layers of representation in sophisticated dialogue systems. The reasons for such a decision come on one hand from the realities of integrating dialogue systems with commodity language technologies, and on the other hand from the belief that one modelling system simply is not capable of being a perfect fit to all computational needs. This layering feature is seen most dramatically in the TRIPS system where a given utterance can be modelled in terms of many different levels of representation, but can also be seen to a lesser extent in models such as SmartKom's discourse representation.

Given the range of design similarities, it may be easy to believe that the fundamental design constraints of dialogue management are now in hand. Such a statement is at best questionable, but even if it were to be true, there still remain a great number of simplifying assumptions made in current systems designs which must be 'unsimplified' before system designs can scale to realistic domains and produce or understand more unconstrained language. Let us consider a number of these simplifying assumptions:

- **Coarse-Grained Linguistic Control:** Dialogue systems such as IBiS, WITAS, and SmartKom use a semantics model which wholly abstracts from linguistic form. The purposes of such abstraction are numerous: (a) facilitating simplified question-answer or statement-frame combination; (b) allowing simple keyword spotting and template-based language analysis and generation systems to be used; and (c) allowing all declarative resources of the system, i.e., linguistic semantics, frame representations, dialogue plans, beliefs, goals, etc., to be modelled within the same propositional framework. While these simplifications follow a general trend in using canonical forms where possible, the sacrifice of linguistic sensitivity has many consequences. The most notable consequence is that it is difficult to shape system-produced language in the ways necessary to produce natural dialogue. Furthermore, simplified analysis techniques may be able to produce some results more often, but they fail to scale to longer utterances and frequently miss subtle meaning differences given by alternative utterance forms.

- **The Generation Gap:** Within dialogue systems, language production is often systematically under-developed since the focus in dialogue systems development is more frequently on language interpretation. This leaves us with a
range of limitations in dialogic production equivalent to what is referred to as
the Generation Gap in complete generation systems [Meteer, 1990]. At the domain level, the relationship between system-specific knowledge and realization
is often ad hoc and thus limited to use with application-specific template-based
generation which cannot hope to scale to the coverage or delicacy required for
natural communication. Significant issues are also present within the dialogue
planning process itself in terms of how the multiple dialogue goals which are
necessary both for grounding processes, and/or mixed initiative multi-threaded
dialogues are handled. While works such as Appelt's [1985] language generation process and PARADIME [Morante et al., 2007] have given considerable
attention to the analysis of multiple dialogue goals as different dimensions of
communicative responsibility, little has been achieved in fluidly integrating distinct goals into single moves.
- **Poor Resource Modularization:** While process modularization, manifest through
distributed dialogue system design, is commonplace, less attention has been
paid to the modularization of knowledge resources. This is typified by the use
of single ontological resources to define knowledge types ranging from linguistic semantics, system capabilities, information state, and domain resources. The
negative consequences of such a perspective is that the development of monolithic knowledge resources is both difficult to manage, and considerably limits
our ability to apply inference mechanisms to reason over declarative knowledge
models. Although more theoretical work in the area of discourse semantics such
as SDRT has begun to address such issues, these trends have not yet made their
way into practical dialogue system construction.
- **Idealised Domain Models:** In all but the most trivial dialogue system designs,
there exists a separation between the dialogue processes themselves and the domain application which is subject to dialogue. From the perspective of dialogue
management the easiest case to handle is one in which this domain application
can be encapsulated as a simple knowledge base that a user either queries or
updates through basic operations. Although early models of dialogue processing did embrace the more complex agent metaphor which extends to a larger
range of domains, more recent work has generally rejected issues of agency and
intentionality both in the modelling of the dialogue processes themselves, and
in the modelling of the domain applications to which those dialogue processes
attach.
- **Idealised Contextualisation:** Related to the issue of idealised domain models, the details of linguistic contextualisation, i.e., what the relevance of a logical surface form representation is to the attitudinal state of the dialogue system, are frequently systematically simplified. This is typified by IBiS's reliance
on a range of declarative domain resource relations, e.g., relevant(A,Q),
resolves(A,Q), combine(Q,A,P), plan(Q,Plan). The use of such
declarative resources rather than the development of equivalent functions obvi-

ously simplifies the construction of simple dialogue theories, but the reliance on case by case declarative statements clearly cannot scale to larger domains.

While each of the above problems pose difficulties to long term goals of naturalistic human-machine dialogue, poor resource modularity, and the idealization of domain application behaviour and contextualization functions, are of particular concern to the development of situated dialogue systems.

As indicated, the nature of contemporary dialogue system domains has allowed relatively simple views of resource modality to be applied. While this assumption holds for many applications, it does not carry over to situated dialogue applications, which can have rich theories of space and action which cannot be easily integrated with linguistic processing resources in monolithic representation theories. Although some spatially-oriented projects such as TRIPS have acknowledged this issue by positing layered models of knowledge representation, there remains much to do in applying more principled ontological and knowledge engineering methods to establish the necessary modularity needed to support meaningful reasoning in a tractable way.

The notion of an idealised database-like domain model is also due to the very nature of projects such as SmartKom and IBiS where users to the most part are simply querying or updating a database with operations of a comparatively quick duration. However, for situated applications this design assumption no longer holds. The nature of situated application domains is fundamentally different from a simple knowledge base of static facts. The domain application itself is capable of performing actions both in response to user commands and pro-actively based on domain-specific decision processes. Moreover, the actions performed by the domain application are typically of considerably longer temporal duration, are often structured into plans and causal relationships, and can hence themselves be the subject of discussion. While some work has been considered in the use of models of explicit action representation in practical dialogue systems, e.g., Lemon's Conversational Intelligence Architecture, this work has not taken account of richer models of agency, nor the practical frameworks which have been developed in software engineering in the last 15 years to support such notions [Pokahr et al., 2003, Collier and O'Hare, 2009].

The issue of idealised domain application is in turn inherently coupled with the idealized contextualization assumptions held by many contemporary dialogue systems. However, for situated applications which have highly complex and dynamic views of their environment which can differ subtly from the models used by or assumed by the agent, the contextualization issue is no longer one which can be trivialized. Although dialogue processing architectures such as SmartKom have developed sophisticated models of context and contextualization, these models have by virtue of the domains of application focused on issues of discourse context and have not needed to consider the potential complications of contextualization in situational models as demonstrated in Chapter 1. Thus, situated language processing will necessarily require more complex models of contextualization than are provided for in current dialogue processing models.

While each of: (a) poor resource modularity, (b) idealization domain application, and (c) idealized contextualization functions, individually pose considerable difficulties, taken as a whole they mean that the development of situated dialogue systems cannot be simply viewed as the process of tacking existing qualitative or quantitative models of spatial reasoning onto existing dialogue architectures. Instead we must bring dialogue system design into alignment not only with spatial reasoning techniques, but also models of rational agency, context-sensitive language interpretation, and detailed spatial modelling resources. While this fact may in itself be unsurprising, the question of how to achieve this in a tractable, modular, and scalable way is a significant research issue.

3.5 Requirements for Situated Dialogue Processing

In light of the shared features of dialogue systems, and the properties of the situated domain, we can now outline a minimal set of requirements for any language processing architecture for a situated domain:

1. **Mixed-Initiative Capable:** The language processing model should support full mixed initiative dialogue as changes in an artificial agent's environment may motivate specific dialogue goals for that agent.
2. **Action-Centric Dialogue:** Rather than focusing on information-seeking dialogues, interaction with situated agents typically concern action-oriented dialogues where user and system negotiate tasks to be performed by either user or agent in their shared environment.
3. **Multi-Thread Handling:** The language processing model should be capable of supporting multi-threaded dialogue since the realities of mixed initiative dialogue, and the general properties of task-oriented interaction, frequently give rise to the temporary suspension of individual dialogue threads.
4. **Untrained User Input:** The nature of situated dialogue domains such as humanoid robotics and embodied conversational agents necessitates that the dialogue architecture work with relatively flexible input and hence not constrain a user to become trained in an artificial command and control vocabulary.
5. **Complex Contextualization Support:** Rather than relying on the assumptions of idealised static contextualization as seen in the previous chapter, the language processing architecture should support contextualization that is sensitive to both physical and task contexts.
6. **Controlled Language Production:** To avoid the canned unnatural nature of most spoken dialogue systems, and to support context-sensitive detailing of spatial information, the language processing model should facilitate highly controlled language production rather than relying on canned textual output.
7. **Communication Management Support:** The decisions made in dialogue processing not only reflect the core task-related content of a dialogue, but also necessarily reflect the uncertainties and processing failures which can occur at the various dialogue processing levels. Particularly in an embodied domain where

perception is problematic and user language unconstrained, the language processing model must incorporate a communication management methodology in the style of the theories of Clark [1996] and Allwood [1997] and the computational realizations of these models by Poesio and Traum [1997] and Larsson [2002].

8. **Tractability:** Computational processes applied in the dialogue architecture should not be so strong as to limit the system's application in real-world domains. System actions, physical or verbal, should be made in a timely fashion with respect to user actions.

9. **Scalability & Re-use:** Dialogue systems are expensive to develop. Therefore, on a highly practical note, both the overall architecture and the individual resources and processes should be designed with scalability and re-use constraints in mind.

10. **Modularity:** The dialogue architecture should be highly modular in terms of interchange of solutions for processing needs. Such modularity is key not only to computational tractability, but also for significantly enhancing the generalizability of individual solutions. Modularity should not however be so strong as to prevent information sharing across components.

While these requirements are difficult to meet, they remain a small subset of the range of features pursued in contemporary dialogue systems research. Other broad requirements for dialogue applications concern multi-modal and multi-lingual communication support, handling barge-in and low-latency communication, emotional and psychological model integration, and mechanisms for dialogue integration with large third-party knowledge sources to name but a few. However, since none of these issues are particularly relevant to the domain of situated dialogue, we do not consider them further here. Nevertheless the methods pursued later do not preclude extension to these research issues, and in Chapter 9 we briefly comment on how the language architecture outlined in the following chapters can be extended in some of these directions.

3.6 Summary

In this chapter we have analysed the current state of computational dialogue to gain a clear picture of the position of relevant theories and technologies with respect to developing appropriate interaction models for autonomous and situated systems. In particular, we saw that spoken dialogue systems are the architectures used to organize language processing and domain components to produce applications with a degree of language competence. While the general layout of these components is clear, the boundaries between them is often fuzzy and vary from project to project. Of all the components within a dialogue system, the dialogue manager, which sits at the centre of such architectures, is the one component which varies most with respect to both responsibilities and design. Several dialogue management models are currently employed within the computational linguistics and engineering communities. These vary from simple finite-state and frame-based designs as used in commercial systems, to information-state and agent-oriented designs which offer considerably more power in theory, but which have yet to realize their intended potential. While implemented dialogue systems are growing in sophistication, limitations in current dialogue systems such as a lack of resource modularity, frequently ad hoc or non-existent situational model sensitive language contextualization models, and a bias of design towards database query domains, mean that the application of dialogue systems to situated and autonomous systems is a non-trivial research goal. We have however been able to outline a set of basic requirements for Situated Dialogue Systems.

Chapter 4
The Situated Dialogue Architecture

Abstract Having established a background in dialogue modelling and management, this chapter motivates and outlines an agent-oriented architecture for situated dialogue processing. We begin in Section 4.1 by setting out a number of key modelling principles in the development of the situated dialogue architecture. Section 4.2 then presents the architecture itself and subsequently describes how this architecture is to be expanded upon in later chapters. The chapter is concluded in Section 4.3 with a brief summary.

4.1 Modelling Principles

In light of the requirements introduced at the end of the last chapter, it is clear that a language architecture based on principles of state-based dialogue management cannot provide the diversity of functionality needed for situated dialogue systems. On the other hand, classical agent-based and plan-based dialogue remains intractable and undeveloped with respect to fine-grained language processing issues. Frame-based and ISU-based dialogue – particularly where processing mechanisms have been extended with probabilistic processing – arguably provide the most flexible basis for any dialogue application. But, as we argued in the last chapter, poor resource modularization, and assumptions of idealized domain models and contextualization functions mean that these models in themselves cannot be directly applied to the development of situated dialogue techniques.

To support dialogue processing in situated systems, we instead require a dialogue management model that acknowledges the agent-like properties of situated applications, as well as the highly context-sensitive nature of situated language. But, it should also account for these complexities in a pragmatic view of dialogue modelling and management that is supported by a complete processing architecture. Achieving these functionalities requires a number of design choices regarding the nature of representation and reasoning within the dialogue management model and supporting architecture. In the following we describe and motivate a number of these choices for the situated dialogue architecture.

4.1.1 Ginzburgean Dialogue Organization

Following Ginzburg and Stalnaker, the situated dialogue architecture assumes a *Dialogue Game Board* style structure as the principle organization of dialogue state at any given time. Thus, rather than being a formal first-order or modal theory as is typical of classical BDI and agent-oriented theories of dialogue, this dialogue game board is structured to provide pragmatic and operationalization transparency. We refer to this structure as the system's *Dialogue State Structure* (DSS), whereas the application's complete knowledge including both domain resources and any intentional state will be referred to as *mental state*.

While the DSS provides the basic characterization of the agent's dialogue state, the content of that state and the processes of dialogue management principally assume a local rather than global organization of dialogue. In other words, while structural phenomena exist in dialogue, the dialogue architecture views these as a result of cognitive organization processes, rather than being a structure that directly drives the language production or understanding processes. While the structural characterization of dialogue is important for empirically building an organization of dialogue potential, it need not necessarily be used directly within processing resources. This places the notion of process rather than structure as central to the dialogue task. As indicated in earlier chapters, this local perspective has been argued for at length in the Conversational Analysis community [Sacks et al., 1974], and is typical of cognitive dialogue theories and information state based dialogue management models.

Moreover, following trends in dialogue management design, a single dialogue management process will be rejected in favour of a number of distinct processing components. These components, when taken together with the DSS, in turn fulfil the core requirements of a dialogue manager. However, unlike Larsson, who draws dialogue management out into two distinct components, the approach taken here has been to further tease apart dialogue management, and develop a number of distinct components which account for various aspects of language contextualization and planning.

4.1.2 Agent-Oriented Domain Modelling

As argued in previous chapters, the nature of spatially-situated domains means that situated applications are often capable of performing actions with spatial and temporal extent, and that a high quotient of the language to be produced or understood by these systems necessarily consists of action descriptions. The agent metaphor (see Ferber [1999] for an introductory text), with its emphasis on action and intentionality, would seem like a sensible design abstraction for dialogue management in situated dialogue domains. However, as we saw in the last chapter, there is a long history of agent-based dialogue management, although this school of dialogue management has generally been found unfruitful for various reasons. Nevertheless, we argue that the agentive nature of situated applications cannot be circumvented. But we also argue that the problems seen in early agent-based dialogue models were due to a single agent design being made responsible for all aspects of application and dialogue management functionality.

Thus, the Situated Dialogue Architecture applies agent-based design, but in a more modular way than was the case for classic agent-based dialogue systems. Specifically, the architecture applies strong agency models to domain organization, but captures dialogue management as meta-behaviours which operate over these cognitive constructs. Thus, the agency model places emphasis on the organization of domain behaviours in a way that is compatible with theories of rational reasoning as applied in cognitive robotics. These agentive constructs are in turn made visible to dialogue processing behaviours, but the dialogue processing behaviours themselves are not driven by agentive constructs.

4.1.3 Move-Oriented Communication Management

Taking an agent-oriented view of a domain application suggests the use of domain-specific action and state definitions as the natural units of content communication between system and user. Namely, by packaging a particular action or state description up within a speech act type, we immediately have an act or *move* that can be used as the primary unit of communication as was the case in both classic agent-based dialogue theories, as well as computational theories of communication and cooperation (e.g., [Wooldridge, 2000]). However, while directly applicable in machine-machine communication, experience with agent-based dialogue models reminds us that such a communicative unit is only the beginning of the story for human-human and human-machine communication.

In reality, communication is complicated by perceptual, production, and cognitive inadequacies, as well as the fact that such communicative acts are in many cases only elements of a larger task-dependent goal or communication protocol. Because of this, the communicative act has been largely subsumed in practical dialogue systems by alternative task-oriented units of modelling – principally frame-like descriptions of task structure as were introduced in Section 3.3.2. It would therefore seem natural to apply the frame directly in the domain of situated dialogue processing to capture the individual actions which a system may perform for a user. However, there are difficulties in applying frames in this way. First, frames are traditionally, but not always, associated with less 'semantic-oriented' views on dialogue and often depend on surface form information – typically literal strings – directly within the frame definition to describe how a question should be put to a user to fill a slot and so forth. In other cases, the frame includes specific database functions used to answer the frame query. While such mechanisms are useful in defining quickly implementable dialogue designs, there is considerable distance between the notion of a frame and the agency-oriented view on the domain application advocated in the last section.

The result of such discrepancies can be seen in the differences between the classical notion of frames and simple task-oriented constructions used by users in situated spatial dialogues. Whereas classical frame structures often assume flattened parametrisations where each slot filler is a single simple entity, in spatial language examples there is often complex information structuring within individual 'slots', e.g., where a user supplies additional attributions to identify an object as in the following example:

(13) pick up the ball beside the lamp

Here while we can view the ball as a directly specified slot within a frame structure, the presence of the additional spatial constraint cannot be accounted for directly within that frame structure. Rather such constraints must be viewed as additional information which helps to resolve the element specified within the frame but which is processed as part of the reference resolution process rather than the direct processing of the frame as such.

More significantly, frames and frame-based dialogue systems were conceived to solve a particular type of query-oriented dialogue. Therefore, there is no guarantee that the metaphor maps directly to other dialogue types. In particular, frame-based models generally tend to be geared towards complex interfaces such as flight booking where a considerable amount of structured information has to be provided by a user. Thus, while dialogue frames, analogous to both paper forms and forms used on web pages, can be of considerably size and complexity, potentially requiring several dialogue exchanges to complete filling the frame, the frame itself is usually assumed to be static, in the sense that arbitrarily structuring frame elements together – as would for example be required to allow a user to dictate plans or route descriptions – is not a natural abstraction. Related to this, the frame filling metaphor itself says nothing about the processes of contextualization – which we already argued are vitally important in situated dialogue – nor does it offer a natural solution to other dialogue phenomena such as indirect speech act handling.

While the frame is a powerful abstraction in dialogue management as an interface unit between domain and dialogue processing, it does not adapt particularly well to dialogue types where complex form filling is not generally required. In action-oriented dialogue types such as the situated application domain, the most natural unit of meaningful update remains the classic *dialogue move* since it maps most easily to notions of action and intention and belief update. What is needed therefore is a natural hybridization of these two constructs.

The adopted dialogue model therefore treats the move, i.e., a combination of an action or state description with a speech act, as the central unit within the dialogue theory, but also views the move as a dynamically constructed unit along the same lines as a frame in contemporary dialogue models. The notion of dialogue move adopted differs from the frame in that it: (a) assumes a relatively simple information complexity; (b) purposefully includes no information on direct surface realizations or database functions; and (c) may be complexed together through the use of relations which reflect both causal and logical connections between individual moves and their contents. Also, critically, the move is not only used for modelling realized expression of user dialogue goals, but also provides means for representing system-initiated dialogue contributions as is required in situated dialogue domains. Moreover, the move will be treated as an analogue to true frames, in that the notion of a *dialogue plan* is rejected outright on the grounds that it undermines the essential usefulness of mental state oriented dialogue management accounts.

While the move is a complete meaningful update of the agent's mental state that indicates the intent of either system or user, moves cannot generally be directly communicated through natural language. The move, like its frame analogue, must at times necessarily exist in varied and incomplete states in addition to a fully specified state. Thus, the dialogue processes do not always operate on moves in the strict sense, but rather on ungrounded, i.e., underspecified or ambiguous, moves. Moreover, following terminology from exchange structure modelling (see Section 2.4.2), we will refer to a finer-grained sub-move unit of communicative exchange as simply *acts*.

4.1.4 Iterative Language Contextualization

Though application of the frame concept as a *staging ground* has advantages, the classical models of frame processing do not carry over well to simple action-oriented dialogues. Specifically, the classical approach to frame construction (for a query, update, or instruction) is through the complete filling of the frame by the dialogue process before the frame is passed to the relevant application knowledge sources. In this *frame-oriented resolution* model, the application knowledge source then resolves form parameters, and, when all parameters are acceptable, processes the frame in a single step. Expected outcomes from the application query can be categorized as follows:

1. All parameters were successfully resolved and the operation could be made (queried information was returned, knowledge base was updated, or an action was performed).
2. Parameters were successfully resolved but the operation could not be made for some reason, i.e, there were no return results, the information is incompatible with the present knowledge base, or an action cannot be performed in the current state.
3. Parameters could not be uniquely resolved but a set of possible parameter resolutions is provided.
4. Parameters could not be resolved because one or more parameters are incompatible with the domain model.

The behaviour of the dialogue manager then varies on the basis of the returned state from the application. For example, in the case of a set of possible parameters being returned, the dialogue process should result in a clarification question.

The advantage to this approach is that it allows dialogue manager functionality to remain highly independent of the task domain. The dialogue manager needs only be aware of the more general interface ontology to the domain and the frame that defines a meaningful update. If the application does return possible interpretations of a given parameter for clarification, then the dialogue manager is only exposed to a small number of domain individuals – but is not exposed to the mechanisms which were used to acquire them. While this separation of concerns is appealing, it is not without complications.

One complication with this model is that while it works well in domains where most slots in the frame structure are mandatory, it works less well in domains where unspecified parameters are easily retrievable within the situational context. To illustrate, consider the case of simple command-oriented dialogue in which the user of a robotic wheelchair attempts to direct the system to make a turn where the situational context makes it abundantly clear just what the direction of turning should be:

(14) a. *User:* turn
          left direction is only obvious direction
      b. *System:* should I turn left or right here?
      c. *User:* left

In such a case the clarification dialogue would be considered unnatural and superfluous. For the classic frame-oriented construction model to work in such cases, it is necessary to add additional machinery to the frame specification language to note that some parameters need not necessarily be explicitly given in the surface form of language.

A second related issue with the frame-oriented construction process is the potential for unnatural interaction in the case of long parameter-dependent interactions as illustrated below:

(15) a. *User:* I want to book a flight to New York
      b. *System:* OK, where do you want to fly from?
      c. *User:* Athens
      d. *System:* OK, when would you like to fly?
      e. *User:* Saturday the 25th
      f. *System:* OK, What class would you like to fly?
      g. *User:* Economy
      h. *System:* OK
          .... Sends information to KB
      i. *System:* Do you mean Athens Greece or Athens Georgia?

Here, a dialogue management policy that collects all frame filling information prior to contextualization results in an unnatural structure in that utterance 15d seemingly confirms grounding of the user's contribution in 15c – although we later see in utterance 15i that this had not in fact been the case. Arguably a more affective strategy would immediately contextualize user contributions and introduced clarification as early as utterance 15d.

Of course, it is not always appropriate to address underspecified parameter information immediately, as in many cases additional information may help to isolate a sensible parameter without the need for explicit dialogic clarification. This is illustrated by the following case where the user's third utterance may in a given spatial setting be sufficient to clarify the underspecified direction of the second utterance:

(16)   a. *User:* go to the junction,
       b. *User:* then turn
       c. *User:* and pass the elevator
       d. *System:* OK

In practice, there may not be any *correct* way to handle when it is appropriate or inappropriate to ask a clarification question. The examples above indicate that it may be neither appropriate to always immediately ask for clarification, nor is it appropriate to always wait. On the other hand the duration of interaction may be an important factor in that immediate resolution may be advantageous in long interactions, while waiting for further information may be appropriate for short interaction.

However, while in the case of multiple possible resolutions there might be differing perspectives on when clarification is appropriate, the case for invalid arguments is far less uncertain:

(17)   a. *User:* I want to book a flight to London
       b. *System:* OK, where do you want to fly from?
       c. *User:* Athens
       d. *System:* OK, when do you want to fly?
       e. *User:* Saturday the 25th
       f. *System:* OK, What class would you like to fly?
       g. *User:* Economy
       h. *System:* OK
           .... Sends information to KB
       i. *System:* There are no flights to London on the 25th

Here, if available to the system, it would seem appropriate that the information regarding no flight availability be conveyed to the user as soon as possible rather than waiting for the complete frame to be assembled.

The issues just discussed are directly related to how tightly the dialogic filling of a frame is coupled to a domain-specific processing of that frame. By keeping the dialogue and knowledge reasoning processes too far apart, the dialogue management process cannot make the most appropriate and context-sensitive decisions on information solicitation from and provision to users. One alternative approach is to have the dialogue manager itself responsible both for populating the parameters of a frame and resolving those parameters against both discourse and situational context through functions defined by the application. This approach thus allows the dialogue process to have complete context-sensitive control over frame parameterisation. But the disadvantage is that the dialogue process requires access to application-specific resolution which violates separation of concerns, and reduces design clarity, scalability, and re-usability.

Rather than adopting either a fill-then-contextualize or merged dialogue management and contextualization process, the move contextualization strategy applied in the Situated Dialogue Architecture keeps the dialogue management and domain reasoning processes separated, but assumes an interleaved processing model where at any processing step all information known to the dialogue process will be factored into the interpretation of the move before context augmentation, guided by the domain application, is applied. This in turn requires a clear distinction within the dialogue management processes between those processes which integrate user-provided information, contextualization and augmentation of that information based on situational context, and dialogue planning processes. Before we see how this affects the structure of the dialogue processing architecture, we consider the final dialogue architecture modelling principle.

4.1.5 Knowledge Modularization

Given a Ginzburgean perspective on dialogue organization, the epistemological foundation of the agent's mental state is of significant importance since both scalability and inter-operability with particular domains is highly dependent on having a well thought out basis for knowledge representation and meaning. Thus, rather than taking an ad hoc approach to the use and definition of types within the agent's mental state, the situated dialogue architecture relies on strongly typed resources provided by type hierarchies and ontologies. This not only requires that well defined existing ontological resources be built upon where necessary, but also that modularity be maintained in the elements of mental state which are accessible by dialogue processes.

With respect to knowledge typing and ontology, the dialogue architecture has been influenced by the desire to separate out linguistic and non-linguistic information within semantics and knowledge representation. In such an approach, referred to as *two-level semantics*, the unit of exchange at the surface semantics interface (i.e., at the interface between semantics and language analysis and generation grammars), is a true linguistic semantic representation that, rather than trying to model utterances directly in terms of the types used for the application's own internal problems solving task, models language in terms of an interface ontology which is better suited to the representation of language. At the domain end of the dialogue system, such a Logical Form may be mapped to a formal model which actually reflects the application's own embodied perspective on the world and which is more suited to the forms of non-linguistic reasoning employed by that agent.

Motivation for a two-level semantics comes from both theoretical and practical perspectives. Farrar and Bateman [2004] extensively reviewed theoretical motivations as backed both by experiences in the development of large-scale language engineering projects, and the more fundamental issues of relating accounts of language form to reasoning theories. The primary conclusions drawn by Farrar & Bateman are that language variability is simply too great to map directly from surface from to conceptual structures.

Practical reasons for favouring a two-level semantics centre on issues of resource re-use and interoperability. Namely, in many cases of dialogue system development, it can be virtually impossible to ensure that the organization of world knowledge used directly within the domain application maps to the organization of world knowledge assumed by grammars of language analysis or production. If either the grammars to be used or the systems's knowledge pre-exist, then bringing these two models into alignment in computational applications necessarily requires the adoption of a two-level semantics approach. Indeed, this perspective has been taken in large-scale computational linguistics projects. For example, as described in Section 3.4.1, the TRIPS project employed a multi-level semantics methodology to separate a general purpose grammar from domain level reasoning.

There is however an alternative perspective, most vocally advocated by Porzel et al. [2003, 2004], that a single ontological account should be given for both the grammar's interface semantics and the back-end knowledge system. This view has a natural appeal in terms of simplicity of representation and development – particularly if the domain application and grammars are not being developed in isolation. While such an approach is typically the privy of smaller dialogue systems, the argument made in favour of such an approach is that if the ontological structure of semantics models is well developed, then the grammars and semantics models constructed will not be tied to limited domains, but can potentially scale naturally to new applications. On the other hand, it is questionable whether a single ontological model can be developed which is optimized both for linguistic semantic representation and domain application reasoning. Thus, there would be a natural trade-off between these two orthogonal requirements. Furthermore, if dialogue systems are being created around domain systems that have been created separately from grammatical resources, then the need for more than one ontological structuring will naturally arise[1].

In the light of the two-level semantics approach, one question which becomes apparent is whether the representation of dialogue state, i.e., the Dialogue State Structure, should be defined in terms of the linguistic ontology of surface information or the application-specific conceptual ontology. Certain elements of any declarative dialogue state model are very close to a linguistic semantics commitment, e.g., representations of analysed user utterances. Thus, it would seems that the types used in these information state elements would be best defined in terms of the same linguistic ontology which interacts directly with a grammar. On the other hand, declarative dialogue states in the style of Ginzburg's dialogue game board also contain types of knowledge which are more domain-oriented in nature. Commitments, intentions, beliefs, but also interpreted dialogue moves, are not linguistic semantic in meaning

---

[1] Two-level semantics are sometimes confused with issues of underspecification as introduced in Section 2.2.3. It should be made clear here however that these are two separate issues, and that it is possible to have a single-level semantics system that employs underspecification methods such as a distinction between quasi-logical and logical representation forms. Thus, two-level semantics is best characterised by the use of two separate ontologies – one for the linguistic semantic categories, and another for the underlying conceptual and domain knowledge held by the agent.

but operate at a different stratum of organization. Thus, it does not seem appropriate that the types used within such attitudinal elements be defined in terms of the linguistic semantics organization. Similarly, it is not appropriate that the linguistic-oriented elements of the information state be modelled in terms of these application-specific semantic constructs.

Therefore, elements of the dialogue state may necessarily be modelled in terms of either the linguistic semantics or application (conceptual) ontologies. Indeed, in practice, several ontologies or ontological modules must be applied together in the expression of either linguistic or conceptual semantics. For example, in later chapters we will see how the linguistic semantics model is composed of a number of different ontological units.

4.2 An Architecture for Situated Dialogue Processing

The remainder of this book develops a dialogue processing architecture which aims to meet the particular needs of situated dialogue systems. The architecture builds on the strengths of declarative dialogue frameworks, but extends these to account for requirements introduced by the agent-oriented perspective, the need for a systematic deep-contextualization model, and a modular approach to linguistic resource development. Some of the basic principles of the model were introduced in the last section. In this section we make the architecture more explicit, and relate it forward to its development in subsequent chapters.

Figure 4.1 presents an outline of the architecture. Rectangles and parallelograms depict processes and data stores respectively, while rounded columns represent short term memories. Solid directed lines reflect the typical direction of control flow within the architecture, while broken lines portray an access relationship between individual modules and mental state containers. While not depicted, it is assumed that processing modules have access where necessary to domain-specific resources that specialize those modules for particular applications.

4.2.1 Processing Modules

The architecture presented in Figure 4.1 can be broken up in various ways to reflect specific collections of operations. The left-hand side of the lower two-thirds of the diagram depicts the core language processing units which together comprise a spoken dialogue system backbone (compare with Figure 3.1). The language processing modules reflect the typical elements of a spoken dialogue system as introduced in Section 3.1 along with additions due to the requirements argued for in this and earlier chapters. Specifically, in addition to the traditional language perception, language analysis, language realization, and language production modules, the following processing modules are assumed:

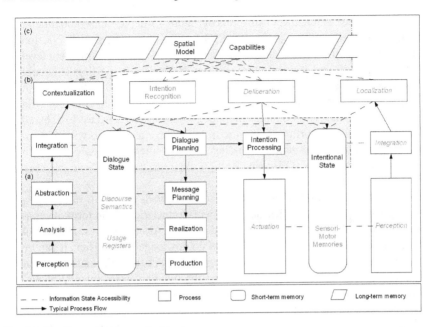

Fig. 4.1 The Situated Dialogue Architecture.

- **Language Abstraction:** Transforms shallow semantics produced at the language interface into domain-specific dialogue acts more suited to dialogic and situated reasoning.
- **Language Integration:** Integrates abstracted language content into the agent's dialogue state through the construction and manipulation of move and act objects which reflect the user's explicit dialogue goals.
- **Contextualization:** Applies domain-specific context to resolve and augment user dialogue moves.
- **Dialogue Planning:** Analyses the dialogue state – including open user moves and system initiated goals – to determine the most appropriate acts to be made by the agent in furthering user move interpretation and system dialogue goals.
- **Message Planning:** Transforms planned system acts expressed in terms of abstract conceptual semantics into a shallow linguistic semantics suited to realization.

When taken together with a model of pragmatic dialogue state, these processing modules constitute a *dialogue manager* in the traditional sense. This pragmatic dialogue state is in turn held within a shared memory of medium and short term language use which, in a fully elaborated language system, may also incorporate theories of discourse semantics as well as registers of grammatical and lexical usage necessary for language shaping and alignment [Pickering and Garrod, 2004, Purver et al., 2006].

The right-hand side of the lower two-thirds of Figure 4.1 depicts the embodiment-relevant units of the cognitive architecture. Following the discussion in Section 4.1.2, this includes both low-level actuators and perceptors, as well as medium level intention management and perception integration. Higher level non-linguistic cognitive processes such as localization, mapping, route planning, and so forth then sit above these basic embodiment functions. It should be noted that in common with the dialogue components, embodiment components may also have access to a shared memory which captures short to medium term knowledge associated with perception and the performance of action including both the intentional state itself and also sensorimotor memories [O'Regan, 2006] – although the latter have not been explored in the present work.

The top tier of the cognitive architecture concerns both persistent information such as long-term spatial memories, action schema knowledge in the form of capability definitions, and other persistent knowledge types. Alongside these sit general reasoning capabilities for spatial and non-spatial knowledge types. Thus, it can be seen that the architecture broadly follows the principles of three tier design in that there is a movement from a sensori-motor layer, through perception integration and action sequencing, through long term memory and deliberative reasoning.

Whereas the distinction between linguistic and non-linguistic functionality can be made figuratively at least in the lower tiers, the distinction breaks down at the long-term memory and deliberation tier since all forms of persistent knowledge can be leveraged off in both linguistic and non-linguistic reasoning. Indeed, looking again to the lower tiers of the architecture, it is clear that linguistic and non-linguistic processing are analogous, and can arguably be seen as two instantiations of the same basic process organization.

4.2.2 Process Control

Although the layout of processing components depicted in Figure 4.1 is clearly modular, there is variability in how the processing architecture as a whole is controlled. Considering just the core dialogue processes themselves, historically the earliest control methodologies have reflected the sense-plan-act algorithms from robotics (see Murphy [2000, ch. 2] for an introduction) where a single processing thread controls execution flow from language perception, through dialogue planning, to language synthesis. While such an approach is highly transparent, language perception, deliberation, and production cannot generally be assumed to be sequential processes treatable in a purely pipelined manner – unless it is acceptable for a system to ignore user input during deliberation or language production phases.

In contrast to applying a strict sense-plan-act control flow, a cognitively-motivated perspective would argue in favour of a highly dynamic structure in which information can be freely exchanged between processing modules. Such an approach was adopted in an earlier version of the work presented in this book where each processing component was cast as an agent within a small multi-agent system [Ross et al.,

2004d]. With agent inter-connection facilitated by a service-based broker, agents passed messages between each other only when necessary. While such a message-based, distributed approach is cognitively appealing, it does introduce considerable timing complications for what is effectively a static system. Moreover, the synchronized maintenance and update of information state becomes considerably complicated, while overall system stability remains open to individual agent failure.

A third control alternative, which is a hybrid between the two models just considered, makes the assumption that most components operate within their own control thread, but that a single sense-plan-act style control algorithm operates over those modules in a continuously repeated 'execution cycle'. This approach, motivated in part by approaches to agent design, and applied for example by Larsson [2002], remains conceptually clear in its relationship to agent control methodologies while guaranteed to operate in a timely fashion. The relative disadvantage of this approach is however that all modules are guaranteed to be called during each execution cycle – regardless of whether earlier processing stages have actually produced results which should be fed into these subsequent modules.

Here, we apply a variant on this last approach with the modification that we assume only modules with a relatively long temporal operation to run in their own threads, and that the central control loop itself factors in the logic of whether or not it is appropriate for a particular module to be called at any given stage in an execution cycle. While complicating the central control loop, the logic of individual components is simplified and the overall processing of the system – even in atypical situations – becomes more transparent. Figure 4.1 depicts the control flow in the idealised situation where user input is detected and successfully passed through all input language modules before the agent composes a response which is channelled through output modules. It should be noted that communication between modules is not direct but rather facilitated by modules reading and writing to the dialogue state structure. One significant exception to this is that dialogue planning interacts with and can directly affect the intention management component – hence the two control flow lines emanating from the Dialogue-Planning component.

4.2.3 Organizational Tiers

In the following three chapters this architecture is expanded upon by providing concrete resource definitions and computational models. Detailing all linguistic and non-linguistic units is however not feasible – nor particularly valuable – in a single book. Instead, we will focus on three distinct collections of functionality which together provide a clear view of the situated dialogue modelling process. These collections of functionality, depicted in Figure 4.1 as shaded segments marked (a), (b), and (c), group dialogue processing functionality into three tiers:

Tier (a) comprises the language interface which defines the semantic resources and grammatical processes which map surface language to units of communicative exchange via a shallow linguistic semantic representation. The language interface will be described in Chapter 5.

Tier (b) comprises the core of the language processing architecture, i.e., an Agent-Oriented Dialogue Management model which includes both non-context-sensitive dialogue processing and the agent's intentional structure and management processes. These resources and processes will be described in Chapter 6.

Tier (c) comprises the deeper application level contextualization processes which integrate user moves with models of situational context. For our purposes we will look specifically at how this level relates to the concrete topic of verbal route interpretation. This will be addressed in Chapter 7.

4.3 Summary

This chapter has set out the basic organizational principles of the situated dialogue processing architecture that is developed in this book. In particular, a set of modelling principles for the dialogue architecture were first argued for. These modelling principles called for an ontologically modularized Ginzburgian view on dialogue organization which, on one hand, accounts for an agent-oriented view on domain application, and on the other hand provides a communication management process which treats the classical move as the primary unit of dialogic exchange while allowing that move to be processed in an analogous manner to classical frames. Based on these principles a specific dialogue processing architecture was then presented in terms of its processing units and the approach taken to process organization.

Chapter 5
The Language Interface

Abstract This chapter details the first tier of the dialogue processing architecture, i.e., the *language interface* that provides mappings between surface language and conceptual units of expression. The chapter begins in Section 5.1 by describing the composition of the language interface. Section 5.2 then describes the representational basis for semantics expressions. Section 5.3 subsequently details the first of two language interfaces semantics layers, i.e., the linguistic semantics which interfaces with grammars of language analysis and production. Section 5.4 then moves up a layer of abstraction from linguistic semantics to the conceptual semantics that couples dialogue knowledge to the application's own representation types. A context-sensitive functional transformation model is then developed to establish the necessary mapping between linguistic and conceptual meanings. In Section 5.6, we draw the chapter towards a close with a discussion.

5.1 Elements of a Language Interface

Amongst the many dimensions under which language process and structure can be analysed, the one which is of critical architectural importance to dialogue system design is the notion of *stratification*. This looks at language as a stack of representational layers where a structure within one layer of representation is expressed or *realized* by units from a lower, more concrete, layer [Halliday and Matthiessen, 2004, p. 24]. This stratification principle permeates both theoretical linguistic frameworks and concrete spoken dialogue system architectures, and is overtly present in the modules of the current language interface depicted in Figure 5.1. The language interface thus reflects a computational view of the stratified language stack, here consisting of four representational layers and three process layers that map between those representation layers. The representation layers are, in increasing order of abstraction: (a) *Signal* - language in a physical medium, e.g., sounds and words on screen; (b) *Wordings* - pre-signal representations of messages to be communicated, e.g., text string outputs of speech recognizers; (c) *Linguistic Semantics* -

descriptions of the wordings in terms of formal categories which reflect the organization of the world in language; and (d) *Conceptual Semantics* - abstractions of linguistic semantics in terms of formal categories which reflect the organization of the world in the agent's own cognitive perspective. The first inter-representation process layer maps between *Signal* and *Wordings*, and in the case of spoken language includes speech synthesis for *Production* and speech recognition for *Perception*. The second process layer maps between *Wordings* and *Linguistic Semantics* and is realized through lan-

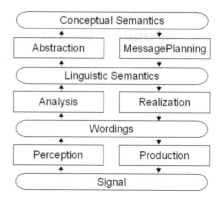

Fig. 5.1 The Language Interface

guage analysis and generation grammars. Finally, the third process layer maps between *Linguistic Semantics* and *Conceptual Semantics* and is achieved by semantics abstraction and message planning.

While significant research potential remains within the processing layers that map between *Signal* and *Wordings*, and *Wordings* and *Linguistic Semantics*, these are not the focus of this book and so will not be considered in any depth here. Instead, we focus on the *Linguistic Semantics*, the *Conceptual Semantics*, and the processes which map between the two. With respect to the lower portions of the language interface, we simply assume the existence of commodity speech perception and production processes, and that suitable analysis and language generation grammars exist which take as meaning interface the linguistic semantics we develop in this chapter. Concrete fillers for these lower layers and processes are briefly discussed in Chapter 8.

Considering the linguistic semantics layer, we saw in earlier chapters that the linguistic semantics serves a pivotal role within any non-trivial dialogue model or system. Principally, it acts as an interface to grammatical resources to facilitate language analysis and realisation tasks; it can form the basis for a discourse semantics structure that enables naturalistic language devices such as anaphora and ellipsis to be processed; and it serves as interface to conceptual resources and deliberative decision processes such as typically embodied within dialogue management and application processing. While developments such as Asher and Lascarides's [2003] SDRT provide a clear perspective on how a linguistic semantics which addresses these diverse responsibilities can be integrated within a logical framework, such a logical framework is however not yet directly employable within practical dialogue systems for the following reasons:

- **Delicacy of Content Description:** Compositional and Discourse Semantics development focuses on the mechanisms of discourse representation without looking at the specifics of world description. While this is acceptable in its own right, application of models to particular domains requires that semantic resources be customized for those domains. Considering the specifics of spatial language modelling and situated communication, the linguistic coverage given by these frameworks are impoverished with respect to our understanding of semantics for many specific domain issues, and, worse still, the categories employed within the semantic theories are frequently ad hoc.

- **Tractability:** A considerably more significant issue is practical model applicability. While the use of glue logic, and logic mixing in general, improve the tractability of models such as SDRT in comparison to both classic agent-based dialogue systems and holistic discourse semantics accounts (see e.g, Hobbs et al. [1993]), reliance on hybrid and modal logics leaves these theories intractable for practical computational purposes. Moreover, many of the assumptions made by these theories with respect to the responsibilities of the conceptual interface mean that such full blown theories are not widely applicable except in a few small cases (see [Schlangen, 2003] for a discussion of the limitations of computational applications of models such as SDRT).

Given the importance of tractability to the development of practical situated dialogue, we therefore forego a full formal linguistic semantics such as SDRT in favour of a light-weight linguistic modelling, which, while being less powerful than a dynamic semantics, provides the level of linguistic expressivity necessary to capture a range of spatial and action-oriented language features. Achieving the necessary expressivity in turn requires that semantics models include a content language sufficiently detailed for the particular domain of application. On the other hand, following from the arguments for *two-level* semantics in previous chapters, the categories used in such a content language should not be based on the types used within a targeted domain application, but should instead take an application independent view while providing a level of detail sufficient for the application. To aid re-usability, it is equally important that the linguistic semantic content language not be defined in an ad hoc way, but should be based on empirical data and well defined knowledge organizational principles. In the last 10 years, the fields of Formal and Applied Ontology have become increasingly important in this regard. The linguistic semantics model developed here therefore aims to provide fine-grained but ontologically well founded accounts of situated language meaning based on existing resources whenever possible.

Considering the conceptual semantics layer, the same issues of expressivity and practical tractability also hold, but unlike in the case of the linguistic semantics, the purpose of the conceptual semantics is not the facilitation of an interface to natural language grammar, or the processing of linguistic devices such as anaphora and ellipsis, but is rather concerned with pragmatic tasks such as the modelling of dialogue goals, the interpretation of answers against questions, and principally the interface to non-linguistic domain-specific reasoning. Thus, while the same principles of ontological organization and weak logic expressivity will be applied to the conceptual semantics, the types used will necessarily draw on application-specific modelling. As a consequence, it should be noted that while the linguistic semantic expression and reasoning is effectively confined within the language interface, the use and definition of conceptual semantics necessarily spills up into higher cognitive processes.

What is relevant within the language interface are the basic units of the conceptual semantics, and how those basic units are mapped to linguistic semantics chunks.

Before considering the details of the linguistic semantics, the conceptual semantics, and the mappings which exist between them, we first consider the representation basis assumed.

5.2 Representation & Modelling Basis

Both linguistic and conceptual semantics models require appropriate languages to both represent knowledge within semantic layers, and to provide a means of defining terms within those knowledge representations. In this section we describe the choices made with respect to these issues.

5.2.1 Description Logics & The Manchester Syntax

As a concept definition language, i.e., a language in which we can define individual concepts and relations which populate the semantics layers, we will apply Description Logic (DL). Description Logics, called terminological systems and concept languages in their formative years, were conceived in part to give a clean semantics to the frames and semantic networks based knowledge representation models that developed through the 1970s and 80s. While not as powerful as first order predicate calculus, Description Logics are sufficiently expressive to describe categories, individuals, and their properties, and in so doing, answer questions on subsumption, classification, and consistency [Nardi and Brachman, 2003].

There are many different variants of Description Logic available, and an equally large number of syntactic formalisms or serializations. All Description Logics of interest are built on the Attribute Language $\mathcal{AL}$ which provides a minimal language of atomic concepts and atomic roles. More interesting Description Logics can then be constructed by adding additional constructors, and it is these constructors that are used to distinguish the different DL variants. Table 5.1 gives a short listing of the most common Description Logic constructors available. It should be noted that in addition to the constructor symbols, the symbol 'D' is often used to describe when a description logic includes data type reasoning, i.e., a reasoner such as Pellet [Sirin et al., 2007], is said to support the Description Logic $\mathcal{SHOIQ}$ with datatype reasoning.

The choice of Description Logic, specifically the variant $\mathcal{SHOIQ}(D)$, as a category definition language is motivated by a number of factors. Significantly, DL is far more computationally favourable than first-order, or higher-order logics, and this has clear significance for its use within practical dialogue systems. Moreover, and on a related note, $\mathcal{SHOIQ}(D)$ is extremely well supported from a tool and reasoner perspective, and is the formal model behind the Semantic Web language, OWL 1.1

Symbol	Description
$\mathcal{Q}$	Qualified Cardinality Restrictions
$\mathcal{N}$	uNqualified Cardinality restrictions
$\mathcal{O}$	nOminals
$\mathcal{F}$	Functionality Axioms
$\mathcal{H}$	role Hierarchies
$\mathcal{J}$	role Inversal
$\mathcal{S}$	$\mathcal{ALC}_{\mathcal{R}+}$
D	Datatype Reasoning

Table 5.1 Description Logic constructor names & descriptions. The construct $\mathcal{S}$ is a shorthand for the popular base language $\mathcal{ALC}_{\mathcal{R}+}$. This base language includes concept negation and intersection, transitive roles, as well as universal restrictions and limited existential quantification.

[Smith et al., 2002]. This amongst other reasons has led to the development of a wide range of reasoning systems for description logics, e.g., Racer Pro [Wessel and Möller, 2005], MSPASS [Hustadt and Schmidt, 2000], and Kaon2 [Motik, 2007]. Moreover, $\mathcal{SHOJQ}$(D) through OWL 1.1 is also well supported by tools; such as, for example, the the Protégé Ontology Editor [Knublauch et al., 2004].

Description Logics can be given many different syntaxes. In this book we will use the Manchester Syntax for OWL 1.1 [Horridge and Patel-Schneider, 2008] for most concept and role definitions. The Manchester Syntax is intended to be a considerably more human-readable notation then provided for in either XML-based or logic-based representations. For example, Manchester Syntax abstracts away from the set theoretic underpinnings of Description Logic and avoids the use of mathematical notation, substituting for example the term `and` for the $\sqcap$ used in the Description Logic Syntax [Baader, 2003]. The reasons for choosing this syntax here, inspite of the relative compactness and model transparency of the still popular Description Logic Syntax outlined by Baader [2003], is its accessibility to non-logicians, and that, at the time of writing, the Manchester syntax seems likely to become a de facto standard. To illustrate a DL concept definition, and the Manchester Syntax in particular, we can define that a man is a male person with the following:

Definition 1. `Class: Man SubClassOf: Person`
`EquivalentTo: Person that hasGender value male`

where all concept, role, and individual names follow the convention that concept and individual names start with an uppercase letter followed by lowercase letters, and that role names start with a lowercase letter.

Manchester Syntax also allows the definition of individuals as will sometimes be used in later examples. To illustrate, we define `Mike` as a `Man` with a particular age as follows:

Example 1. `Individual: Mike`
`Types: Man`
`Facts: hasAge 41`

Description Logic, and in particular $\mathcal{SHOJQ}$(D) (OWL 1.1) gives us a representation formalism which is powerful enough to characterize the ontological constraints on a frame language for the representations of fragments of linguistic and conceptual meaning, while yet retaining all important aspects of tractability and decidability. Naturally, Description Logics are clearly limited in their own right; for example we can not express the sort of ontological knowledge required for representing many facets of general knowledge, i.e., knowledge beyond category definitions, hierarchies, and properties. However, we argue that a simple meaning representation language need not include such information, and that if such information is in fact needed for some element of reasoning in the Dialogue Manager in general, it can be provided though a secondary reasoning system based on something approaching a first order language as necessary, e.g., RuleML [Wagner et al., 2004].

5.2.2 Frame Object Structures

Given the use of Description Logic, and OWL in particular, as a concept definition language across both semantics layers, it is appealing to cast parts of the agent's mental state, and in particular the individual semantics specifications and other mental state content directly into DL instance models which can be reasoned over by off-the-shelf systems. However, such a direct use of the Description Logic formalism in dialogue, as for example investigated by [Buecher et al., 2001], or indeed any conventional formal logic, belies the structural and highly dynamic nature of both representations of spatial information and the Stalnakerian view of dialogue state as pursued here.

Instead, we require a flexible model of semantic and mental state representation. A number of options are available for providing a basis for such a model. One option is to make use of Description Logic reasoners enhanced through appropriate use of ontology modularization [Kutz et al., 2004] and rule-based Description Logic extensions to provide a basis for a representation system. While recent work towards this goal looks promising [Hois and Kutz, 2008], a complete and tractable theory integrating heterogeneous information state descriptions, application knowledge, and the algorithms and operations over those knowledge types is not yet realizable. On the other hand, a formal dynamic logic approach can be used to capture some of the dynamic aspects of both theories of discourse representation [Asher and Lascarides, 2003] and agency [Wooldridge, 2000]. Unfortunately, at the present time the direct application of rich dynamic logics to discourse modelling is intractable except in very small cases [Schlangen, 2003]. In contrast to highly formal accounts, an alternative view is to pursue an *engineering-oriented* representation based on domain needs and making light use of traditional symbolic representation. Although such an approach both benefits rapid prototyping, and simplifies transformations between symbolic and non-symbolic representations, the advantages of symbolic reasoning, the explicit declarative presence of information state representation, and the clean separation of domain resource from generalized processing mechanisms are lost.

As a compromise between these perspectives, in this book we will assume that individual processing components are subject to their own internal modelling constraints, but that a common knowledge representation language can convey information between these diverse processing components. This approach ensures that semantic and mental state instances can be easily mapped to a number of different models as used within language and domain processing components. Specifically, the representation model consists of typed, frame-like structures based on the modelling potential of Description Logic, but which also includes variable constructs more typical of query languages such as the SPARQL query language for RDF [Prud'Hommeaux and Seaborne, 2008]. The frame language expressions, termed *Frame Object Structures*[1], are used to capture instances of linguistic and conceptual semantic expressions exchanged between architectural components, and as the basic expression language for contents of the dialogic mental state developed later. Frame object structures consist of the following primitives:

- **Types:** All objects, roles, and propositions within a Frame Object Structure are assumed to be typed. We assume role and type hierarchies to be defined in terms of an appropriate ontology. Following conventions in RDF and the web ontology language (OWL), types start with a capital letter, and are assumed to exist within a name-space, and that in the case of possible ambiguity, that namespace is expressed as a colon delimited prefix.

- **Objects:** Objects denote concrete entities or individuals in DL notation. Objects must be typed according to some ontology, and in notation are linked to their type via a # delimiter.

- **Propositions:** A proposition[2] is used to assert features on an object without introducing a new object. Like objects, propositions are typed, but do not introduce new discourse referents in a linguistic semantics. Since the proposition is inherently typed, only the proposition itself need be represented explicitly, e.g., Female is a proposition definition assumed to be defined as a type within an accessible ontology.

- **Datatypes:** The primitive data types string, float, integer and boolean may be used in the assertion of object properties.

- **Variables:** Variables identify unknown objects or proposition values. Variables are written as terms where the first character is a question mark, and the remainder is a nominal or proposition expression. For example, ?Ball#x1 represents an unknown object referentially accessible as x1 of type Ball, while ?Gender represents an unknown Gender type. Variables are commonly used

---

[1] The Frame Object Structure (FOS) used here can be traced genealogically to the representation formalism used by Collier [2001] and later [Ross et al., 2004c] in the modelling of agent programming languages. In particular the Frame Object Structure retains elements of the basic syntactic formalism such as the bracketing conventions. However, whereas the original formalism was based on untyped predicates, variables, and composite functions, the Frame Object Structure considerably extends the original with features of frame languages and Description Logic in particular, e.g., strong typing, role relations, objects, and propositions. Thus, the term Frame Object Structure is a retrospective re-terming of Collier's [2001] First Order Structure.

[2] In DL terminology, this notion of a proposition is roughly equivalent to a nominal.

in expressing queries at a conceptual level, as well as expressing variables in
plan definitions within the application interface.

- **Roles:** Roles explicitly express relationships holding between objects, or between objects and propositions. Role types are assumed to be defined in terms
  of an applicable ontology, and are written according to the conventions of Description Logic and OWL.

By convention we write representation expressions as frame structures. To illustrate the format's use, a number of simple examples are now given. The most
basic statements allow objects to be linked, or properties to be asserted on individual objects. For example, we might express that a red apple is being eaten with the
following form:

Example 2.   `Eating#e1(actee(Apple#a1(color(Red))))`

where `e1` is an object of type `Eating`, `actee` is an object role, `a1` is a object of
type `Apple`, `color` is a proposition role, and `Red` is a proposition.

Within a semantics-oriented account of dialogue management, the representation
of questions and unknown information is almost as important as representing known
information. Historically the semantics of a question has been captured in terms of
lambda abstractions and the set of propositions which can act as answers to that
question [Hamblim, 1973]. Instead of using lambdas here, we make use of existentially scoped variables and sets to extend the FOS language towards handling a range
of standard questions which can when necessary be resolved against an underlying
model through unification. Here we introduce the syntax for representing questions
but return to the topics of question-answer combination and model querying as required in later chapters. Specifically, we directly apply the FOS mechanisms for
elicitation questions. For example, the following expression can be used to identify
an object of type `Person` which is assumed to have played the `actor` role in an
`Eating` object:

Example 3. `Eating#e1(actee(Apple#a1(color(Red))),`
                         `actor(?Person#p1)))`

Similarly we assume polarity questions to be expressible as a variable over event
or process types, e.g., to capture "Will I drive you to the kitchen" at the conceptual
level we write:

Example 4. `?Drive#e1(actee(Person#h1),`
                         `destination(Kitchen#k1))`

Choice questions on the other hand can be expressed as a set of polarity questions
when fully expanded. For example, to express, "should I drive to the kitchen or the
office?" at the conceptual level, we extend the above representation through the use
of an in-built logical operator OR as follows:

Example 5. `OR(alpha(?Drive:e1(actee(Person:h1),`
                         `destination(Kitchen:k1))),`
                `beta(?Drive:e1(actee(Person:h1),`
                         `destination(Office:o1))))`

In practice however it can be useful to merge the options into a single event which
takes a set of variable types as parameter as follows:

Example 6. `Drive:e1(actee(Person:h1),`
`                    destination(Set(setElement(?Kitchen:k1),`
`                                    setElement(?Office:o1)))`

While this treatment through the use of sets requires that a role can be filled by a generic set rather than an object of a specific type, the advantages of such a representation, for example in the construction of transform rules in language production as described later in Section 5.5.1, outweigh the disadvantage of such a modelling choice.

5.3 The Linguistic Semantics Layer

This section details the linguistic semantics model that operates at the syntax/semantics interface. The section thus begins by considering the role of ontologies in providing a foundation for linguistic semantic categories.

5.3.1 The Meaning of Things

A variety of 'upper ontologies' and lexical resources have been developed in recent years which could in principle be used as a modelling basis for the content of linguistic semantics. DOLCE [Masolo et al., 2003], OpenCyc [Cycorp, 2004], and SUMO [Niles and Pease, 2001] are, for example, popular foundational ontologies that provide a cognitively inspired upper description of the world. The BFO [Bittner and Smith, 2003] on the other hand is a 'realist ontology' which attempts to provide a more accurate reflection of how the world really is, rather than our conceptualisation of it. Such ontologies provide both a sound formalised foundation for semantics content, and are well suited to the organization of application-oriented knowledge. However, their cognitive and realist stances place them at too far a distance from the organization of meaning conveyed by conventional grammatical resources. At the other end of the scale, there exist a number of lexically-oriented resources which have been applied as descriptions of linguistic semantic content. In particular, both the WordNet [Fellbaum and Miller, 1998] and FrameNet [Baker et al., 1998] classifications offer considerably more fine-grained and linguistically motivated views on meaning potential than those offered by the upper ontologies in isolation. Unfortunately, the organization of these resources lies so close to surface language that it is extremely difficult to make direct generalizations from surface language to a level of meaning suited to knowledge representation processes.

Rather than making use of either conventional upper ontologies or lexically-oriented linguistic resources, here we will anchor propositional meaning within a so-called *linguistic ontology*. Unlike conceptual or realistic ontologies, a linguistic ontology provides a formal theory of the world based on categories motivated by lexicogrammatical evidence. This places the ontology at a level that is well suited to describing meaning at the syntax/semantics interface, yet allowing abstraction in many cases over common organizational phenomena in language. In particular, we apply Bateman et al.'s [2008] Generalised Upper Model (GUM) version 3, a linguistic ontology rooted in the school of Systemic Functional Linguistics [Halliday and Matthiessen, 2004, Eggins, 2004, Thompson, 1996]. Although the reasons for selecting GUM include its lineage as a semantics description resource in computational linguistics, the most important reason for selecting GUM as a propositional content potential is that the current Generalized Upper Model includes a detailed view of the semantics of action and spatial language as required for situated dialogue.

We proceed by first introducing the upper categories and relations within the Generalised Upper Model. We then look to the lower-level organization of language within the linguistic semantics model applied here. Finally we provide a small number of linguistic semantics examples for spatial language.

5.3.1.1 Bateman's Generalised Upper Model

The modelling potential provided by GUM is most sharply conveyed in terms of the type and relation hierarchies which form the backbone of the ontology. Figure 5.2 depicts the upper categories of GUM's type hierarchy. The top category GUMThing has three subclasses:

- The Configuration, or *figure* in Halliday & Matthiessen's terminology, can be viewed as the basic fragment of experience that embodies one quantum of change [Halliday and Matthiessen, 1999, p. 128]. More concretely, a configuration is the central modelling unit within GUM, and as such classifies groupings of meaning commonly expressed in natural language. Thus, a configuration groups together a nuclear process, zero or more central participants within that process, and any number of peripheral circumstances for the process. Types of configurations thus reflect grammatically licensed collections of participants in language.
- The Element is the super type of all types which can participate within a Configuration. Thus, from a grammatical perspective, whereas a configuration typically captures the meaning of whole clauses, instances of the Element type encode the meaning of simple clause constituents. Elements are in turn broken into Process, SimpleThing, SimpleQuality, and Circumstance which map roughly to, or are realized by, the grammatical categories of verbs, noun phrases, adjectives, adverbs, and prepositional phrases.
- The MultiConfiguration is the super type for all types which can be used to assemble individual Configuration instances into larger scale complexes of ideational meaning. Thus, from a grammatical perspective, the MultiConfiguration defines the semantic reflex of super-clauses structures such as what we might call multi-clause *sentences* or *texts*. Two subclasses of MultiConfiguration are provided: Expansion, which describes a

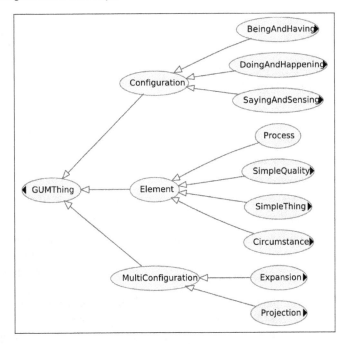

Fig. 5.2 The upper taxonomy of concepts in GUM 3

material extension, elaboration, or enhancement of one configuration by another; and Projection, which accounts for constructions such as quotes or
mental processes where one configuration is expressed as living on a different
content plane to another.

Participation of elements within configurations, and configurations within multi-
configurations, are expressed through elements of a separate role hierarchy rather
than relying on valency – thus positioning GUM as a neo-Davidsonian or dependency semantics. The uppermost tiers of the relation hierarchy are depicted in
Figure 5.3. Three of the four upper relation types are used to identify the roles
played by individual Element types within Configuration types. One upper role type participantInMultiConfiguration is the super type for all
relations which describe the role played by a given configuration within a multiconfiguration.

As indicated, GUM 3 has been specifically extended to provide a comprehensive
account of the natural language semantics of spatial language. These extensions, described in detail by Hois et al. [2008], are rooted in the traditions of formal spatial
language semantics (e.g., [Eschenbach, 1999]), theories of cognitive spatial semantics [Talmy, 2000], and more descriptive accounts of spatial phenomena in language
(e.g., Levinson [2003b]). The approach thus results in category types that are motivated by the distinctions made by language in its construal of space. We will make

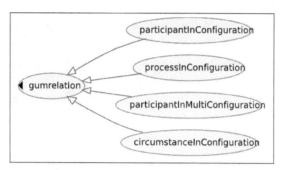

Fig. 5.3 The upper taxonomy of roles in GUM 3

no attempt to replicate the detail of the spatial semantics model here. Instead, we
highlight some of the more salient modelling points so as to introduce the overall
approach, and detail how specific instances of situated language constructions are
handled by GUM.

Of the configurations, those of most relevance to the linguistic description of spatial language are the classes of `SpatialLocating`, `NonAffectingSpatial-`
`Doing` and `AffectingSpatialDoing`. The `SpatialLocating` category
captures the semantics of *locating expressions*, while the latter two configuration
types capture the semantics of a wide range of motion expressions. Such motion
expressions range from simple turnings to more complex transitive expressions
which express motion along a path. Configurations define typical dependent semantic participants, e.g., the performer of the action, the direction towards which
the action is to be made, and so forth. Of particular relevance to the specification
of spatial information in language are the categories of `GeneralizedRoute` and
`GeneralizedLocation`. These categories licence the semantics of surface language which expresses route elements and prepositional phrases respectively. Based
on grammatical evidence, a generalized route is defined as consisting minimally of
one of: (a) a source, (b) destination, or (c) a path placement role. Roles such as
source and destination are in turn filled by so called `GeneralizedLocation`
entities which comprise both the `relatum` of an expression, and the semantic spatial relation expressed by a dynamic spatial preposition.

5.3.1.2 Lower Model Organization

Before considering illustrative examples of GUM modelling, and its description of
spatial language in particular, it is useful to also consider how GUM, as an upper ontology, can be extended to account for low-level modelling phenomena. The
most common methodology for lower-model extension of linguistic ontologies at
the syntax/semantics interface has been to associate lexical items with ontological
categories. Such an approach is exemplified in the Sentence Planning Language

(SPL) which is used as input to the KPML language generation system [Bateman, 1997], e.g., :

Example 7. The weather is bad.

```
(PA / PROPERTY-ASCRIPTION :TENSE PRESENT
    :DOMAIN (WE / OBJECT :LEX WEATHER :DETERMINER THE)
    :RANGE (B / QUALITY :LEX BAD))
```

where PROPERTY-ASCRIPTION and OBJECT are ontological types taken from a predecessor of GUM, while a LEX macro is used to select a lexical stem feature for the upper ontology object. This mixture of ontological and grammatical types is common to many shallow semantics models. Although this approach is highly pragmatic in that it allows language analysis and generation to operate without detailed lower ontologies, it is difficult to see how the approach can be used directly within semantic and pragmatic reasoning resources.

An alternative approach is to develop lower model ontologies which formally extend the upper model definition with domain-specific types and relations. This approach is appealing since it gives a more uniform mechanism for semantic type definition, and hence allows us to develop clean ontological descriptions for many of the entities described in our semantics. However, whether or not lower model types can be created for all relevant lexical choices is questionable. For example, in the case of near-synonyms, it is unclear whether two distinctive lower model categories should be postulated. What is more, the development and maintenance of a meaningfully axiomatized and relatively complete lower-model is far from trivial.

Rather than adhering strictly to either of these alternative, let us take a pragmatic middle-ground to the lower model organization problem. Namely, we assume that, where reasonably possible, the GUM upper ontology is extended by populating lower layers of modelling; however this delicacy of distinction is not required to extend all the way to lexical choice. For the current work, the population of lower-model entities has been based on the requirements of the application domain which will be described in Chapters 7 and 8. Nevertheless, a more principled population of the lower model for a given language may be possible through an alignment of GUM categories with those of existing lexical resources such as WordNet [Fellbaum and Miller, 1998] or FrameNet [Baker et al., 1998].

5.3.1.3 Sample Linguistic Semantics

To illustrate the form of GUM semantics in general, and in particular the semantics of spatial language, consider the following sentence along with its propositional semantics licenced by GUM and expressed as a set of OWL statements:

Example 8. The box is on the table

```
Individual: B1  Types: Being

Individual: B2  Types: Box

Individual: S1  Types: Support
```

```
Individual: T1  Types: Table

Individual: SL1 Types: SpatialLocating
                Facts: process B1,
                       locatum B2,
                       placement GL1

Individual: GL1 Types: GeneralizedLocation
                Facts: relatum T1,
                       hasSpatialModality S1
```

Here the types SpatialLocating, GeneralisedLocation, Support, and the relations process, locatum, placement, hasSpatialModality, and relatum are supplied by GUM, and the types Box and Table are lower model categories which subclass the GUM category Simplething. Although the Manchester Syntax used in Example 8 is compact, it is sometimes useful to re-write the model as a traditional dependency structure in FOS notation as follows:

Example 9. The box is on the table

```
(gum:SpatialLocating#SL1(
    gum:process     (gum:Being#B1),
    gum:locatum     (lm:Box#B2),
    gum:placement (gum:GeneralisedLocation#GL1(
        gum:hasSpatialModality (gum:Support#S1),
        gum:relatum     (lm:Table#T1))))
```

where the prefixes gum and lm denote categories as members of the GUM ontology itself or the lower model respectively.

The GUM based semantics of a single clause can frequently require the presence of more than one configuration. To illustrate, consider the following example clause and its propositional semantics which includes three configurations:

Example 10. The red box is on the large table

```
(gum:SpatialLocating#SL1(
    gum:process     (gum:Being#B1),
    gum:locatum     (lm:Box#B2),
    gum:placement (gum:GeneralisedLocation#GL1(
        gum:hasSpatialModality (gum:Support#S1),
        gum:relatum     (lm:Table#T1)))
(gum:ColorAttributeAscription#CAA1(
    gum:domain      (B2),
    gum:range       (lm:Red#B1)))
(gum:SizeAttributeAscription#SAA1(
    gum:domain      (T1),
    gum:range       (lm:Large#L1)))
```

Here one configuration expresses the spatial arrangement, with the other two expressing colour and size attribution to the box and table objects.

It should be noted that the presence of multiple configurations in the semantics above is distinct from the treatment of the semantics for clause complexes. As noted earlier, the Generalized Upper Model provides the MultiConfiguration class to model the reified relationship between two instances of either configurations or multi-configurations. Since in this book we are focused on action and plan descriptions, the expression of logical relationships between action descriptions is of most importance to us. Thus, we assume the existence of three domain-specific `MultiConfiguration` subclasses to handle logical relationships between clauses. These subclasses, `SequenceMC`, `DisjunctionMC`, and `ConjunctionMC`, may be given common-sense interpretations.

5.3.2 Interpersonal & Textual Meaning

Strictly speaking, the semantics given in example 10 is also the propositional meaning licensed by the Generalized Upper Model for variants such as "the box on the large table is red", "is the red box on the large table?", and many more. Within a Systemic Functional Linguistics framework, the semantic differences between these sentences concern so called *interpersonal* distinctions such as those typical of speech act theory, and *textual* distinctions typical of information structure accounts. GUM on the other hand is, in Systemic Functional terms, an *ideational* model of meaning which explicitly excludes issues of *interpersonal* or *textual* meaning. We must therefore extend the linguistic semantics beyond the ontological organization offered by GUM. Before detailing these extensions, we first briefly introduce the Systemic Functional Linguistics framework so as to better describe both what is missing from GUM, and thus what is required of the extension.

5.3.2.1 (Meta-)Functionality in Meaning

Systemic Functional Linguistics addresses the function of language rather than simply its structure; thus contrasting significantly with the approach taken by Chomsky's Formal Linguistics. While acknowledging the importance of structure in language, functional linguistics focuses in equal measure on functional and structural aspects. Thus allowing us to investigate the processes that go into play in understanding and creating language.

Although Functional Linguistics provides an integrated view of language within a multi-dimensional framework, here we are concerned with just one dimension of analysis, i.e., the *meta-functional* dimension which characterizes language in terms of the highly generalized functions language has evolved to serve [Halliday and Matthiessen, 2004]. In particular, we say that language is split into three distinctive meta-functions: the **ideational**, the **textual**, and the **interpersonal** meta-functions. These three different meta-functions of language in turn are manifest at all strata of the language system, i.e., grammar, semantics, and context. We can best explain the meaning of these meta-functions in terms of how they partition the semantics stratum into three distinct units:

- **The Ideation Base:** is the part of the semantics stratum which can be viewed as a resource potential for the structuring of meanings [Halliday and Matthiessen, 1999]. Thus, the Ideation Base corresponds most closely – but not identically – to propositional semantics within formal linguistics.

- **The Text Base:** contains those aspects of meaning construction and interpretation which involve the identification of focus spaces over the ideational content. The text base also contains the mechanisms which describe how these focus spaces evolve over the course of a complete text or dialogue. Thus, from a language production perspective, the text base allows us to put the ideational content of one message in the context of the conversation.
- **The Interaction Base:** captures the meaning of language from the perspective of social exchange. Thus, whereas the ideation base sets out the resources which allow us to express situations within a message, the interaction base sets out the resources which allow us to express our social roles, and also our attitude towards those contents. Such attitude is expressed in terms of whether we question, agree with, request, or commit to these contents and so forth.

The definitions of both interpersonal and ideational semantics are broad and can hence encompass considerable descriptive breadth. Since interpersonal semantics should cover the meaning of a text or utterance as used in a communicative exchange, a true interpersonal semantics account should cover a substantial portion of speech act theory and discourse structure introduced in Chapter 2. Analogously, textual meaning subsumes theories of information structure and other *aesthetics-oriented* patterning in language.

While GUM does not consider these aspects of meaning, there have been attempts within the functional linguistics community to populate the models of interpersonal and textual semantics (e.g., [Martin, 1992]). Unfortunately these accounts are not easily accessible to compositional theories of grammar, and though we consider comprehensive accounts of both interpersonal and textual semantics essential to developing a controllable and fine-grained interface to grammars – something which is indispensable in natural communication – it is not possible to build and apply such detailed accounts here. Therefore, we will take a highly pragmatic approach to the question of interpersonal and textual semantics and simply extend the semantics provided by the Generalized Upper Model with a minimal account of interpersonal and textual meaning which is sufficient for current purposes.

With respect to textual semantics, Halliday and Matthiessen [1999] define the core unit of textual information, i.e., the *message*, as a view on the ideational content of a complete discourse as captured within a single language contribution. In other words, ideational semantics is assumed to be relatively static like a knowledge base, and the primary responsibility of textual semantics is then to pick out the elements of ideational semantics conveyed in a single utterance. In practice however, the Generalized Upper Model is used most often as the basis for a *micro-semantics* that reflects the meaning of individual statements at the syntax/semantics interface. Thus, ideational semantics as used in GUM already picks out its own semantic content, and therefore the primary responsibility of the textual semantics message is defunct for our current purposes. Textual semantics however also involves structural mechanisms such as *theme and rheme* to signal new and important information.

Although the integration of such information structuring mechanisms with computational grammars has been previously investigated (see Kruijff-Korbayova and Steedman [2003] for an overview of issues, or Kruijff-Korbayova et al. [2003] for a specific application of information state theory in a semantics-rich grammar development), these factors are not central to the task of deep contextualization considered here. Thus, for the current linguistic semantics model, we assume the configuration to serve the same purpose as a message, and subsequently assign to it the same responsibilities. It is clear however that such an assumption would have to be removed if the current semantics model was to be extended towards a full discourse semantics.

5.3.2.2 Modelling Interpersonal Meaning

Although we apply a minimal model of textual semantics, interpersonal meaning is of considerably more importance to dialogic exchange, and, thus, we require a more complete account of this meta-function of semantics. To develop an account of interpersonal linguistic meaning, three core issues must be addressed: (a) how delicate should the account of interpersonal meaning be with respect to units of lexicogrammatical expression? (b) what potential of interpersonal expression do we make available for describing interpersonal meaning? and, (c) to what units of meaning expression do we assert features of interpersonal semantics? These three issues, and the approaches taken to their treatment are addressed in the following.

Units of Interpersonal Expression
As described in Chapter 2, 'speech act'-like interpersonal analysis is frequently applied to surface form units of varying granularity. That is, speech act analysis is often applied to units ranging from minor clauses, through major clauses, sentences, and conversational turns. The choice of analysis unit depends on one hand on the goals of analysis, but also on the commitment made to whether a speech act is to be viewed as a surface form oriented or cognitive entity. For example, within the Dialogue Act Taxonomy community, analysis units are typically utterances or turns since cognitive classification choices are allowed based on the overall context of a conversation. On the other hand, a classical surface speech act theory operates at a considerably finer grained level of analysis.

For our model of linguistic semantics, we are concerned not with deep contextual analysis, but with the provision of a formulation of interpersonal meaning at the syntax/semantics interface. It is well known that interpersonal meaning has a direct reflex in the grammatical mood of the major clause in English. Thus, the interpersonal semantics should be capable of modelling that clause-oriented interpersonal meaning. Moreover, units smaller than the major clause can carry interpersonal meaning, i.e., sub-clause discourse particles such as "yes", "no", and "okay". Thus, the interpersonal meaning model must be capable of attributing the semantics of such discourse particles consistently. In light of this, the current model eschews the assignment of interpersonal meaning to coarse units such as utterances and turns

in favour of a linguistic semantics model centred around the analysis of clause and
sub-clause units.

Speech Function Types

Following SFL conventions, we will use the term Surface Speech Function to refer
to the unit of interpersonal meaning assigned to the semantic reflex of a major or
minor clause.[3] A number of detailed sources are available to inform the categorisa-
tion of surface speech function types. However, most are either too cognitive or too
detailed to be applied as a simple model of interpersonal meaning at the syntax/se-
mantics interface. Thus, here we make use of a simple categorization of surface
speech functions. Following the SFL methodology, this categorization is motivated
by grammatical evidence. Grammatical mood is the obvious realization of interpersonal meaning, and therefore it is appropriate to organize interpersonal meaning in
the linguistic semantics based on systematic distinctions in the grammatical mood
system. While necessary, such a mood-driven model is not sufficient since it does
not account for interpersonal meaning supplied by discourse particles such as "yes",
"no", and "okay" which cannot be represented at the syntax/semantics interface in
terms of grammatical mood. In light of these motivations, Table 5.2 depicts the set of
speech function types to be assumed along with typical realizations of those speech
function features.

The speech function types listed are terminal category types from the speech
function portion of an Upper Interaction Ontology UIO which has been developed
to complement the ideational organization offered by the Generalized Upper Model.
The UIO is however not a strict linguistic ontology in the same manner as GUM
since, as we will see later, the UIO also includes categories and relations involved
in the conceptual level description of interpersonal meaning.

Speech Function Type	Example Utterance
StatementSSF	The box is on the table
QuestionSSF	Where is Prof. Kane's office?
CommandSSF	Take this to the kitchen
OfferSSF	I'll drive to the kitchen
AcceptSSF	okay
RejectSSF	no
ConfirmSSF	yes
DenySSF	no
GreetingSSF	hello
TakeLeaveInitSSF	bye
SignalUnderstandingSSF	uhuh
SignalNonUnderstandingSSF	pardon?

Table 5.2 Surface Speech Function types with example realizations.

---

[3] See Section 2.3.3.3 for a discussion of the relationship between speech function and speech acts
and illocutionary force

Units of Interpersonal Semantics

Within the SFL view of meta-functionality, the surface speech function semantic property is assigned to an organization unit termed *move*, which is the interpersonal analogue of the ideational configuration or textual message. However, in the strictest interpretation of meta-functional analysis, the ideational, textual, and interpersonal meaning of a text should be analysed as three distinct aspects of meaning; thus resulting in disparate semantic structures for the same textual realization. While such an analysis, performed manually, provides a rich description of a given text, an automated three-way analysis of utterances is beyond the scope of the syntax/semantics interface possible here.

Rather than requiring a semantics with distinct interpersonal and ideational content sides, an alternative approach is to view the ideational content as a single aspect of a complete dialogue move. This view, for example taken by O'Donnell [1994] is highly appealing because it places the interpersonal move as the primary unit of expression in dialogic exchange. Moreover, the approach corresponds tightly with a traditional pragmatics view on speech act theory, where propositional meaning is the content of a complete speech act. To illustrate how such a model would be applied here, consider the following revised semantics for example 8 earlier:

Example 11. The box is on the table

```
(uio:Move#M3(
    uio:hasSurfaceFunction {uio:StatementSF#S2),
    uio:content (gum:SpatialLocating#SL1(
        gum:process     (gum:Being#B1),
        gum:locatum     (lm:Box#B2),
        gum:placement (gum:GeneralisedLocation#GL1(
            gum:hasSpatialModality (gum:Support#S1),
            gum:relatum    (lm:Table#T1)))
```

While such an approach is appealing in its clarity, compositionality considerations make this modelling approach difficult to achieve in commodity grammars. For this reason, we follow the approach taken earlier to the textual meaning unit, i.e., the *message*, and weaken the strict meta-functional distinctions of systemic functional semantics by conflating the unit of interpersonal exchange, i.e., the *move*, with the ideational configuration. Hence, we can now re-write Example 10 with a speech function as follows:

Example 12. The red box is on the large table

```
(gum:SpatialLocating#SL1(
    uio:hasSurfaceFunction (uio:StatementSSF#S2),
    gum:process     (gum:Being#B1),
    gum:locatum     (lm:Box#B2),
    gum:placement (gum:GeneralisedLocation#GL1(
        gum:hasSpatialModality (gum:Support#S1),
        gum:relatum    (lm:Table#T1)))
(gum:ColorAttributeAscription#CAA1(
    gum:domain     (B2),
    gum:range      (lm:Red#R3)))
(gum:SizeAttributeAscription#SAA1(
    gum:domain     (T1),
```

```
gum:range      (lm:Large#L1)))
```

where the relation `hasSurfaceFunction` is defined within the Upper Interaction Ontology, and relates a speech function to a configuration.

Given that discourse particles such as "okay" do not explicitly capture ideational meaning, the linguistic semantics for such an utterance consists of only the base configuration category and an assigned surface speech function, e.g., :

Example 13. "okay"

```
(gum:Configuration#C1(
    uio:hasSurfaceFunction {uio:AcceptSF#A1))
```

In Section 5.2.2 we described how questions are modelled at the conceptual level in terms of existentially quantified variables across FOS statements. While such a semantics formalism can be produced by commodity grammars, the GUM ontology which we build upon, and the grammatical resources made use of in Chapter 8, favour a treatment of questions which uses roles to explicitly identify missing information from within a configuration. Thus, following O'Donnell's [1992] micro-semantics model, we introduce the upper interaction role `required` to flag a discourse entity as being the subject of a question as illustrated by the following example:

Example 14. What is on the table?

```
(gum:SpatialLocating#SL1(
    uio:hasSurfaceFunction   {uio:QuestionSF#Q1),
    uio:required             (X1),
    gum:process              (gum:Being#B1),
    gum:locatum              (gum:SimpleThing#X1),
    gum:placement            (gum:GeneralisedLocation#GL1(
    gum:hasSpatialModality   (gum:Support#S1),
    gum:relatum   (lm:Table#T1))))
```

where `X1` is a reference to the same entity defined as the filler of the locatum role.

5.3.2.3 A Note on Terminology

Having conflated the interpersonal, textual, and ideational meaning units, GUM's `Configuration` class has become central within the linguistic semantics. However, referring to Example 12, we see that the linguistic semantics for a given utterance may nevertheless consist of a set of configurations, or even fragmentary contributions; thus, it does not make sense to equate the configuration with our primary unit of linguistic semantics. Instead, we re-introduce the concept of textual *Message* to refer to units of linguistic semantics as used at the syntax/semantics interface. This use not only simplifies discussions of meaningful chunks of linguistic semantics, i.e., we can now make reference to semantic messages, but also serves to recognize that the conflation of textual, interpersonal, and ideational units is a simplification within the context of an SFL based semantics framework.

5.4 The Conceptual Semantics Layer

Having introduced the linguistic semantics model, this section moves up to the Language Interface's conceptual semantics layer. Unlike the linguistic semantics layer which captures meaning in terms most suited to a reusable grammar interface, the conceptual semantics is a domain-specific layer of modelling that is more suited to application reasoning. In the strictest sense, a conceptual semantics layer extends at least to an interlocutors's dialogue state model. Such a dialogue state model will be developed in the next chapter. Here, however, we first focus on the unit of conceptual semantics for individual utterances, i.e., the Dialogue Act.

5.4.1 The Dialogue Act

The *Dialogue Act* is a conceptual-level description of a dialogue contribution made by an interlocutor. Thus, the dialogue act is the conceptual semantics corollary of the linguistic semantics *Message*. However, whereas a Message consists of a set of configurations – one of which is assumed to have an attributed surface speech function – the act reflects a traditional pragmatic view of communicative function. Namely, a dialogue act is defined as an entity which: (a) is performed by some agent; (b) potentially takes a propositional content defined in terms of the agent's domain ontology; (c) is performed at a particular time; and (d) has an associated speech function type.

Concretely, the dialogue act is defined by the UIO as follows:

Definition 2. `Class: DialogueAct`
`                    SubClassOf: UIOThing`
`                         and (function exactly 1 SpeechFunction)`
`                         and (performer exactly 1 Agent)`
`                         and (content some Thing)`
`                         and (timeStamp exactly 1 String)`
`                         and (parent some Move)`

where `function`, `content`, `performer`, `timeStamp`, and `parent` are concept properties defined in the Upper Interaction Ontology; `SpeechFunction`, `Agent`, and `Move` are concrete classes defined in the Upper Interaction Ontology; `String` is a data type which we use to encode a time stamp; and `Thing` is a placeholder for domain-specific propositional content. The nature of the `parent` role and `Move` will be discussed in the next chapter. In the following however some of the other aspects of the dialogue act are described in more detail.

5.4.1.1 Speech Function Types

The dialogue act is most frequently characterised in terms of an associated speech function type. This speech function type denotes the act's functional contribution to dialogue progression. Table 5.3 depicts available speech function types. The range of types reflect the needs of basic task-oriented dialogue, as well as a minimal coverage of communication management types.

Speech Function Type	Description
Assert	Act performer asserts a piece of information.
Inform	Act performer provides act receiver with information.
Accept	Act performer signals acceptance of provided information.
Reject	Act performer signals rejection of provided information.
Question	Act performer requests information from act receiver.
Answer	Act performer provides act receiver with requested information.
Instruct	Act performer requires act hearer to perform action.
AcceptInstruct	Act performer signals intention to perform requested action.
RejectInstruct	Act performer signals rejection of action request.
Offer	Act performer offers to perform action.
AcceptOffer	Act performer accepts offer to perform action.
RejectOffer	Act performer rejects offer to perform action.
GreetingInit	Act performer initiates interaction.
GreetingResponse	Act performer acknowledges initiation of interaction.
TakingLeaveInit	Act performer initiates termination of interaction.
TakingLeaveResponse	Act performer acknowledges termination of interaction.
SignalNonUnderstanding	Act performer signals that a dialogue contribution was not interpreted.
SignalPartUnderstanding	Act performer signals that a dialogue contribution has been partly interpreted.

Table 5.3 Speech function types

It should be noted that though the range of function types available is similar to the set of surface speech functions defined earlier, the speech function types defined here are not purely surface form oriented. Thus, for example, an utterance modelled with a Statement SSF surface speech function, may, dependent on the relevance of the statement to dialogic progression, be attributed either an Inform or Answer speech function type at the conceptual level.

Of the speech functions depicted in Table 5.3, we refer to the last six as purely conversational speech functions. Whereas other speech functions are typically co-present in acts which require contextualization as part of a language integration process, purely conversation speech functions are assumed to have no associated propositional content and thus do not require contextualization. This allows a slight simplification of the agent-oriented dialogue processing model developed in the next chapter.

5.4.1.2 Propositional Content

Acts may have an associated propositional content which is the object of negotiation or exchange in dialogue. Whereas all acts have an associated speech function, not all acts need have an associated content. An act's content is defined in terms of the agent's application ontology. An example of such an ontology will be introduced in Chapter 7 for a simulated robot environment. Here, it is sufficient to know that a small number of constraints are placed on the interface to domain knowledge in order for that domain knowledge to be applicable within the dialogue architecture developed in this book. Minimally, the interface to the conceptual semantics is required to be expressible in terms of the knowledge representation structures introduced in Section 5.2.

5.4.1.3 Dialogue Act Performer

Dialogue acts are defined as being performed by some interlocutor. While an interlocutor is most correctly a named agent within the application's own domain model, it is convenient in two-party dialogue to abstract interlocutors to simply 'Self' for the dialogue system and 'User' for the human interlocutor.

5.4.2 The Dialogue Act Complex

Just as linguistic semantic messages are complexed, i.e., joined together by logical relations, to capture the semantics of clause conjunctions at the surface form, dialogue acts can also be complexed together via conceptual semantic relations. Such modelling is necessary to capture the conjunction, disjunction, or sequencing of instructions and statements as seen frequently in task-oriented dialogues such as route instructions. We thus introduce the `DialogueActComplex` as a conceptual semantic complex of dialogue acts.

The dialogue act complex is defined as follows in the UIO:

Definition 3. `Class: DialogueActComplex`
`                SubClassOf: UIOThing AND`
`                    alpha EXACTLY 1 (DialogueAct OR`
`                                        DialogueActComplex) AND`
`                    beta  EXACTLY 1 (DialogueAct OR`
`                                        DialogueActComplex)`

where `alpha` and `beta` are properties of the `DialogueActComplex` which mark the primary and secondary participants in the complex. These properties are in turn filled be either an atomic `DialogueAct` or a `DialogueActComplex`.

5.4.3 Dialogue Act Examples

To illustrate the structure of the dialogue act, and contrast the conceptual semantics organization with that of the linguistic semantics, consider the following dialogue act instance which restates Example 12 at the conceptual level:

Example 15. The red box is on the large table

```
(uio:Act#A1(
          uio:function      (uio:Inform),
          uio:performer     (uio:System),
          uio:timestamp     ''10:01:01'',
          uio:content       (ns:Location#O1(
                                  dm:locatum(ns:Box#B1(
                                      ns:colour(ns:Red))),
                                  ns:Place#P1(
                                      ns:modality(ns:Support),
                                      ns:relatum(ns:Table#T1(
                                          ns:size(ns:Big)))))))))
```

where the prefix ns denotes content defined within the domain application ontology used in Chapter 7, and the prefix uio denotes content defined within the Upper Interaction Ontology.

5.5 Inter-Layer Mapping

Over the last two sections, the semantic representation of utterances at different layers of abstraction has been outlined. While the use of two distinct representation models may seem overly complex at first, this two-level approach reflects real-world grammar re-use requirements across application domains. However, for this two-level modelling approach to succeed, we also require processes to align the linguistic and conceptual semantics during language analysis and production. To that end, this section details a rule-based transformation system which has been applied to this problem.

5.5.1 Functional Transforms: A Guided Tour

We will make use of a transform-based approach to mapping between the strata of linguistic and conceptual semantics. The transform approach makes use of an on-tologically sensitive functional transform system which takes as input a set of FOS structures defined in terms of some input ontology set, and maps these to an out-put set of FOS structures defined in terms of a second ontology set. The general process is illustrated in Figure 5.4. As an instance of function-based transforms, the approach developed here has similarities with, and indeed has been partially

motivated by, the XSLT transform language [Kay, 2007]. It should be noted however that while XSLT provides mechanisms for the rewriting of XML documents,
the approach developed here goes well beyond XSLT by necessarily incorporating
ontological sensitivity in transform content selection and production. In the following, we informally introduce the transform rule methodology before providing more
formal details in Section 5.5.2.

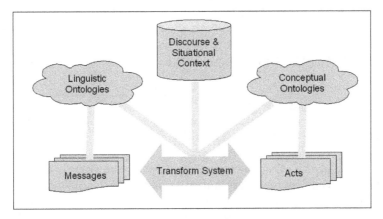

Fig. 5.4 Semantic layer mapping with functional transforms

5.5.1.1 Elementary Transformations

At the heart of the model is the notion of a transform rule which describes how a
given element of an input structure is to be transformed into an output structure. All
transform rules consist of a *signature* and *body*. The signature minimally determines
whether, in a particular context, the rule is applicable to a given input structure. The
body on the other hand defines a structure of transformation functions which are
then applied to the semantic input.

The most basic transform rule type simply creates a new proposition in the output
structure if triggered by a specific input category. For example, if the same colour
space interval is captured by the symbol 'Red' in both the input and output ontologies, then an appropriate mapping rule is written in an XML syntax as follows:

Example 16.
```
<transform content="Red">
      <create_prop type="Red"/>
</transform>
```

where the transform rule's content selection constraint requires the rule only be
applied to an element of type Red and the transform rule's body consists of a single
application of the create_prop function which will create a proposition of type
Red.

In fact, the rule's signature above will not only be applicable to any symbol which is directly of type `Red`, but will also accept any symbol whose type is a subclass of `Red`. Thus, if an input ontology is more specific than the output ontology, such a rule generalizes to map a wide variety of *reds* to a single output category.

By introducing a key-term `$type`, which is replaced at runtime with a symbol which is lexically identical to the input symbols terminal type, but which is assumed to live within the namespace defined for the default output ontology, we can make use of the type generalization above to abstract over sets of transforms where there is inherently high alignment between input and output ontologies. For example, if the symbols used to represent colour space intervals across input and output models match, we can generalized Example 16 as follows:

Example 17.
```
<transform content="Colour">
    <create_prop type="$type"/>
</transform>
```

Within either input or output models, there is however rarely a single all-encompassing ontology which covers all definitions within a given semantic layer. For example, the linguistic semantics layer does not make use of a single ontology, but instead builds upon a set of ontologies including the core GUM ontology, the Upper Interaction Ontology, and domain level ontologies. Following the Web Ontology Language's modularization conventions, ontological modules are assumed to be assigned specific namespaces. Transform rules are in turn sensitive to those namespaces in both constraint selection and object creation functions. Thus, we can re-write Example 17 with appropriate namespacing as follows:

Example 18.
```
<transform content="lm:Colour">
    <create_prop type="ns:$type"/>
</transform>
```

where `lm` is the shortform namespace identifier for the *lower model* module of our linguistic ontology, and `ns` is the shortform identifier for the Navspace application ontology defined in later chapters.

Of course, it is necessary to create more than individual propositions in a transformed structure. In accordance with the properties of the FOS structures introduced in Section 5.2.2, transform rules must also support the arbitrary creation of objects and relations in output structures. Thus, in addition to the `create_prop` function, a transform rule body may also make use of the `create_object` and `create_role` functions which operate analogously to the `create_prop` function. The `create_object` function creates an object within the output structure with a specified type and optionally specified instance identifier (`id`) as follows:

Example 19.
```
<transform content="lm:Cup">
    <create_object type="gum:SimpleThing"
                   objectID=''x101''/>
</transform>
```

Similarly, the `create_role` function create a new role of a specified type that is asserted to hold between a parent object and some object or proposition defined within the body of the `create_role` function. To illustrate the `create_role` function's use alongside the `create_object` and `create_prop` function, consider the case of transforming from conceptual to linguistic semantics in a case where the specificity of an object in the application model is not matched directly by the specificity of the linguistic semantics domain – thus requiring the assertion of a specific lexical item to be used. We may write an appropriate transform rule as follows:

Example 20.   `<transform content="ns:Cup">`
                          `<create_object type="gum:SimpleThing"/>`
                          `<create_role role="lex">`
                             `<create_prop type="$type"/>`
                          `</create_role>`
                       `</transform>`

Although not possible in the case just given, it should be noted that if the execution of the body of the `create_role` function fails to result in the construction of a valid object or proposition, then the `create_role` function itself fails and no role is created.

5.5.1.2 Content Selection Constraints

If input and output models differ in structural organization – which they are likely to – it can be useful to allow a transform signature's selection constraint to look for a specific feature within an input structure rather than selecting based on the head of the semantics chunk. For example, within the linguistic semantics model, a speech function is treated as a property of a configuration, and any given linguistic semantics message can consist of multiple configurations which do not all necessarily have speech function properties. However, on the other hand, within the conceptual semantics, propositional content is treated as a semantic feature of a dialogue act. In making a transformation from linguistic to conceptual semantics the fact of whether or not a given configuration has an associated speech function is an important determining factor in the selection of an appropriate transformation rule. To support the appropriate choice of transform rules, the argument given to a selection constraint is treated as a structural query of the input term. To illustrate, consider the following partial example which limits a configuration transform rule to be applicable only to Configurations which have a `uio:hasSurfaceFunction` role asserted on a `gum:Configuration`:

Example 21.   `<transform`
                          `content="gs:Configuration/uio:hasSurfaceFunction">`
                          `...`
                       `</transform>`

By default if the selection constraint is met by the input term then the entire term remains in scope for subsequent functions within the transform's body. However, it can be useful to use a selection constraint to isolate a portion of the input term which will be in scope for subsequent functions rather than the entire input term. The selection constraint specification language supports this functionality through the use of the selective path operator (!), which unlike the default path operator used above (/), isolates the specific sub-structure of the input term which is to be in scope for the transform body. To illustrate, we can extend the above example to

limit the rule's application to configurations with a specific speech function, and
have only that speech function in scope for the transform body as follows:

Example 22. <transform
content="gs:Configuration/uio:hasSurfaceFunction!uio:GreetingSSF">
                        ...
                        </transform>

5.5.1.3 Maintaining Referential Equivalence

As indicated, new objects are by default created by the create_object function
with an arbitrary object id. However, by default this behaviour will cause an object
referenced multiple times in an input model to be cloned into multiple objects with
distinct identities in the output term. Unchecked, this behaviour would for example introduce multiple discourse referents at the conceptual layer where there was
only one such referent in the linguistic semantics layer. To avoid this problematic
behaviour we must ensure that, despite the number of distinct references to an input
object, only one output object is instantiated. Thus, we may make use of an input
object's unique identifier as a template for the identifier given to a new output object.
This is achieved through the use of the objectID attribute of a create_object
function, along with two new constructs:

- termType: an additional transform rule selection constraint which limits
  transform application to input terms which are headed by either a object,
  role, or proposition, and
- $objectID: a keyterm, which, when applied within a create_object's
  objectID attribute, is replaced at runtime with the identifier of the matching
  input term.

We can illustrate the use of these features together within a single transformation
rule as follows:

Example 23.   <transform content="gum:SimpleThing"
                              termType="object">
              <create_object type="ns:$type" objectID="$objectID"/>
              </transform>

While the application of this function does result in the creation of multiple objects, each object is of the same type and identifier; thus ensuring referential equality.

5.5.1.4 Context Sensitive Transformations

In Chapter 1, we saw how the meaning of language can be tied to situational contexts in numerous ways. While reference resolution is the most obviously influenced
context dependency, context also affects the behavioural interpretations we give to
certain constructions.

To illustrate, consider again the interpretation of the simple command "turn".
Within the linguistic semantics layer, the meaning of this command is expressed

in terms of a `NonAffectingOrientationChange` instance regardless of the context in which this utterance is performed. However, within the domain model developed in Chapter 7 for example, the same utterance can be given one of two distinct deep interpretations depending on the movement state of the agent. Namely, when stationary, the utterance is interpreted as a `Reorient` behaviour which entails rotational movement only. However, when moving, the same utterance is interpreted as a `Redirect` behaviour which entails both rotational and transpositional movements.

While plan inference at the application level can be used to interpret such context-sensitive meaning, it can be useful to handle context sensitivity directly at the semantic abstraction layer when full plan inference is deemed too costly. To support such context-dependent variance in mapping, we introduce an additional rule selection constraint, i.e., `context`, which defines a query applicable directly to the agent's mental state model. To illustrate, consider the following transform rule which can be used to effect the appropriate interpretation of a "turn" command:

Example 24.
```
<transform
    content="gs:NonAffectingOrientationChange"
    context="BEL(state(Stationary))">
<create_object type="ns:Reorient"/>
    ...
</create_object>
</transform>
```

where the statement `BEL(state(Stationary))` is a query of the agent's non-dialogic mental state.

5.5.1.5 Controlling Rule Selection

Rather than selecting and applying a transform to each term in an input FOS, we make use of the functional aspects of the transform to control transform application. This control is effected through the use of an `apply_rules` function which selects a portion of the FOS structure in scope and attempts of find and apply a transform for the selected portion. The rule selection process itself, invoked both for `apply_rules` function application, as well as for the head of each input structure at the start of transform processing, then chooses the transform rule whose selection constraints, defined in the rule's signature, are most compatible with the input FOS structure passed to the rule selection process. To illustrate, consider a transform rule that reorganizes the linguistic semantic assertion of a colour property ascription, captured in GUM as a `ColorPropertyAscription` instance, into a conceptual semantics structure which asserts a colour property on a given object:

Example 25.
```
<transform content="gum:ColorPropertyAscription">
<apply_rules select="*/gum:domain!gum:SimpleThing"/>
<create_role role="ns:hasColor">
    <apply_rules select="*/gum:attribute!gum:Color"/>
</create_role>
</transform>
```

where the selection constraints passed to the `apply_rules` function also include a wildcard $*$ which allows any type within the given portion of a search query.

5.5.2 Functional Transforms: Model Definitions

Having introduced the available functional transforms, here we provide an implementation oriented perspective on the Functional Transform Rule (FTR) model. Although a formal semantics would provide the most precise account of the model, the development of a full formal semantics for the FTR model is beyond the scope of this book.[4] Instead, we will proceed with a pseudo-formal description of the model components, as well as the algorithms used in selecting and applying transform rules. Any readers who require only a cursory overview of the language interface can skip this section and move directly to the discussion.

5.5.2.1 Transform Rules & The Transform Process

From a model perspective, we define a transform rule as a 2-tuple:

$$FTR = \{FTH, FTB\} \tag{5.1}$$

where FTH is the head of the transform rule which is used to determine if a rule is applicable to a given input structure, and FTB is the body of the transform rule which defines the set of operations to be performed in constructing the output structure.

A transform head is defined as follows:

$$FTH = \{CntSC, CxtSC\} \tag{5.2}$$

where $CntSC$ defines a content selection constraint which limits the rule's applicability to a particular input pattern, and $CxtSC$ defines an optional context state constraint which limits the rule's applicability to a particular state holding in the agent's mental state.

A transform body is defined as an ordered set of transformation operations as follows:

$$FTB = List(Op) \tag{5.3}$$

where each Op is a Transform Operator as will be defined in Section 5.5.2.3.

The transform system may be defined in terms of a transform function as follows:

---

[4] However, for guidance on how a formal semantics could be approached, the reader is directed to the introduction to term-rewriting by Baader and Nipkow [1998], and in particular to Wadler's [1999] note on developing a formal semantics for the path selection portion of the XML transformation language XSLT.

```
Set(FOS)out transform(Set(FOS)in, Set(FTR)in, Oin_a, Oin_b){
    Set(FOS)out = new Set(FOS);
    foreach(FOSi : Set(FOS)in){
        FTRBest = select(FOSi, Set(FTR)in, Oin_a);
        Set(FOS)out.add(apply(FTRBest,FOSi,Oin_a,Oin_b));
    }
    return Set(FOS)out;
}
```

Fig. 5.5 The transform process.

$$Set(FOS)_{out} = transform(Set(FOS)_{in}, Set(FTR)_{in}, O_{in\_a}, O_{in\_b}); \qquad (5.4)$$

where: $Set(FOS)_{in}$ is the input set of FOS structures to be transformed; $Set(FOS)_{out}$ is the resultant set of FOS structures following transformation; $Set(FTR)_{in}$ is a set of functional transform rules; O_{in_a} defines an ontology which provides category definitions for $Set(FOS)_{in}$; and O_{in_b} defines an ontology which provides category definitions for $Set(FOS)_{out}$.[5]

Figure 5.5 gives a pseudo-code description of the transform system's application process. At this high-level of abstraction, the process is very simple. Given an input set of semantic structures, along with supporting ontologies, and the set of functional transform rules, a transform rule is selected for each of those input structure. The transform rule is then applied to the input structure, and the results of that application added to the resultant semantic structure set. The heavy lifting of the model is achieved by the transform selection and application functions. These are addressed in subsequent sections.

5.5.2.2 Transform Selection

Given an input semantic structure (FOS), ground in terms of an input ontology (O_i), the transform selection function (select) returns a transform rule judged most applicable to the input structure. For any given input structure, a number of transform rules may be applicable at any given time. While a 'first-best' selection can work well as long as rules have been given an appropriate ordering, here we favour a more controlled application of transform rules. Namely, the transform selection process returns the rule which best fits the given input structure. As depicted in Figure 5.6, the select function thus iterates over all transforms, and maintains a best transform at any given time. The transform selection function makes use of two additional functions: isValid, to evaluate if a transform can be applied to a given

---

[5] In this and future chapters we apply the convention that normal-script terms indicate types, whereas subscripts are used to denote instance names.

```
FTRBest select(FOSin, Set(FTR)in, Oin){
   FTRBest = null;
   foreach(FTRi : Set(FTR)in){
      if(isValid(FTRi, FOSin, Oin){
         if(FTRBest == null){
            FTRBest = FTRi;}
         else if(compare(FTRBest,FTRi,FOSin,Oin) ≥ 0){
            FTRBest = FTRi;}}}
      return FTRBest;
}

bool isValid(FTRin, Oin, FOSin){
   if(!((FTRin.CntSC != null) && holds(FTRin.CntSC))){
      return false; }
   else if(! search(FTRin.CntSC,Oin, FOSin){
      return false; }
   else return true;
}
```

Fig. 5.6 The transform selection process.

input; and `compare`, to evaluate the specificity of two transform rule heads.

The transform validation function (`isValid`) is also sketched in pseudo-code in Figure 5.6. This function makes use of the transform rule content and context selection constraints, i.e., *CntSC* and *CxtSC*. Context selection constraints reduce to knowledge base queries on the agent's mental state. Content selection constraints, on the other hand, make use of a structural query mechanism (`search`) which is specifically tailored to the transform process. In the following, we look at the properties of these content selection constraints, and describe their application in both the structural query (`search`) and rule head comparison functions (`compare`).

Content Selection Constraints
We define the content selection constraint (*CntSC*) as a list of path descriptor elements. Each path descriptor element is composed of: (a) a type constraint that limits the path descriptor element's applicability to objects of a certain ontological class; and (b) a path operator which indicates which section of a searched semantic object is to be returned by the content selection constraint. From a denotational perspective, we can model this as follows:

$$CntSC = List(PD) \tag{5.5}$$

where *PD* is a Path Descriptor:

$$PD = \{PD\_Type, PD\_Op\} \tag{5.6}$$

composed of a path descriptor type (*PD_Type*) and path descriptor operation (*PD_Op*). These two types are in turn defined as follows:

$$PD\_Type \in (* \cup O_i) \tag{5.7}$$

$$PD\_Op \in (PD\_Op_! \cup PD\_Op_/) \tag{5.8}$$

where PD_Type is an element of the input ontology (O_i) or the wildcard type ($*$) which subsumes all category and property types in the ontology; and PD_Op is one of either the selective path operator ($PD\_Op_!$) or the default path operator ($PD\_Op_/$).

Each content selection constraint is defined as having one, and only one, path descriptor that has been defined as including the selective path operator. However, as we saw in Section 5.5.1, the serialization of content path constraints need not necessarily include a selective path operator. In this case, a pre-processing step first analyses a path selection constraint; if no path selection constraint is included in the constraint beyond the 0th path description element, then the 0th element is assigned a path selection constraint. This pre-processing step simply means that unless the rule writer has specified otherwise, the entire content of a matching FOS chunk will be in scope for the function transform rule body.

Finally, as a list structure, we also assume that the *CntSC* can be accessed and manipulated in the usual ways, i.e., list elements can be iterated through, and there exist functions *head* and *tail* which respectively return the first element of the list, and the remainder of the list (possibly null), respectively.

Content Selection Constraint Querying

As discussed, we require content selection processes which: (a) decide whether a particular transform rule is applicable to a given semantic input; and (b) determine which part of that semantic content is to be made available to the rule construction process. Both these requirements can be met by a single operation. One way this operation could be achieved is through a modified form of unification which rewrites the path selection constraint as a frame structure to be unified with the input semantics. Here we take a more straightforward approach to the problem by applying a modified path search algorithm.

Figure 5.7 depicts the modified path search function. As indicated, the search function takes as input the transform rule's content selection constraint, the input semantic structure, and an ontology definition which is assumed to ground the types both within the selection constraint and the semantic input. The function, which operates recursively, returns a 2-tuple consisting of a boolean which indicates if the search has been successful, and a FOS which captures the semantic structure which is later to be made available to the transform rule body. The function operates by comparing the first element of that content selection constraint with the functor of the FOS input. If neither the selection constraint is a wildcard, nor is the selection constraint a super class or super property of the FOS functor, then the search fails at the current element and returns. Otherwise, the search function proceeds based on whether there remain any elements within the content selection constraint. If no elements remain, the function returns true, with the semantic content returned varying on whether the current content selection constraint has a default or selection

```
{bool,FOS} search(CntSC_in, O_in, FOS_in){
    if((head(CntSC_in) == *) ||
        (isSubordinate(head(CntSC_in).PD_Type,FOS_in.functor,O_in))){
        if(tail(CntSC_in) == null){
            if(head(CntSC_in).PD_Op == PD_Op/){
                return {true,null};}
            else {
                return {true,FOS_i};}}
        else{
            foreach(FOS_j : FOS_in.args){
                {bool_j,FOS_j} = search(tail(CntSC_in)),O_in,FOS_j);
                if(bool_j == true){
                    if(head(CntSC_in).PD_Op == PD_Op/){
                        return {true,FOS_j};}
                    else {
                        return {true,FOS_i};}}}}}
    return {false,null};
}
```

Fig. 5.7 The search algorithm.

operator. On the other hand, if further elements remain in the content selection constraint, then the algorithm proceeds by iterating through each of FOS's arguments and performing a recursive analysis of each of those children.

Content Selection Constraint Comparison

As indicated, we also require a function which given two content selection constraints selects the more detailed of the two. The definition of *detailed* here is dependent on two criteria. First, given two selection constraints, the selection constraint which contains the first path descriptor element which is judged to be more specific than the path descriptor at the same position in the other selection constraint is judged to be more specific. Second, if two selection constraints have been judged equivalent up to the length of the minimum selection constraint, then the longer of the two selection constraints is judged to be more specific. If based on these two criteria both selection constraints are equivalent, then we fall back and assume the ordering of selection rules to be significant.

Although the metrics above work perfectly well in most content selection constraint comparisons, there are a number of features of constraint formalization which complicate the comparison process. First, different selection constraints can correspond to two alternative paths taken through a semantic input, and therefore two different path constraints may apply to a semantic chunk but represent completely different semantic structures which cannot be compared by simply performing a subsumption check. Second, the ontologies of linguistic and conceptual representation assumed in this work support multiple inheritance in keeping with the properties of Description Logic. Thus it is entirely possible that even single-entity content selection constraints can apply to a single semantic element, but yet neither category

```
CntSC_Best compare(CntSC_a, CntSC_b, O_in){
    if(CntSC_a longer than CntSC_b{
        CntSC_small = CntSC_b;
        CntSC_other = CntSC_a;}
    else{
        CntSC_small = CntSC_a;
        CntSC_other = CntSC_b;}
    for(int i = 0; i < CntSC_small; i++){
        if(compareDP(CntSC_small[i], CntSC_other[i]) == CntSC_small[i]){
            return CntSC_small;
        else if(compareDP(CntSC_small[i], CntSC_other[i])
                                == CntSC_other[i]){
            return CntSC_other;
        else {} }
    return CntSC_a;
}

DP_out compareDP(DP_a, DP_b, O_in){
    if(subsumes(DP_a, DP_b, O_in)) return DP_b;
    else if(subsumes(DP_b, DP_a, O_in)) return DP_a;
    else if(equivalent(DP_a, DP_b, O_in)) return null;
    else if(minDepth(DP_b, O_in) > minDepth(DP_a, O_in)) return DP_b;
    else if(minDepth(DP_a, O_in) > minDepth(DP_b, O_in)) return DP_a;
    else return null;
}
```

Fig. 5.8 Content search constraint comparison function.

subsumes the other. To handle both these cases we appeal to the depth features of taxonomic structures, and assume that a category or property whose minimum ancestry length to the most generic class is greatest is the more specific of the two compared features.

Figure 5.8 illustrates an algorithmic model that captures these comparison properties for content selection constraints. As indicated, the first function (compare) takes two content search constraints and iterates over these two constraints in accordance with the length of the smaller of the two constraints. If a segment in either constraint is judged to be more specific than the same indexed segment of the other constraint, then that first constraint is judged to be more specific and returned. This decision is made on the basis of the (compareDP) function which evaluates two individual constraint types based first on a relative comparison, i.e., subsumption, and then on an absolute measurement in the case that neither of the compared entities subsumes the other.

```
FOS_out apply(FTR_i,FOS_i){
   Table_variables = new Table();
   Table_variables.put(TYPE,FTR_i.FTH.get(TYPE));
   Table_variables.put(LEX,FTR_i.FTH.get(LEX));
   Table_variables.put(ID,FTR_i.FTH.get(ID));
   return transform(FTR_i.FTB,null,Table_variables);
}

FOS transform(Set(Op)_in,FOS_Head,Table_variables){
   foreach(Op_i : Set(FOS)_in){
      Operator_i = Table_operators.getOperator(Op_i);
      FOS_Head = Operator.apply(FOS_Head);
   }
   return FOS_Head;
}
```

Fig. 5.9 The transform application algorithm.

5.5.2.3 Transform Application

Once a transform rule has been selected for a given input FOS, the transform body, consisting of a number of transform operations, is applied to that input. The input is not structurally altered, as a non-destructive transform process resulting in the creation of a new output FOS structure is applied. The transform application process can thus be seen as a tree construction algorithm where transform operators perform traversal steps over the input FOS and contribute, where applicable, elements of the output FOS structure.

Figure 5.9 presents pseudo-code for the outer transform application algorithm. The algorithm takes as input: (a) a transform rule (FTR) that was identified as most applicable for the input semantics unit; and (b) (FOS_i), a fragment of that semantics unit that is in scope for the transform's application. Although omitted here for the sake of simplifying the description of the application process, it can also be assumed that the transform process is passed reference to the input and output ontology definitions, and that the process makes these definitions available to transform operators as well as subsequent semantics selection and construction function calls.

This application function itself operates by first making rule signature variables globally available so that they may be later referenced by transform operators. The application function then applies transform operators in sequence to construct the output FOS. Moreover, the application process maintains: (a) a reference to the head of a portion of the input FOS that is in scope; (b) a reference to the next element in a sequence of transform operators to be applied; and (c) a pointer to a branch of that output FOS which dictates where individual transform operators attach constructed semantics.

Transform operators are supplied with the specification defined parameters for that operator, along with the pointer to the current focus of the output FOS.

Constructive transform operators such as `create_object`, `create_role`, and `create_term` then attach new semantics to that referenced output FOS element. While some operators attach no semantics to this local focus, e.g., `print`, and some attach only single semantic elements, e.g., `create_object` and `create_-term`, other operators such as `apply_rules` and `create_role` lead directly to the calling of nested transform operators which result in a more complex structure being attached to the local focus of the output FOS.

In the following, the most salient features of three of the more important transform operators are briefly described to further illustrate the transform process.

- **The `create_object` operator:** The `create_object` operator constructs a new FOS object with a type and object ID. Critically, the `create_object` operator can only be applied successfully if the local context of the output FOS element is null or an open role definition. In other words, following the Description Logic basis of the FOS semantic structures, an object cannot be asserted as an argument of another object or a proposition. Assuming the applicability constraints are met, a `create_object` operator then proceeds to gather information on the type and object ID for the FOS object to be created. Type and object ID are assumed to be supplied in the operator specification passed into the `create_object` instance, but default object IDs can be applied, and the input specifications may also make use of key terms which are matched against the input variable table.

- **The `create_role` operator:** The `create_role` operator constructs a role relation between the local focus of the output FOS and a child FOS object or FOS proposition. Like the `create_object`, the `create_role` object has applicability constraints which restrict its use. Specifically, the `create_role` operator can only be applied to a FOS object. Application to either a FOS proposition or other FOS role will fail. This constraint, along with the applicability constraints for both `create_object` and `create_term` are summarized in Table 5.4. Assuming input constraints are met, `create_role` proceeds in a similar manner to `create_object` by assembling the role type from the input specification and variable table. However, unlike the `create_object` operator, the `create_role` input specification typically includes a set of child transform operators. These operators are first applied in an analogous manner to the original application of the `transform` operator, thus resulting in the creation of the child FOS object or predicate to be attached to the created role. If the call to child transform operators does not result in the creation of either a FOS object or FOS proposition, then the application of the `create_role` transform is considered to have failed, and thus no element is attached to the local focus of the output FOS structure.

- **The `apply_rules` operator:** The `apply_rules` operator selects a section of the input FOS element and applies the complete transformation process to that segment. The resulting transformed elements are then attached to the local focus of the current output FOS. The `apply_rules` operator is thus a new application of the complete transform application process which includes the use of the rule selection process. If the selection constraints specified for the `apply_rules` instance do not select a section of the input FOS, or if no rule is found that results in the creation of an output FOS structure, the `apply_-rules` operator application does not result in any element being attached to the local focus of the output FOS element.

Operator	Target Focus	Outcome
create_object	null	object
	object	fail
	role	role-object
	prop	fail
create_role	null	fail
	object	object-role
	role	fail
	prop	fail
create_prop	null	prop
	object	fail
	role	role-prop
	prop	fail

Table 5.4 Local context application conditions for primary transform operators.

To illustrate the transform application process, Figures 5.10, 5.11 and 5.12 present a worked example of transform application for an input semantics unit. For each in a series of transformation processing steps, we see: (a) the transform selected for application along with the current point in that transform's application as illustrated by an arrow; (b) the initial input semantics unit along with the local focus on that input FOS as indicated by an arrow; and (c) the output semantics unit along with the local focus on that output FOS as also indicated by an arrow.

5.6 Discussion

The dialogue architecture's *Language Interface* layer has been designed to provide a practical and extensible approach to the development of reusable linguistic resources for situated dialogue applications. This has involved, on one hand, the development of appropriately expressive models of linguistic semantics for spatial and action-oriented language, and, on the other hand, the recognition that real-world applications often make use of ontological models that are not identical to those used in the definition of semantic interfaces to grammatical resources.

Although this chapter has focused on the semantics layers and the transform model that operates between these layers, the Language Interface's grammar levels have been instantiated with concrete grammar resources for the situated dialogue ap-

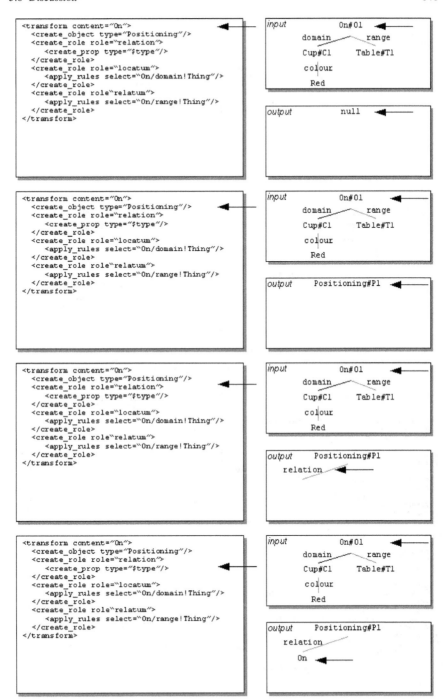

Fig. 5.10 Transform Application Examples. Stages 1-4.

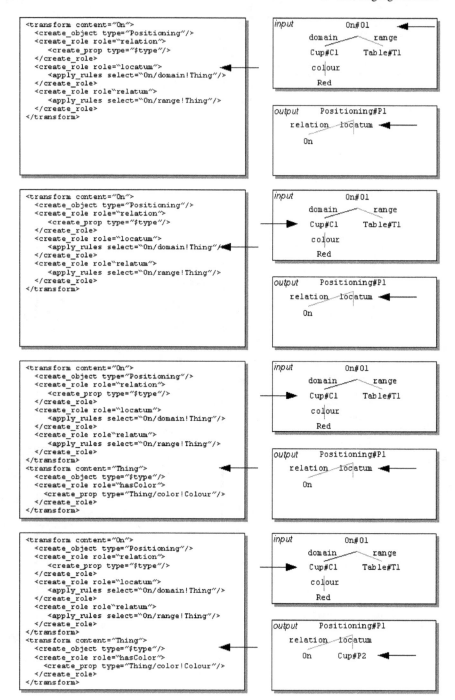

Fig. 5.11 Transform Application Examples. Stages 5-8.

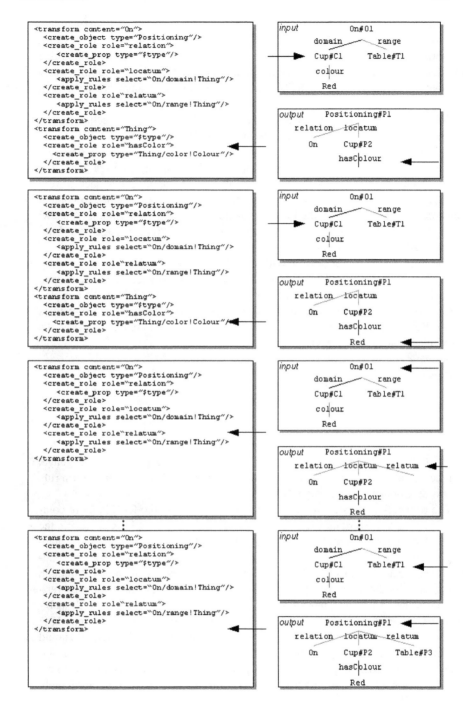

Fig. 5.12 Transform Application Examples. Stages 9-11 and final stage.

plication described in Chapter 8. Namely, linguistic analysis for both German and
English has been provided through grammars based on Steedman's [2000] Combinatorial Categorial Grammar formalism via the OpenCCG framework [Baldridge
and Kruijff, 2002]. For language generation, Systemic Functional Grammars created within Bateman's [1997] KPML language production framework have been
applied.

With respect to the Functional Transform Rule (FTR) system, its mechanism can
be viewed as a special case of term-rewriting (see Baader and Nipkow [1998] for
an introduction). Classical term-rewriting systems are however typically designed
to provide generic solutions to algebraic specification analysis, and are hence too
computationally expensive to be applied in time-critical applications such as dialogue systems. Less generic, but more domain applicable, approaches to content
transformation have been studied in fields including machine translation [Alshawi
et al., 1992], information presentation [Kay, 2007], and dialogue management [Lau-
ria et al., 2002, Dizikovska et al., 2007].

The basic principles of term-rewriting in the language engineering community
are probably best illustrated by the rule-based machine-translation model developed
by Alshawi et al. [1992] for the Core Language Engine (CLE). Within their model,
sets of declarative transfer rules are used to map between the linguistic semantics
of two base languages, namely English and Swedish. To illustrate Alshawi et al's
approach, consider the following two rules:

Example 26. `trans(carl <=> bill)`

Example 27. `trans([call_name,tr(ev),`

              `qterm(<t=quant,n=sing,l=ex>,A,[entity(A)]),`

              `tr(ag),`

              `tr(name)]`

    `<=>`

    `[heta,tr(ev),tr(ag),tr(name)])`

where each side of a rule is a Quasi-Logical Form (QLF) expression of language,
and the operator between the sides indicates the directionality of the transform rule
(in this case both are bi-directional).

The declarative transfer rule-based approach as used in the CLE is more compact and straightforward than the FTR approach, and, unlike the FTR approach, the
CLE's declarative rules theoretically allow the development of bi-directional mapping resources. In practice however, declarative rule-based approaches are best applied where rule domains are of similar ontological and epistemic bases. Although
this is partially the case in mapping between the CLE's linguistic semantics variants, even the CLE had to introduce the capacity for uni-directional as well as bi-
directional rules. Indeed, in the CLE's English-Swedish transfer system, some 10%
of rules were not bi-directional. While this number seems relativly small, it should
also be considered that approximately 70% of rules were atomic rules (i.e., translations between individual predicates usually corresponding to words); thus, the proportion of non-atomic uni-directional transfer rules was presumably considerably
larger than 10%.

A second issue to note with the declarative transfer rule approach as typified by
the CLE, is that vanilla declarative rules do not easily handle variability in category

constituency. This however is the case in the notion of Configuration used in our linguistic semantic model. While specific configuration types denote the range of mandatory constituent roles to be filled, extra *circumstances* can be used to augment a given configuration, e.g., the location circumstance is commonly used to augment a host of motion process specifications. A declarative rule-based approach such as that employed by the CLE would require the enumeration of multiple rule instances to handle the different circumstantial qualifications which can be added to a configuration. Through the optional applicability constraints built into functions such as apply_rules, the FTR approach needs no such rule repetition.

Although outside of the computational linguistics domain, one approach to content transformation which has attempted to provide considerably more power than these declarative rules is the W3C's recommended XML transformation technology, i.e., XSLT [Kay, 2007]. XSLT is a truly functional approach to model transformation in that rule consequents can include a host of selection and manipulation constructs which allow rule writers greater flexibility in the rendering of information. Also, rule antecedents, simply declared as template declarations in XSLT, do not require that the arguments for given terms be declared in the template definition, hence avoiding the problem of circumstance variability in configuration types. To illustrate the XSLT methodology, Example 28 below depicts a simple XSLT transform which when applied to XML content headed by a records element, attempts to create a HTML page and subsequently apply further transforms to child elements of records to complete production of the HTML page.

Example 28.
```
<xsl:template match="/records">
  <html>
    <head> <title>Student Data Records</title> </head>
    <body>
       <h1>Student</h1>
       <ul>
       <xsl:apply-templates select="student">
          <xsl:sort select="family-name" />
       </xsl:apply-templates>
       </ul>
    </body>
  </html>
</xsl:template>
```

In this example we see that XSLT freely mixes standard HTML document content, e.g., html, head, ul, etc., with explicit XSLT commands, i.e., those beginning with the xsl prefix. Of these, the xsl:template tag is used to define the start of the transform, while the xsl:apply-templates tag invokes a function that selects child content from the input record, and subsequently applies that content to further transformations.

Similarities between the FTR and XSLT approaches are the use of simple structure driven selection mechanism on the rule antecedent side, and the explicit inclusion of functions in rule consequences. Differences between FTR and XSLT approaches centre on the logical underpinnings rather than structural underpinnings of the FTR approach. The most significant difference between the XSLT and FTR models is that XSLT has no notion of typing and ontological sorting as used in the

FTR approach. XSLT is thus only sensitive to string patterns in both rule antecedent and rule consequent descriptions. Thus, while XSLT is a useful structure re-writing system, XSLT cannot work with strongly typed input data as is typical of linguistic semantics. On the other hand, XSLT has advantages over the FTR model. On one hand, XSLT offers a considerably richer set of transformation constructs and document query functions than are currently provided by the FTR model, while on the other hand, XSLT allows the direct authoring of content in rule consequents as seen in the HTML outline of Example 28.

Back within the computational linguistics domain, the content mapping model that is most similar in motivation, though not execution, to the FTR model is that of Dizikovska et al. [2007]. In her model, Dizikovska develops an approach to mapping from linguistic semantics to knowledge representation for the TRIPS dialogue engine. Unlike XSLT and many classical rewriting systems such as those used by the CLE, Dizikovska's rule system is ontologically sensitive in rule selection, and is geared towards a neo-Davidsonian semantics rather than a valency based model. As indicated, the development of Dizikovska's model has been motivated by the same basic needs as the FTR model. Despite this, there exist a number of subtle but important differences between the models. On an operational level, Dizikovska's model is a classically declarative approach, and, thus, does not allow the use of functions in rule consequents – a feature found useful in both the FTR model and in contemporary transform methodologies such as XSLT. Moreover, Dizikovska's model does not provide global context sensitivity in rule application, nor does it allow local context sensitivity due to the lack of function support in rule consequents. On the other hand, whereas the functions used in the FTR allow a conditionality based approach to variability in configuration constituents, Dizikovska's model provides an elegant solution to the problem through the use of *abstract transfer rules* that can be used to implement a form of transfer rule property inheritance.

5.7 Summary

In this chapter, we developed the language modelling levels of the agent-oriented dialogue processing architecture. These levels, referred to as a whole as the Language Interface, account for the structures and processes which map between surface language and high-level representation tiers. As such, the elements of the Language Interface are the foundations on which the larger dialogue processing model is built, and it is thus important that the nature of the representation layers were made clear, even if it is self-evident that such models are necessary.

The chapter began with an introduction to the Language Interface as a whole, and detailed that Language Interface in terms of its processes and representational layers. Of those processes and representational layers, this chapter focused on two semantics representational models, i.e., the linguistic and conceptual semantics, as well as the processes which map between them. After sketching the formal modelling basis for semantics expression, an ontologically well founded model of linguistic semantics was then argued for. This model built upon Bateman's Generalized Upper Model, an ontological account of linguistic meaning that offers a rich model of action and spatial semantics. The linguistic semantics model built on this GUM core by accounting for interpersonal semantic features not covered by GUM. A model of conceptual semantics for utterances was then developed. Although the conceptual semantics layer will not be fully developed until the dialogue model and domain application interface are outlined, the dialogue act model provided is sufficient for handling the conceptual semantics of individual utterances. Finally, following the development of the linguistic and conceptual semantics layers, a functional transform system was proposed to provide the mapping between the layers.

Chapter 6
Agent-Oriented Dialogue Management

Abstract This chapter moves up from the natural language interface to the *Agent-Oriented Dialogue Management* (AODM) model. The AODM model couples a theory of rational agency with an information-state derived account of dialogue management that applies an incremental process of situational context application. Section 6.1 details the first part of this model, i.e., the agency theory which we apply to the modelling of domain state. Section 6.2 then details the AODM model's dialogue state backbone in terms of types, structures, and basic operations. Next, Section 6.3 develops the mechanisms by which language is integrated into dialogue state. Section 6.4 then develops a generalized iterative contextualization process. Thereafter, Section 6.5 describes the dialogue planning process which takes advantage of the intentionality model described at the head of the chapter. Finally, the presented dialogue model is briefly compared to some existing works in Section 6.6.

6.1 The Agency Model

In this section, we outline the agency component of the AODM model. The basic principles of agency, agent-oriented design, and the relationship between theories of agency and dialogue have already been introduced in Section 1.2. As noted in that section, the mental state elements of agency models can be broadly compartmentalized into two areas: (a) models of action and intentionality; and (b) models of belief and knowledge. In the AODM model, the belief-centric elements of mental state are captured in terms of the model's Dialogue State Structure (developed in the next section) and the domain knowledge of a given application (developed in the next chapter). In this section, we concern ourselves chiefly with the intentional aspects of the mental state.

The view of action and intentionality applied here is derived from work in the area of agent-oriented programming. In particular, we make use of an agent-oriented programming language in the style of Shoham's [1993] AgentO. The language, AL-PHA (A Language for Programming Hybrid Agents), was originally developed as a middleware solution for dialogue system integration [Ross et al., 2004c,b]. AL-PHA extends Collier's [2001] Agent Factory Programming Language which provides constructs for the management of a temporal logic based mental state for individual agents. The ALPHA model also includes a Stalnakerian view of information state, intention structures (termed commitments), and reactive behaviour control. Means-end reasoning support is also provided in the ALPHA language, and has been back-ported to its predecessor. Furthermore, ALPHA also incorporates more advanced constructs including role specification and design inheritance.

Although ALPHA provides a rich general-purpose agent theory that has been applied to rudimentary dialogue interaction for service robots [Ross, 2004, Ross et al., 2004a], for our current purposes we draw on a sub-set of the ALPHA model. The primary reason for this is that, as a full agent-oriented programming language, ALPHA requires that the majority of the agent model be defined explicitly in custom high-level programming constructs. In practice however, it is often more convenient and efficient to apply the core theory directly with derived models. Thus, we take the general model of actions, plans, and the management of these objects at the intentional level from ALPHA, but do not require that their description be made explicit in the context of a stand-alone agent-oriented program specification.

In terms of its principle features, the AODM model assumes an agent to be endowed with an inventory of capabilities, i.e., a *capability library*, each of which describes, at the intentional level, some action or construction of actions, i.e., a plan, that the agent is capable of performing. Whereas a behaviour implementation or actuator provides a sub-symbolic implementation, the action itself describes that low-level implementation at the symbolic level, and hence in a manner accessible to dialogic and deliberative reasoning. Plans are in turn simply constructions of these action descriptions via logical operators. The importance of the capability to the dialogue process is that it describes what the agent is capable of doing at a human cognition oriented level, and is hence a natural basis for communication between system and user. It should be noted however that whereas the granularity of capability descriptions can be arbitrary in some agent models, here we assume that the capabilities available to a dialogue enabled agent are relatively coarse and corpus motivated in the sense of Lauria et al.'s [2002] IBL inventory. We return to the issue of capability types for particular domains in Chapter 7.

Whereas capabilities define the potential to act, the agent must instantiate a capability to perform an action. Within the AODM model, the notion of an instantiated capability will be referred to as an intentional action or simply *Intention*. The intention-oriented equivalent of a plan is a tree-like *intention structure*. An intention is a fundamental unit within the mental model and can be reasoned on and queried just like other mental state notions such as spatial models or dialogic state. This model of intentionality is based on the formal analysis of commitment and commitment management made by Collier [2001] and extended through the ALPHA language, but has been simplified for the more modest needs of the dialogic agent model.

As we will see later, the benefit of the intention model to the dialogue process is that it enables the performance of a variety of communicative behaviours such as confirming the initiation and completion of actions in an automated, domain independent way. On that note, the needs of situated language processing require that the AODM model's notion of intentionality go beyond the original properties of the ALPHA language design. In particular, the usage scenarios introduced in Chapter 1 require that the AODM model be capable of resolving the contextualized meaning of a simple command like "stop" against what might be multiple actions being performed by the agent. Having explicit commitment and action structures brings us part of the way to addressing that requirement in that we may develop formal models of individual actions and their internal states, and use these to determine whether a generic command might apply to the action in that state. However, in the case of multiple active actions, which can all 'be stopped' or altered in some way, the intention models must be augmented with a notion of salience.

6.1.1 Model Components

Models of agency and intentionality are most formally expressed in logical theories which combine modal, temporal and dynamic logic components, e.g., Wooldridge [2000]. Since our goal here is to apply a portion of a complete agency model to the task of dialogue management, we pursue an implementation-oriented description. This description will be provided in two parts. In this first section, we describe the basic building blocks of the intentionality model. Then, in Section 6.1.2, we expand on this description by looking at the processes responsible for maintaining the agent's intentional state over time.

6.1.1.1 Actions, Plans & Capabilities

The basic element of the agency model is the notion of an *Action*. An action is an intentional level representation of a primitive activity that the agent can perform to alter its environment or own mental state. Actions are primitive within the agency model, but outside of the model, an action can be split into more basic primitives. These primitives are implementation specific and cannot be reasoned over at the intentional level. Primitives are encapsulated within an *Actuator* which models the embodiment of a given action. Following the Situation Calculus [McCarthy, 1968] and derived models of planning and logic programming, e.g., [Fikes and Nilsson, 1971, Levesque et al., 1997], an action is principally characterized in terms of a *Signature*, and *Pre-Condition*. The signature of an action defines the action type along with all parameters required to instantiate that action. Both the signature and parameter types are assumed to be defined in terms of the dialogue architecture's conceptual rather than linguistic ontology. The *Pre-Condition* for an action is a state description that must hold for an action to be successfully initialised. Pre-conditions can draw on any elements of the mental state, but are assumed to draw primarily on non-dialogic content. It should be noted that the current model does not include an explicit static post-condition as means-end reasoning is not necessary for current purposes, and that the actuator actually best models the potential effects of an action in a given situation.

Since the performance of actions in the real world are neither instantaneous nor guaranteed to succeed, an action also has an associated state that provides an abstraction over the status of the underlying actuator. Allowed action states include: *NEW*, the action has just been instantiated with necessary parameters; *ACTIVE*, the action is currently being performed by the agent; *PAUSED*, the action was active but is now temporarily suspended, *FAILED*, the performance of the action failed, *SUCCEEDED*, the action was performed successfully, and *TERMINATED*, the action is now defunct. Transitions between these states are self-explanatory with the exception that *TERMINATED* is the only terminal state, and actions can move from the *PAUSED*, *FAILED*, or *SUCCEEDED* states directly to the *TERMINATED* state.

In summary, an action is denoted as a 4-tuple as follows:

$$Action = \{Signature, Actuator, PreCondition, State\} \qquad (6.1)$$

Actions may be complexed into *Plans* in the usual way through the use of plan operators to allow the agent to perform and reason over more complex actions at the intentional level. A *Plan* is a composite, which, like an *Action*, is defined in terms of a *Signature* that defines the plan type and arguments required to instantiate the plan; a plan *Pre-Condition* that defines the necessary state conditions for a plan to be applied; and a plan *State*.

Unlike an *Action*, the operation of a *Plan* is not defined in terms of an actuator, but in terms of a *Body* that defines a construction of actions or plans connected via logical operators. Although the ALPHA model defines a rich set of plan operators (see Ross [2004]), here we limit ourselves to three operation types, i.e., logical sequencing via the *SEQ* operators, logical conjunction via the *AND* operator, and logical disjunction via the *OR* operator. These three operators are sufficient for the situated language processing scenarios pursued in this book. When needed, additional operators including conditionality and recursion constructs can be added in a straightforward manner following the models developed for the more complete ALPHA language.

While the state of an action is directly coupled to the state of the underlying actuator, the state of a plan is coupled to the semantics of the plan operator which heads the plan body. The semantics of the three plan operators are straightforward and linked to the manipulation of their arguments. Arguments of the *SEQ* operator are moved to the *ACTIVE* state in exclusive sequential order such that the *SEQ* operator is *ACTIVE* if any of its arguments are in an *ACTIVE* or *PAUSED* state. If any argument enters a *FAILED* state, then the *SEQ* operator also enters a *FAILED* state; while, if all arguments enter a *SUCCEEDED* state, then the *SEQ* operator enters the *SUCCEEDED* state. Arguments to the *AND* operator are moved to the *ACTIVE* state in parallel. If any child enters a *FAILED* state, then the *AND* operation also enters the *FAILED* state; while, if all children eventually enter the *SUCCEEDED* state, then the state of the *AND* operation is also *SUCCEEDED*. Like *SEQ*, the arguments of the *OR* operator are moved to the *ACTIVE* state in sequence until either one argument enters the *SUCCEEDED* state; in such a case the *OR* operation enters the *SUCCEEDED* state

In summary, a plan is defined by the following 4-tuple:

$$Plan = \{Signature, Body, PreCondition, State\} \tag{6.2}$$

The AODM agent is assumed to be endowed with one or more action definitions and zero or more plan definitions. We use the term *Capability* to generalize over actions and plans, and thus assume the agent to have a *Capability Library* that defines an inventory of available plans and actions. It should be noted that plan bodies can be composed dynamically outside the scope of named plan types, thus allowing a user to conjoin action and plan types arbitrarily.

We define the signatures of all capabilities, i.e., actions and plans, as having certain shared properties. First, we assume all capabilities to be performed by an agent – in our case either the dialogue agent itself or the user. Second, we assume that all capabilities have a certain earliest time at which a parameterised capability may be invoked. We may express these constructs from an ontological perspective as:

Definition 4. `Class: Capability`
        `SubClassOf: Thing`
                `and (actor exactly 1 Agent)`
                `and (earliestStartTime exactly 1 Time)`

where `Agent` and `Time`, `and`, `actor` and `earliestStartTime`, are respectively concepts and roles defined within the agent's conceptual ontology.

The AODM capability library is defined in terms of the agent's conceptual ontology both in terms of the range of capabilities available and the parameter types associated with each capability. A concrete inventory of capabilities will be defined for the verbal route processing domain in Chapter 7. Here, we take one of those capabilities, `Reorient`, and describe its properties so as to give a basis for the illustration of the connection between specific actions and the operation of the language processing system in later sections.

A `Reorient` is the simplest type of `SpatialAction` assumed for the route processing domain. The properties of the capability are that, given the agent is in a stationary state at a given location, then the agent will turn on its rotational axis towards a specified direction through a specified extent. For our simple model, no change in pose other than orientation change is necessary. The directional movement is modelled as a `GeneralDirection` which is parameterized with the direction modality and the direction extent. We thus define the signature of the `Reorient` capability as follows:

Definition 5. `Class: Reorient`
                `SubClassOf: SpatialAction`
                        `and (direction exactly 1 GeneralDirection)`

where

Definition 6. `Class: GeneralDirection`
                `SubClassOf: ComplexSpatialEntity`

```
and (modality exactly 1 DirectionalModality)
and (extent exactly 1 QuantitativeExtent)
```

Rather than assuming that a direction of movement is expressed with respect to the moving agent's ego-centric frame of reference – as is typically the case in a low-level robotic controller – it is useful to allow the direction to be expressed with respect to alternative frames of reference. The reason for doing so is that a user has relative freedom in reference frame selection, and the conceptual interface should reflect such flexibility (even if low-level behaviours necessarily make a mapping to an ego-centric reference frame). Thus, we assume that within the agent's cognitive layer, a direction is defined with respect to a particular frame-of-reference as follows:

Definition 7. `Class: DirectionalModality`
```
            SubClassOf: SpatialModality
                and (ref-frame exactly 1 ReferenceFrame)
```

The `Reorient` behaviour signature also inherits a number of properties through subsumption within the concept type hierarchy. Specifically, from `Capability` the properties of having an actor and earliest start time are gained, while from `SpatialAction` the properties of having a particular abstract speed and placement are gained. We summarize these properties along with the asserted properties for the `Reorient` class below:

Definition 8. `Class: Reorient`
```
            SubClassOf: SpatialAction
                and (actor exactly 1 Agent)
                and (earliestStartTime exactly 1 Time)
                and (placement exactly 1 Place)
                and (speed exactly 1 AbstractSpeed)
                and (direction exactly 1 GeneralDirection)
```

6.1.1.2 Intentions

Intuitively, an intention can be thought of as a commitment that is made by an agent to perform some capability. From a model perspective, we define an intention as a 7-tuple:

$$Intention = \{Capability, Requester, Actor, StartTime, State, \qquad (6.3)$$
$$Children, Parent\}$$

where *Capability* is the action or plan that is intended; *Requester* is the name of the agent who wishes the capability to be performed; *Actor* is the name of the agent who is to perform the capability; *StartTime* is the earliest time at which the agent will attempt to fulfil the intention; *Children* is an ordered set of child intentions (potentially empty); *State* is the state of the intention; *Parent* is the intention's parent (potentially null); and *Outcome* is the set of likely outcomes. The notion of *Capability* has been defined above, and we can give common sense interpretations to the notions of *Requester*, *Actor*, and *StartTime*.

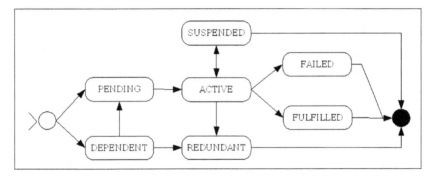

Fig. 6.1 Intention states with allowed transitions.

The state of an intention signifies at what stage of fulfilment the intention is at. An intention can have seven non-trivial states: DEPENDENT, an intention that has been adopted, but is not due to be actively fulfilled until some fruition constraint has been achieved; PENDING, an intention that the agent will begin to actively fulfil in its next execution cycle; ACTIVE, an intention that is being actively pursued by an agent; SUSPENDED, an intention that is still held but which is temporarily not being pursued by the agent; REDUNDANT, an intention that will no longer be pursued; FULFILLED, an intention that has been successfully realized; and FAILED, an intention which the agent judges to be no longer achievable. These seven states, the start and terminal states, and allowed state transitions are depicted in Figure 6.1.

Each intention can have a number of child intentions, where each child is an intention to perform some capability, which if successful, partially achieve the parent intention. For example, an intention to perform two actions can be refined to two intentions to perform more primitive actions. Thus, an intention structure is formed, where a coarse grained intention is broken down into sub-structures of finer grained intentions. Following Collier's proposal, we refer to the topmost intention in an intention structure as a primary intention. The agent's intention set (*Intentions*) contains a number of these intention structures, where each structure was added through the initial adoption of a primary intention through the application of some deliberative process. Intention set elements may be in any allowed state. However, for reasons of practical efficiency, the intention set is a leaky set in that intentions that have reached the terminal state are removed. The semantics of the plan operators, and the general processing of intention structures are addressed shortly, but first, we consider the salience structure that abstracts over the intention set.

The *Intention Salience List* is a stack of atomic intentions used to explicitly track the most prominent intentions within the agent's mental state. We define an atomic intention to be most salient if, and only if, it has just moved to either the *ACTIVE* or *PAUSED* states. After movement to the *ACTIVE* or *PAUSED* state, the intention is pushed onto the top of the salience list where it remains until the intention enters a terminal state, or until the intention is pushed down the intention stack through the pushing of a new atomic intention to the top of the stack. Unlike other aspects of the intention state, the Intentional Salience List is exclusively used for dialogue management concerns. Specifically, the Intention Salience List facilitates process resolution as required for interpreting highly elliptical process resolving commands such as "stop".

6.1.2 Model Dynamics

Having detailed the intentionality components of the agency model, in this section we describe how these components change over time. Such changes involve the adoption of intentions, as well as the maintenance of the agent's intentional structures. We begin however with a description of the abstract agent execution model that underpins the intention management processes.

6.1.2.1 Agent Execution

The AODM model assumes that the agent operates over a number of execution cycles. During each execution cycle, basic execution steps are performed to query perceptors, update mental state, plan, deliberate on actions to be performed, and perform those actions. For illustration of the general principles, Figures 6.2 and 6.3 respectively depict the life cycle and execution cycles for the agent model. Each execution cycle can be viewed as a three step process. First, the agent's mental state is updated on the basis of any new percepts or temporal constraints on existing knowledge. Second, system goals are updated by: (a) removing goals that have been achieved; (b) adopting new goals on the basis of goal adoption rules; and (c) performing means-end reasoning to adopt intentions towards achieving those goals. Third, agent intention structure is refined through: (a) the adoption of new primary intentions; (b) the refinement of existing intention structures; and (c)

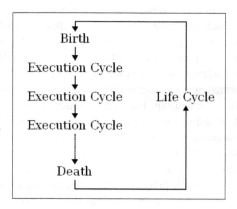

Fig. 6.2 Agent Life Cycle

the firing of actuators. The complete execution model was developed in Ross et al. [2004a]. However, as indicated in Chapter 4, we are not concerned with the entire agent design here. Rather, the following focuses on the intention management process by considering the adoption and refinement of intention structures.

6.1.2.2 Intention Management

Intention Adoption is the process by which an agent decides to add an intention to its set of maintained intentions. Intentions may be adopted by an agent for many different reasons, e.g. as a result of external physical stimuli, or in response to a communicative move made by another agent. In practice, many different computational mechanisms may be applied to deciding upon new intentions. The classical approach, as for example explored by Haddadi [1996], is to apply means-end reasoning over the systems goals and abilities to decide upon intentions for adoption. A less-powerful but more computationally tractable approach is to make direct use of rational rules as for example applied by Collier [2001]. Both the means-end reasoning based and rational rule-based approaches to intention adoption were implemented for the ALPHA language interpreter [Ross et al., 2004a]. In the AODM model we view the specification of rules for the adoption of intentions in response to non-communicative states as an application-specific design consideration. However, the adoption of intentions in response to user communicative moves is a core part of the dialogue management strategy. In Section 6.5 we thus develop an intention adoption strategy in the context of the dialogue planning process.

Regardless of how intentions come to be adopted, once per execution cycle each of the agent's intention structures are updated. The update of intention structures is managed by a two-pass algorithm. This algorithm prunes and expands the intention structure in response to changes in the state of individual intentions and broader changes in the agent's mental state. A pseudo-code representation of the intention update algorithm is presented in Figure 6.4.

During the first update phase, the algorithm descends to the leaves of an intention structure, and examines the state of each intended capability. If a capability has

```
Mental State Management
    Perceptor Management
        Empty Perceptor Queues into Mental State
        Fire Each Perceptor in its own Thread
    Mental State Cleanup

Deliberation / Goal Management
    Drop any goals which have been met
        Drop any subsequent intentions
    Apply means-end analysis rules once to each goal.

Intention Management
        Adopt new primary intentions
    Take each primary intention in turn
        Update intention structures
            Fire Actions
```

Fig. 6.3 Pseudo-code for the execution cycle of a full ALPHA agent

```
update_Intentions(){
    foreach(Intention_i : Set(Intention)_Primary){
      prune(Intention_i);
      expand(Intention_i);
    }
}

 prune(Intention_i){
   if(Intention_i.Children.size == 0){
      if(Intention_i.Capability.state == SUCCEEDED){
          Intention_i.State = FULFILLED;
          Salience.remove(i);}
      else if(Intention_i.Capability.state == FAILED){
          Intention_i.State = FAILED;
          Salience.remove(Intention_i);}
   }
   else{
       foreach(Intention_child : Intention_i.Children){
         prune(child);
         if(Intention_i.Capability.state == UPDATED){
            Intention_i.Children.update();}
         else if(Intention_i.Capability.state == SUCCEEDED){
            Intention_i.State = FULFILLED;}
         else if(Intention_i.Capability.state == FAILED){
             if(Intention_i.StartTime){
               c.State = PENDING;}
             else Intention_i.State = REDUNDANT;}}}
 }
  expand(Intention_i){
   if(Intention_i.State == ACTIVE){
     foreach(Intention_child : Intention_i.Children){
        expand(child);}}
   else if(Intention_i.State == PENDING){
       if(Intention_i.Parent){
          Intention_i.State = ACTIVE;
          Intention_i.Capability.state = ACTIVE;
          foreach(Capability_i : Intention_i.Activity.Children){
            Intention_new = new Intention(Capability_i);
            add(Intention_i.Children,Intention_new);}}}
 }
```

Fig. 6.4 Pseudo-code for Intention Structure Updating

succeeded or failed, then the intention state is updated to FULFILLED or FAILED respectively, along with the removal of the atomic activity from the salience list and the update of a parent intention if present. The checking of capability states percolates up through the intention structure until all intentions have been checked. As parent intentions are notified of the change of state of child intentions, other previously DEPENDENT child intentions are moved to the PENDING state. Such transitions are dependent on the semantics of the plan operation committed to.

During the second update phase, all atomic PENDING intentions are moved to the ACTIVE state, with a subsequent transition of the intentions's activity from the start state to the ACTIVE state. Similarly, all intentions with child structures are examined until all plan operations have been expanded. The consequence of moving the intentions's activity to the ACTIVE state is to create any new secondary intentions, marking these new intentions as DEPENDENT, ACTIVE, or PENDING as appropriate. The exact actions that are performed during the setting of an activity to ACTIVE are dependent on the semantics of the plan constructs introduced earlier. A detailing of these semantics is considerably beyond the scope of interest here; but, in summary, if the activity resolves to an actuator, then activation involves the triggering of the actuator in its own thread. However, if the activity corresponds to a plan identifier, then a child intention towards the plan body is adopted. If the activity corresponds to a plan operator (e.g. SEQ), then the creation of child intentions is dependent on the semantics of the operator in question.

For illustration, Figure 6.5 illustrates the update of intention structures and the intention salience structure. Here, the left column depicts an intention set consisting of one true intention structure and one atomic intention. The right column depicts the intention salience list which can be thought of as a view on the underlying intention structure. Rows *a* through *e* depict changes in intention structure and salience lists over time as intentions become active, are fulfilled, and suspended. Note the suspension of intention towards `act_c` causes that intention to be promoted through the salience structure.

6.1.3 Discussion

In this section, we outlined a rudimentary account of action and plan structure to provide a basis for the intentionality aspects of the AODM model. While an account based on a well-established framework of action and goal management such as the Procedural Reasoning System [Georgeff and Lansky, 1987] or GOLOG [Levesque et al., 1997] could be employed for the same purposes, the advantage of using a minimal custom account is that it is ideal to highlight just those aspects of intentionality that have consequences for integration with dialogue processes in subsequent sections – without getting into too much detail on the nature of the complete goal-oriented or agent-oriented architecture. That said, it would be interesting, both from a theoretical and practical perspective, to apply such well developed frameworks directly as a basis for the AODM model. This issue is considered further as future

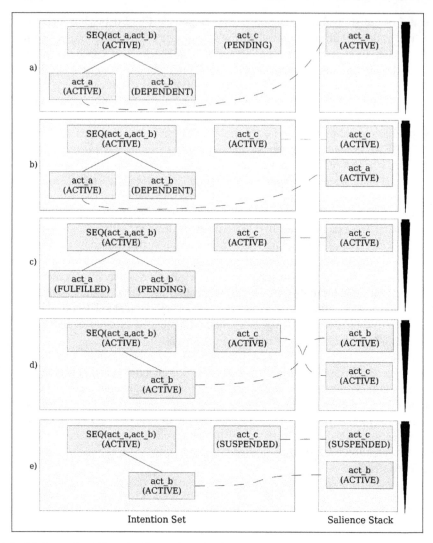

Fig. 6.5 Dynamics of Intention Set and Intention Stack update.

work in Chapter 9.

The presence of explicit models of action, capability, and intentional structure have beneficial consequences for the development of dialogic behaviours. Capabilities can be used as a primary input to a language understanding process in that they define what the agent is capable of doing, and hence what the user is likely to request of an agent in a task-oriented environment. Also importantly, dialogue planning can take advantage of transitions in intentional state to generate appropriate descriptions of intentional transitions for users, thus ensuring that user and system models of intentions and activities are aligned over the course of a dialogue. And finally, as already indicated, an intention salience list can be used to identify the elided parameters to many meta-process requests such as stopping, pausing, or restarting a capability. In the remainder of this chapter and in the next chapter we will see how the intentional structures developed here are applied to just these dialogue behavioural goals.

6.2 The Dialogue State Model

Having developed an account of intentional action as the basis of domain modelling, in this section we detail the dialogue state model. The dialogue state model's description will be given chiefly in terms of the dialogue state types (Section 6.2.2), the dialogue state structure itself (Section 6.2.3), and the modules and control processes that link together the dialogue state model with dialogue processing (Section 6.2.4). We begin however with an informal introduction to the locus of this dialogue structure account, i.e., the dialogue move.

6.2.1 Dialogue Moves: An Analysis

In Chapter 4, a frame-like modification of the dialogue move was proposed as the unit of exchange between dialogue management and domain-specific reasoning. This proposal was motivated by our need to take an agent-oriented perspective on dialogue processing, yet builds on the frame metaphor as a staging ground for meaningful unit composition.

To illustrate the relationships between a dialogue move and other units of the agent's mental state, Figure 6.6 depicts a dialogue example along with a partial discourse structure that emphasises the inter-stratal relationships between representation units. The discourse structure relates dialogue moves to units of representation from both the language interface and non-communicative mental state. Full lines denote realization relationships between units, i.e., within the language interface, an utterance realizes a message, and a message is a realization of an act. Coarse dashed lines on the other hand express a constituency relationship between dialogue moves and dialogue acts; however, reading these lines as a *concerns* relationship is useful

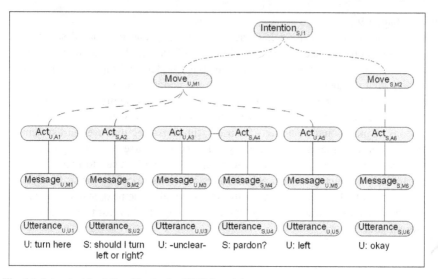

Fig. 6.6 Inter-stratal relationships in the AODM model.

here since it takes emphasis away from an overly mereological analysis. Finally, the fine dashed lines neither express realization nor constituency, but instead express the loose causal relationships which exist between dialogue moves and other units of the agent's mental state – in this case an intention.

While the distinction between the dialogue move and the dialogue act used here is in keeping with both the terminology and intent of Exchange Structure style analysis (see Section 2.4.2), the present use of the two distinct notions is motivated by pragmatic concerns rather than the desire for a constituency-based analysis of dialogue structure. As a result, there are a number of distinctions between the notions of dialogue act and dialogue move that are of note:

- First, whereas dialogue acts can be instantiated for any of the speech function types outlined in Table 5.3 in the last chapter, dialogue moves in the AODM model are only instantiated for task-relevant speech function types, i.e., not purely conversational types. This distinction is due to the level of non-task exchange elements being handled by the dialogue management processes without any need for explicit domain contextualization.
- Second, while the contents of both dialogue acts and dialogue moves are defined in terms of the agent's conceptual ontology, the content of a dialogue act can be any consistent selection from this ontology, whereas the content of a dialogue move must be headed by an application state or capability. Thus, a move is assumed to constitute a 'meaningful' update of the agent's state rather than a fragmentary piece of information. It is then the responsibility of the dialogue process as a whole to make the mapping from fragmentary acts to complete moves.

Role	Type	Filler	Solution A	Solution B
actor	Agent	nil	1.0, system	1.0, system
placement	Place	nil	1.0, Place	1.0, Place
			modality Containment	modality Containment
			relatum Kitchen1	relatum Kitchen1
earliestTime	Time	nil	1.0, now	1.0, now
direction	GenDir	GenDir	0.5, GenDir	0.5, GenDir
		modality Left	modality Left$_{Ego}$	modality Left$_{Allo}$
			extent 90	extent 90
speed	Speed	nil	1.0, normalSpeed	1.0, normalSpeed

Table 6.1 Instance of an Instruct-Reorient move with frame description and two possible solutions. For this example we assume that the move was constructed following the contextualization of information provided in the utterance "turn left" when the actor is situated in a kitchen.

- Third, dialogue acts and dialogue moves evolve very differently over the course of a dialogue. The content of a dialogue act is essentially immutable in that once it is instantiated within the language interface for either language analysis or production it is generally unaltered. The content of a dialogue move, however, is a dynamic structure in language understanding. In other words, we assume a dialogue move can exist in states ranging from a skeletal description of move potential, through states with partially or ambiguously defined content, to the fully contextualized instances which can be subsequently integrated into the agent's mental state.

From Figure 6.6, it should also be clear that not all dialogue acts have a direct relationship with a dialogue move. Uncertainty in language processing can give rise to non-integratable dialogue acts to which system acts are immediately composed as clarification questions. In the AODM model, such clarification questions are not assumed to be directly related to a dialogue move.

The licensed content of a dialogue move is directly coupled to the agent's range of capabilities and potential mental states. More specifically, user dialogue moves and the intentions an agent may adopt are coupled in the usual way in terms of classical illocutionary logic rules which dictate that if the system is requested to perform some activity, and the system is capable of performing that activity in the current state, then the system should adopt the intention to perform that activity. Similarly, if the user informs the system of some knowledge, and that knowledge is not incompatible with the agent's current knowledge, then the agent should add that knowledge to its mental state. These core mental state adoption rules are modelled as one aspect of the dialogue planning process described in Section 6.5.

Due to the dialogue move's role as a frame-like construct that sits between the language interface and the agent's intentional state, we must model the dialogue move as something considerably more structured and dynamic than a simple predicate structure, or even the FOS structures introduced in Section 5.2. Therefore, in the case of language understanding we model the dialogue move as having three components:

- **The Move Template:** defines the move type and content potential in terms of concept and role definitions extracted from the agent's application ontology.
- **The Move Filler:** is the shallow description provided by the user to fill out the roles in the move template. Move filler content reflects precisely the information raised from individual user dialogue acts – hence the move filler is also defined in terms of the agent's application ontology.
- **The Solution Set:** is the set of possible interpretations of the move filler following domain contextualization. While solution contents are defined in terms of the agent's application ontology, they are also referentially-transparent with respect to the application model, have associated interpretation probabilities, and typically include content which was not directly provided by the speaker.

For illustration, Table 6.1 depicts a move instance which includes the Move Prototype, Move Filler, and Solution Set information for an instruction to perform a Reorientation. For this particular example we see that the speaker only provided direction information explicitly, and that all other parameters in the presented solution were filled through contextualization.

At any given time in the dialogue, the choices made in both language integration and dialogue planning are due in part to the properties of such dialogue moves. We generalize over these properties into a number of *move states*:

- **New:** The move has been initialised with the content of a user dialogue act, but no contextualization has yet been performed, and thus the solution set is empty.
- **Open-Parameter:** The move contains solutions which include at least one parameter field that is empty and mandatory. Information is required from the interlocutor to fill this parameter which cannot be filled though contextualization.
- **Multiple-Solutions:** The move contains at least one parameter which has multiple possible interpretations with respect to the domain model. A clarification question is required to determine which of the possibilities is intended.
- **Rejected-Parameter:** The move contains a parameter for which the provided filling value is incompatible with the domain model.
- **Open-Question:** A question has been raised by the system with respect to a move.
- **Complete:** The move contains exactly one valid solution.

In task-oriented dialogues such as those we are concerned with in this book, speakers often request that two or more activities be performed as structured by a logical operator, e.g., a user might request that two actions A and B be performed in sequence. In Section 6.1, we saw how this is handled at the agency level in terms of the complexing of actions through plan operators and resultant intention strictures. In simplified terms, the resultant intentional structure for the given example would be:

$$Intention(SEQ(A,B)) \tag{6.4}$$

In terms of the dialogue move stratum, this same construction is handled in terms of two distinct moves that are related via a reified relation that holds between the two move instances. This *move complex* treatment stands in contrast with a model where a single user move instance takes the complex contents SEQ(A,B). Although the move complex treatment is arguably not an optimal modelling solution since it asserts a relationship between two moves rather than a relationship between the contents of those moves, it does have one advantage. Namely, the complexing approach allows the assertion of direct sequential relations between capabilities to be treated at a domain-specific level rather than introducing over-commitment in the earlier language integration process.

In addition to modelling user contributions, the dialogue move can also be applied to the modelling of system dialogue moves, albeit with some differences to account for the initial certainty in system rather than user dialogue moves. Specifically, unlike user dialogue moves, system dialogue moves only have a single contextualized interpretation as there is no ambiguity in system generated content. System dialogue moves are expressed in terms of the same application ontology content as user dialogue moves. User dialogue moves are however typically generated by domain-specific logic which determines when it is appropriate to convey some information to the user or to have the user answer some question. System dialogue moves can also be generated by the dialogue management process in response to user dialogue moves. In either case the content of both moves is the same. There are then subsequent processes which extract the information from a dialogue move which is to be conveyed to a user. This is performed in an initial act planning process which takes the moves and composes some content. All dialogue planning issues will be dealt with in Section 6.5.

6.2.2 Dialogue State Types

The dialogue move provides a natural unit of representation for the conceptual modelling of user and system contributions in the dialogue management model. It is however only one element of the complete discourse model. Messages, acts, and intentions also play important roles alongside other cognitive modelling units. In the remainder of this section, we make some of these other aspects of the dialogue state more concrete, before looking to the language integration, contextualization, and planning processes in subsequent sections. In the following we detail the data types which underpin the dialogue state representation to be outlined in Section 6.2.3.

6.2.2.1 Basic Data Types

In keeping with a Stalnakerian view of mental state, the dialogue state is not a 'bag' of beliefs, but rather a structured model. The FOS data model introduced in Section 5.2 acts as the base representation for all knowledge represented within the dialogue state. In particular, that data model provides us with a typed model of objects, terms, and relationships, as well as the traditional atomic data types of `String`, `Float`, `Integer`, and `Boolean`. The dialogue state structure however also requires a number of state container types and operators to manipulate and query these containers. Namely, we also introduce the structural containers `Set`, `List`, `Stack`, and `Atoms` along with standard querying and manipulation operators. Out of these basic data types we compose more structured types as needed to represent language at various stages of interpretation. Two of the most important representation types are the dialogue act and dialogue message. Since those types have already been introduced in some detail in Chapter 5, we will not repeat them here. Instead, we turn immediately to the details of the dialogue move.

6.2.2.2 Dialogue Moves and the Dialogue Move Template Library

A *move template* is a potentially interpretable move that is defined in terms of a task relevant speech function type and an application-specific state or capability type. While the application state or capability type can be defined by reference to a named class within the application ontology, it is useful to make explicit the content of such a type in a structure optimized for the needs of dialogue processing. Thus, we define a move template more concretely as follows:

$$MoveTemplate = \{SpeechFunction, TaskType, Parameters\} \qquad (6.5)$$

where SpeechFunction is a task relevant speech function type, i.e., one of the task relevant types given in Table 5.3, TaskType is an application defined action or state, and Parameters defines the set of parameters required to instantiate an instance of the given TaskType as defined by the agent's application ontology. Concretely we define the Parameters as a set of Parameter objects where a parameter object captures a single role which must be filled for a parameter, along with a property which indicates if the parameter must be filled before contextualization of the move type is feasible, i.e.,

$$Parameter = \{Role, FillerType, Mandatory\} \qquad (6.6)$$

The set of available move templates is defined within a *Move Template Library*. The move template library has a relationship to the agent's capability library, but they are not one and the same. Whereas a capability can be thought of as some action that the agent is capable of performing, a move can be thought of as such an action plus a dialogue move type concerning that action. Moreover, moves need not only concern capabilities. Moves may also concern conceptual state types in the case of querying or updating mental states.

For a given user act, it is necessary to determine the most relevant move to which the act can be attributed. Move selection will be addressed in Section 6.3, but for current purposes we need only consider the existence of an instantiation operator, which given a user dialogue act and a move template defined as relevant to the act, returns a move instance with all content information within the act transposed to the relevant parameters within the move, i.e.,

$$Move = f_{inst}(Act, MoveTemplate) \tag{6.7}$$

A move instance can be defined in turn as an extension of the move template with the set of specific parameters provided by a user, along with a set of disambiguated interpretations for those parameters:

$$Move = \{SpeechFunction, TaskType, Parameters, Fillers, Solutions\} \tag{6.8}$$

Table 6.1 provided an explicit example of what such a move instance in turn looks like.

Language interpretation and planning mechanisms also define a number of operations which update and attach information to the dialogue move. We return to these as needed, but decisions on which operation to perform often depend on a generalization over the state of a dialogue move. We define the generalized dialogue move states, introduced in Section 6.2.1, as follows:

$$MoveState = \{New, OpenParameter, MultipleSolutions, \tag{6.9}$$
$$RejectedParameter, OpenQuestion, Complete\}$$

and introduce a function that returns the state for a given move:

$$MoveState_m = f_{state}(Move_m); \tag{6.10}$$

6.2.2.3 Processing Error Types

In both human language competence and computational models of language processing, errors can be introduced at a number of processing stages. These errors may be due either to uncertainty in information, or because a speaker used constructions, concepts, or domain knowledge that a hearer is not familiar with. Within the AODM model, we explicitly capture the occurrence of errors in the language processing backbone so that an appropriately informative message can be planned and conveyed to the user to inform them of the error when necessary. Table 6.2 thus defines the set of error types that can be used to describe the deviations from processing norms which can occur.

As can be seen in Table 6.2, an error type may be complex rather than purely atomic in that a content may be associated with the type. Whereas no content type is appropriate for a `PerceptionPoor` error, an `UnknownWords` error takes a set of strings as content where each string in the set defines a single world that is out of lexicon in the input utterance. The content of a `ParseFailure` on the other hand is the complete string that failed to be parsed. An `AbstractionFailure` takes the actual linguistic semantic content for a message that could not be abstracted to an application-specific level, while the `IntegrationFailure` error takes the

Error Type	Content Type	Description
PerceptionPoor	n/a	Speech recognition detected a user attempting to speak, but either volume levels were too low for perception, or the perceived result was below a minimal quality threshold.
UnknownWords	Set(String)	Language analysis could not process an input because one or more words were outside of the lexicon.
ParseFailure	String	Language analysis could not process an input because one or more constructions were outside of the grammar.
AbstractionFailure	Message	A conceptual semantics could not be generated from a linguistic semantics. Thus, while the grammar covers the input, it is out of range for the current domain application.
IntegrationFailure	Act	A user dialogue act could not be integrated into the Dialogue State Structure.

Table 6.2 Enumeration of dialogue processing error types

act which could not be integrated into the dialogue state.

Within any computational system, a great range of distinct states can be defined to enumerate possible erroneous processing outcomes. The collection of states provided here reflects a balance between the range of true possible problems in language processing, and what information can be conveyed usefully to a user. The set can thus be seen to reflect the states of communicative action proposed by Allwood [1997], Clark [1996], and applied by Larsson [2002], but customized to the processing architecture considered here. Moreover, whereas Larsson conflated all notions of errors in dialogue processing directly into an abstract type of dialogue move, i.e., his interactive communication management moves, here we model errors in processing as particular states rather than moves – although, as we will see in later sections, these states may result in the dialogue planning process adopting dialogue goals which are reflexes of these states.

6.2.3 The Dialogue State Structure

The Dialogue State Structure (DSS) is a record-like model of the dialogue state in terms of the types introduced in the last section as well as the propositional content potential defined by the agent's application and linguistic ontologies. Table 6.3 summarizes the DSS grouped according to function within the language processing architecture. In the following, each of these groupings are detailed in terms of the

Slot	Type	Grouping
Latest-User-Utterance	{String,float}	Input Abstractions
Latest-User-Message	Message	
Latest-User-Act	Act	
Non-Integrated-User-Acts	Set(Act)	User Act Containers
Open-User-Moves	Stack(Move)	User Move Containers
Closed-User-Moves	Stack(Move)	
Planned-System-Moves	Stack(Move)	System Move Containers
Raised-System-Moves	Stack(Move)	
Closed-System-Moves	Stack(Move)	
Planned-System-Acts	Set(Act)	System Act Containers
Open-System-Acts	Set(Act)	
Next-System-Act	Act	Output Abstractions
Next-System-Message	Message	
Next-System-Contribution	String	
Input-Error	ErrorType	Error types

Table 6.3 The Dialogue State Structure

relevant DSS entries.

6.2.3.1 Input Abstractions

Driven by the two-level semantics methodology, but also by the realities of pipelined processing architectures, any given user contribution is modelled at different levels of abstraction. The Input Abstraction entries provide the fields for representing a user contribution at these different levels of abstraction.

The Latest-User-Utterance captures the surface information for a user contribution. Here we assume a simple input surface model since our focus is on the deep contextualization issues. Specifically, Latest-User-Utterance is modelled as a 2-tuple consisting of: (a) a representation of what the user most likely typed or said – modelled as a String; and (b) an associated confidence score for that interpretation – modelled as a float with values ranging from 0 (very low confidence) to 1 (very high confidence).

The Latest-User-Message field captures the linguistic semantics for a given user contribution following shallow analysis. Shallow analysis frequently results in multiple analyses for a given utterance. Here, we assume that the results of analysis can be filtered though the use of ontological constraints as well as the application of simple heuristics based on likelihood of interpretation to obtain a single most likely surface interpretation[1]. Thus, we model Latest-User-Message as

---

[1] This is the case for the language analysis models introduced in Chapter 8.

a single instance of a `Message` object as was described in Chapter 5

The `Latest-User-Act` provides the second, more conceptual, representation of a given user contribution expressed in terms of an `Act` type as introduced in Section 5.4. We assume a one-to-one mapping between `Latest-User-Message` and `Latest-User-Act`.

6.2.3.2 User Act Containers

Mixed initiative dialogue is not always strictly sequential and based upon single isolated contributions. Thus, not all user acts can be immediately integrated into the dialogue state. This may for example be due to: (a) the system asking a confirmation question before integrating a user act; (b) a user having contributed an act complex rather than an atomic act; or (c) because the system must attend to system-initiated dialogue goals before addressing a new user contribution. The `Non-Integrated-User-Acts` field, defined as a stack of `Act` objects, captures all user acts which have not yet been integrated into the dialogue state.

6.2.3.3 User Move Containers

As argued earlier, the AODM model views the move as the principle unit of exchange in dialogue, but acknowledges that user moves cannot always be interpreted from a single communicative act in one step. Instead, we require a 'staging ground' for user moves. The `Open-User-Moves` field captures this staging ground as a set of moves that the agent believes the user is trying to make, but for which the agent has not yet derived a unique interpretation through contextualization or dialogic means. More concretely, `Open-User-Moves` is defined as a stack of `Move` elements where each `Move` can be in any move state other than the `Complete` state.

If a successful interpretation of a move is composed, the agent will act on that move in accordance with the user move handling rules introduced in Section 6.3. Regardless of the agent's stance on a given user move from the dialogue perspective, we can consider the interpretation of the move to now be complete. The field `Closed-User-Moves` provides a long term store of moves for which a single unique interpretation has been composed. The use of a separate field allows moves to persist for reference, but removes them from the set of moves that are actively considered in act interpretation, thus simplifying computational operations.

6.2.3.4 System Move Containers

Just as user moves cannot always be interpreted in a single act, system moves can potentially not be communicated to a user in a single unambiguous act. We thus also need a staging ground for moves that are to be made, and also a means of distinguishing between the moves that the system wants to make, and those moves that the system is actively trying to make. We introduce three fields to account for the different global stages of system dialogue goal processing:

- `Unraised-System-Goals` defines those dialogue goals that the agent wishes to have made but which the dialogue engine is not yet processing;
- `Raised-System-Goals` defines those dialogue goals that the agent is actively attempting to convey to the user; and
- `Closed-System-Moves` the set of dialogue goals that the agent is no longer trying to convey to the user.

All three fields are defined in terms of a stack of `Move` objects.

6.2.3.5 System Act Containers

Within the DSS we introduce two fields for modelling system dialogue acts. `Planned-System-Acts` is the set of acts which the agent has planned to further user move interpretation or its own dialogue goals but which the agent has not yet attempted to convey to the user. `Open-System-Acts` on the other hand are those acts that the agent has attempted to realize but which have not yet necessarily been acknowledged by the user. Both `Planned-System-Acts` and `Open-System-Acts` are modelled in terms of a set of `Act` objects.

6.2.3.6 Output Abstractions

At any given point during an interaction, the agent may select content from its set of planned acts and attempt to contribute them to the dialogue. We label the selected content as the `Next-System-Act` which is of type `Act`. If the given act is to be performed explicitly as a verbal contribution, then a linguistic semantic message is composed. Within the DSS, we label the planned message as `Next-System-Message` of type `Message`. Following realization, a representation of the verbal contribution is available for production. This value, labelled `Next-System-Contribution` in the dialogue state structure, is of type `String`. Although the `String` is a very generic type, it should be noted that this can act as place-holder for anything from a literal string representation of what is to be said or displayed by the agent to an XML document with precise prosody and timing information if such features are available within the language realization and production components. This is the case for the DIASIE framework introduced in Chapter 8, where an encoding of the utterance to be produced is described by a SABLE document [Sproat et al., 1998] encoded as a string.

6.2.3.7 Error Types

One DSS field is introduced to account for possible failures at different stages of language processing. The `Input-Error` field, of type `ErrorType`, captures failures in language processing during an execution cycle which may be due to competence inadequacies in the language processing resources. The range of available input error and output error types was defined in Section 6.2.2.3.

6.2.4 Dialogue Architecture Control

In addition to the Dialogue State Structure, the AODM model defines the processes by which system dialogue moves and acts are planned, and user dialogue contributions are integrated into the dialogue state. As part of this functionality, we need mechanisms by which the DSS model is updated and information is passed between processing components within the complete architecture. In Section 4.2 we anchored the introduction of the dialogue architecture by providing a brief overview of the processing components and control methodology. Here we give detail to the module interfaces and the control algorithm that connects these modules together.

6.2.4.1 Module Interfaces

The following defines the interfaces assumed for each of the dialogue processing modules.

Perception: The perception module generalises over typed and written language input and provides an interface method (`perceive`) which writes the most recent user utterance (if any) along with a confidence score to the `Latest-User-Utterance` field of the dialogue state. One error case is also assumed for this module, i.e., that if perception levels are poor, an error is written to the `Input-Error` field. Since the production of language, i.e., speaking or typing words, takes a relatively long time in comparison to the other steps of language understanding or dialogue planning, the perception module also includes the boolean function `is-PerceptionActive`, which returns true if the perception module is actively processing user speech or typed text, or false otherwise. This property is made use of by the dialogue planning process to prevent the system from making contributions while a user is speaking.

Analysis: The analysis module defines the `analyse` method which looks to the `Latest-User-Utterance` field for input. If input is available, the module attempts to compose a linguistic semantic description of that content. If semantics could be derived, they are written to the `Latest-User-Message` field. If no semantics are produced for the input, either because one or more words were out of lexicon, or because the input utterance cannot be parsed with the available analysis grammars, then the relevant error is written to the DSS's `Input-Error` field.

Abstraction: The abstraction module defines the `abstract` method that looks to the `Latest-User-Message` field for input. If input is available, the message is transformed into an act type expressed in terms of the conceptual ontology. The result of that transformation is in turn written to both the `Latest-User-Act` and `Non-Integrated-User-Acts` fields. If no transformation can be made, an error is written to the `Input-Error` field.

Integration: The integration module defines an `integrate` method which looks to the `Latest-User-Act` field for input. If an input act is present the integration module attempts to integrate that act into the dialogue state by using it to update an existing open move, or to create a new move interpretation for the act. If a suitable integration can be made, either a new move will be added to, or an existing move will be altered within, the `Open-User-Moves` DSS field. If no integration could be found, an error move is written to the `Input-Error` field.

Contextualization: The contextualization module defines a `contextualize` method that looks to the `Open-User-Moves` DSS field for input. The contextualization function takes any open moves present and applies context to disambiguate and enhance the information present in any open moves and thus manipulate the move's solution set. Any updated move structures are written back to the `Open-User-Moves` field.

Dialogue Planning: The dialogue planning module defines a `plan` method that looks to a number of elements of the dialogue and broader mental state for input and determines what, if any, dialogue moves and dialogue acts should be performed by the dialogue system. Any planned content is in turn written to the system move containers and system act containers.

Message-Planning: The message-planning module defines a `compose` method that looks to the `Next-System-Act` DSS field for input and composes a linguistic semantic message to express this act. The output of this composition is written to the `Next-System-Message` field.

Realization: The realization module defines a `realize` method which looks to the `Next-System-Message` for input and uses an appropriate language grammar to compose a surface realization for that message which is in turn written back to the `Next-System-Contribution` field of the dialogue state structure.

Production: The production module defines a `produce` method which looks to the `Next-System-Contribution` for input, and produces verbal and/or textual output for that content. As with language perception, the language production module defines a function used by the dialogue planning module to determine if it is appropriate for the system to immediately perform a new dialogue act. The function, `isProductionActive`, returns true if language production is currently presenting information to the user via spoken or textual communication channels, false otherwise.

In addition to the above, the Dialogue State Structure itself includes a `clear` method which, when invoked, clears the set of Input Abstraction and Output Abstraction slots in the Dialogue State Structure.

6.2.4.2 The Control Process

With the description of both the dialogue state structure and processing modules in hand, we can now detail the concrete architecture control algorithm that was first introduced back in Section 4.2.2.

The AODM control algorithm is given in pseudo-code in Figure 6.7. The control algorithm consists of a continuously repeating loop or *execution cycle*. During each execution cycle, calls are made to each of the language processing components in order from language input through integration and contextualization, to dialogue planning, output, and clearing of transient DSS slots. In an idealised execution cycle, each of the language processing modules is called upon in turn with information read from and written to the dialogue state structure as appropriate. In practice, not each component is necessarily called per execution cycle. If no language input is detected, or if a competency error occurs during language processing, control moves im-

```
while(true){
  perceive();
  if((Latest-User-Utterance == null) ||
     (Input-Error != null)) goto plan;
  analyse();
  if(Input-Error != null) goto plan;
  abstract();
  if(Input-Error != null) goto plan;
  integrate();
  if(Input-Error != null) goto plan;
  contextualize();
plan:
  plan();
  compose();
  realize();
  produce();
  clear();
}
```

Fig. 6.7 The AODM control algorithm.

mediately to the dialogue planning steps to either decide on an appropriate system initiated dialogue contribution, or to inform the user of the language understanding problem. It should be noted that we assume language perception and production to operate in their own execution threads and that the `perceive()` and `produce()` methods operate on message queues that link the central execution cycle with these modules.

6.3 Language Integration

Having described the dialogue state structure and the system control algorithm in the previous section, this section focuses on the processes by which conceptual semantic representations of user dialogue contributions, i.e., dialogue acts, are integrated into the dialogue state structure. As with other processes described in this book, we eschew an implementation-oriented description of the operations made on mental state, in favour of a description which captures the most salient points of the model. We begin the section with an overview of the language integration process for individual dialogue acts.

6.3.1 Integrating Dialogue Acts

Figure 6.8 gives pseudo-code for the integration process for single dialogue acts.[2] The key points of the algorithm are as follows:

- If there is content on the LATEST-USER-ACT slot of the Dialogue State Structure (DSS), the integrate function first removes that content. If no content was present, no further integration is required.
- If the latest user act, Act_{LUA}, is a communication management act, e.g., GreetingInit, then the integrate method places the act on the Non--Integrated-User-Acts field of the DSS where it will be handled by the dialogue planning process.
- If Act_{LUA} is not a simple communication management act, the integrate process attempts to find a task-specific move for the dialogue act. Two separate integration cases are considered:
  - If Act_{LUA} is relevant to an open system dialogue act, then Act_{LUA} is combined with that act to produce a proposition that is used to update the user or system move that motivated that open act.
  - If Act_{LUA} is not relevant to any open dialogue acts, the integration process attempts to initialize a new dialogue move based on a relevant template from the dialogue move library. The instantiated move is then added to the DSS's OPEN_USER_MOVE field.
- If no task-specific move was identified for Act_{LUA}, then the INPUT-ERROR slot of the DSS is updated with a new integration error.

Ignoring error cases and the handling of basic communication acts, the integration of task-specific user acts in the model just described can be distilled into a two step

---

[2] The integration process, like all processes in the language processing architecture, is described in a procedural style rather than in the declarative update rule style of TrindiKit and derived Information State Update dialogue models. Although the procedural approach abstracts away from the declarative methods more often associated with agent design methodologies, the procedural style is considerably clearer since single method specifications can replace multiple declarative rules.

```
void integrate(){
  if(DSS.LATEST-USER-ACT == null) return;
  ActLUA = pop(DSS.LATEST-USER-ACT);
  if(ActLUA.Type subclassOf CM-Act){
    push(ActLUA,DSS.Non-Integrated-User-Acts);
    return;
  }

  MoveSelect = null;
  foreach(Acti : DSS.OPEN-SYSTEM-ACTS){
      if(isRelevant(ActLUA,Acti)){
          FOSresult = combine(ActLUA,Acti);
          MoveSelect = Acti.getMove();
          remove(DSS.OPEN-SYSTEM-ACTS,Acti);
          apply(FOSresult,MoveSelect);
          return;
      }
  }
  if(MoveSelect == null){
      MoveSelect = findMove(ActLUA);
      if(MoveSelect != null){
          apply(Acti,MoveSelect);
          add(DSS.OPEN-USER-MOVES,MoveSelect);
          return;
      }
  }

  if(MoveSelect == null){
      IntegrationFailureNew = new IntegrationFailure(ActLUA);
      push(IntegrationFailureNew,DSS.INPUT-ERROR);
      return;
  }
}
```

Fig. 6.8 Outline of the Dialogue Act Integration Process.

process. First, determine what move (or open act) the given user act was attempting to further. Second, update that move (potentially via an open act) with the contents of the user act. In the following we examine the mechanisms of relevance measurement and move update.

6.3.2 Measuring Relevance

For a given task-specific input dialogue act, the language integration process must determine which open move that act is relevant to. Here we consider two distinct cases of relevance. The relevance of an act to an existing open act and the relevance

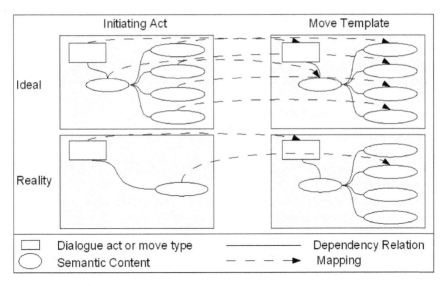

Fig. 6.9 Ideals and realities of dialogue act to dialogue move mapping.

of an act to a move template. We begin with the case of move template relevance.

6.3.2.1 Move Template Relevance

As introduced earlier in the chapter, the *move template library* is an inventory of dialogue moves that the dialogue system is capable of interpreting. If a user has performed an act to initiate a new move, then the move selection algorithm should return a move from this library. Although this process is similar in scope to dialogue plan and frame selection in information state update based and frame-based dialogue management models, these approaches often depend on a user making a highly declarative initial act such as "I would like to book a flight". However, in the situated systems domain where small grained but regular actions are typical, we are more likely to encounter an initial act such as the single word, "left", rather than the complete phrase, "I would like to make a turning". A mechanism of move selection is thus required which can match up highly underspecified descriptions against potential move types.

Since dialogue act and dialogue move contents are expressed in terms of the same conceptual ontology, the contents of dialogue acts can in general be transposed directly to dialogue move definitions. This gives us the basis of a way in which moves can be selected for a given user dialogue act. However, dialogue acts are often highly underspecified in comparison to dialogue moves. Not only may a speaker omit several semantic dependencies, but, in the case of a fragmentary contribution, the process or event type can be entirely omitted.

This discrepancy between the ideal and the reality of dialogue act to dialogue move mapping is depicted by Figure 6.9. The result of this discrepancy is that straight-forward unification cannot be relied upon to determine act/move-template compatibility. Instead, the approach taken here makes use of a relevance function that provides a quantitative measure of act/move-template compatibility. Broadly speaking, the function builds a relevance measure between an act and move template in accordance with the following three principles:

1. Increase the relevance measure for all act content that is present in the move template;
2. Decrease the relevance measure for any content in the move template that is not present in the given act;
3. Decrease the relevance measure for any content in the act that is not present in the move template definition.

In terms of the weighting given to increases and decreases in move applicability, it is often the case that dialogue move content is not present in a given dialogue act. Thus, the relevance increase for a chunk of available information should outweigh the penalty for information not present in the act. On the other hand, if a dialogue act provides more information than is accounted for in a move definition, then the relevance measure should be heavily penalised.

Based on these general principles, Figure 6.10 depicts a pseudo-code outline of the AODM model's move relevance evaluation function. The function makes use of two subsidiary functions defined below:

- `isSubclass(Term a, Term b)` which compares two types via the associated domain ontology and returns true if a is of the same type or a subtype of b, or false otherwise.
- `attach(FOS, Move)` which takes a fragment of dialogue act content and attempts to assign it to a unique field in the move template definition via a breadth-first search through the template's slot definitions.

6.3.2.2 Open Act Relevance

To determine the potential relevance of an input answer act to an open question act, the AODM model checks whether the answer resolves the question through a structural question-answer combination mechanism. The question-answer combination mechanism used is similar to that applied in many semantics-oriented dialogue management models including Larsson's [2002] IBiS system. Thus, we limit ourselves here to a short overview of the mechanism.

The potential relevance of an answer act to a given question act is checked by attempting combination of the question act and answer act. If the combination of the two acts into a predicate is possible, then the answer is judged to resolve the question. The combination process depends on the type of question examined. As described in Section 5.2, the AODM model handles three question types: polar questions, elicitation questions, and choice questions. The combination of answers with

```
int calcRelevance(act,mt){
  // int rating = 0;
  int rating = 0;
  // compare dialogue act and move types
  if((act.type == Statement) ||
     isSubclass(act.type,mt.type)){
    rating += 5;
  }
  else rating += -20;
  // attach all content fragments to move template
  foreach(content : act.contents){
    boolean attached = attach(content,mt);
    if(!attached) return -100;
  }
  // rate move completeness
  foreach(filler : mt.content.filler){
    if(filler != null)
      rating += 5;
    }
    else rating += -2;
  }
  return rating;
}
```

Fig. 6.10 Measuring move template relevance.

polar and elicitation questions depends on a simple binary operation described below. For choice questions, the conceptual semantics representation of a question is first flattened into a set of polar questions which can then be individually analysed as per the standard polar question answer combination technique. The low-level combination of answer acts with elicitation questions first attempts a straightforward unification-based combination approach if the content of the answer act is headed by a valid conceptual capability or state type. If not, i.e., if the answer is a short answer in the Information State Update terminology, the combination mechanism iterates over the semantic roles of a question and attempts to fill an uncertain slot through sortal type constraints.

Question-answer combination results in a single predicate. Before a move can be contextualized in the situational context, that predicate must be used to update the contents of that open move. Move update and move filling will now be described.

6.3.3 Content Application

Following the identification of a relevant open act or move template, input act content is applied to the selected move template or the open move which precipitated an identified open act.

Filling a move template with dialogue act content is a straightforward process since the dialogue act and dialogue moves share the same conceptual ontology. Specifically, we can make use of a unification-like process to copy all content from the dialogue act specification to the newly created dialogue move. If the input dialogue act is headed by a capability or event type, then all dependent conceptual semantic roles are applied to the move specification. If the act is not headed by a configuration or event type, i.e., if it is a short statement or short answer, then the content of that short act is copied over to the move role to which it is ontologically most applicable. Following initialization of the move with act contents, the move enters a *New* state with no solutions, and remain so until the contextualization process addresses the move.

In the case of applying an input act to an already open dialogue move, the process is somewhat more involved. Having identified the open system act resolved by the input user act, the question act combination function `combine` returns a conceptual semantic element, FOS_{Result}, which is the unification of that question and answer act. The integration process then applies this predicate to the open user or system move that gave rise to the system question being raised. This move is directly accessed from the open dialogue act's non-content specific `parent` role.

The application of FOS_{Result} to the identified move requires a reduction in the open move's solution set. The application process thus iterates over each of the new predicate's child roles. A pruning function associated with the solution set is then called for each role in FOS_{Result}. The prune function itself iterates over each solution in the solution set. If the value of a given solution role value does not match the corresponding role value from FOS_{Result}, then this solution is removed from the solution set. In this way, the application of an answer via question-answer combination is used to reduce the size of a solution set. As is the case for applying the content of an input dialogue act to a new move template, the reduced open move is then left for another round of contextualization before the dialogue planning process determines what actions the dialogue manager should take with respect to the open move.

6.3.4 Integrating Act Complexes

During any given dialogue turn, the result of natural language analysis and abstraction can be a dialogue act complex rather than an isolated dialogue act. As described in the last chapter, a dialogue act complex is a construction of dialogue acts that indicates a relation between individual dialogue acts through a reified structure. The act integration process described in Section 6.3.1 is capable of handling the inte-

```
void integrate(){
  if(DSS.LATEST-USER-ACT == null) return;
  Act_LUA = pop(DSS.LATEST-USER-ACT);
  integratePrivate(Act_LUA);
}

Move_Result integratePrivate(Act_in){
  if(Act_in is complex){
    Move_alpha = integratePrivate(Act_in.alpha)
    Move_beta = integratePrivate(Act_in.beta)
    if((Move_alpha != null) &&
       (Move_beta != null)){
        remove(DSS.OPEN-USER-MOVES,Move_alpha);
        remove(DSS.OPEN-USER-MOVES,Move_beta);
        Move_Result = complex(Move_alpha,Move_beta);
        add(DSS.OPEN-USER-MOVES,Move_Result);
        return Move_Result;
    }
    else if(Move_alpha != null){
        return Move_alpha;
    }
    else if(Move_beta != null){
        return Move_beta;
    }
    else return null;
  }
  else{
    if(Act_in.Type subclassOf CM-Act){
        push(Act_in,DSS.Non-Integrated-User-Acts);
        return null; // no move contextualization required
    }

    Move_Select = null;
    foreach(Act_i : DSS.OPEN-SYSTEM-ACTS){
        if(isRelevant(Act_LUA,Act_i)){
            FOS_result = combine(Act_LUA,Act_i);
            Move_Select = Act_i.getMove();
            remove(DSS.OPEN-SYSTEM-ACTS,Act_i);
            apply(FOS_result,Move_Select);
            return Move_Select; // returns the move that should DSS
        }
    }
    if(Move_Select == null){
        Move_Select = findMove(Act_LUA);
        if(Move_Select != null){
            apply(Act_i,Move_Select);
            add(DSS.OPEN-USER-MOVES,Move_Select);
            return Move_Select; // returns a move that it just DSSed.
    }}

    if(Move_Select == null){
        IntegrationFailure_New = new IntegrationFailure(Act_LUA);
        push(IntegrationFailure_New,DSS.INPUT-ERROR);
        return null; // could not integrate.
}}}
```

Fig. 6.11 Outline of the Dialogue Act Complex Integration Process.

gration of individual dialogue acts, but not the integration of these act complexes. With an explanation of the basic integration processes in hand, we now update that integration algorithm.

A dialogue act complex is a binary tree structure where parent nodes are act complex definitions, and child nodes can be either act complexes or individual dialogue acts. Therefore, we make use of a recursive algorithm to traverse the act complex during integration. Figure 6.11 presents the revised dialogue act integration process which consists of two functions, the second of which is the recursive part. The first function is an initial condition check which returns immediately if there are no acts to integrate, but calls on the recursive function to integrate otherwise. This recursive function first determines if the act to be integrated is simple or complex. If simple, the act is integrated in a manner similar to the original simplex integration process. If complex, the child nodes of the complex act are first called to be integrated. If the result of both integration attempts are selected dialogue moves, then a dialogue move complex is created around those moves and added to the OPEN-USER-MOVES slot of the DSS in place of the individual selected moves. Otherwise, if integration of either child node failed to return a selected move (either due to an act being conversational, or not-relevant to any open or known move types), then no complex move is created. Thus, the integration of a given act complex may result in the revision of either dialogue move complexes, or individual dialogue moves, in the dialogue state structure.

6.3.5 Discussion

At the end of the dialogue integration process, any new dialogue acts – be they atomic or complex acts – will have been integrated into the information state by either updating or creating a new move instance in the ideal case, or leaving the act for direct processing by dialogue planning in the case of an unexpected act or a non-task act. Here we have focused on the most salient aspects of this integration process with respect to the relationship between dialogue acts and dialogue moves, but have said little about the mechanisms by which communication and grounding errors are handled by the AODM model. In practice such issues are handled in a manner similar to Larsson's [2002] IBiS, and thus we omit a discussion of these aspects of the integration process here.

For our current purposes however, we have argued that one of the great problems to be overcome in applying classical information-state based systems to the situated domain is the relationship between the information state itself and domain models, and in particular how units of communicative exchange are systematically contextualized against the situational state. In the AODM model this issue is handled by dialogue move contextualization – a process called after the integration of dialogue acts. In the next section we turn to this contextualization process.

6.4 Generalized Language Contextualization

As argued earlier, the dynamic nature of situated knowledge renders a unification or theorem proving based model of move contextualization inappropriate. Simply put, while unification might suffice for resolving a simple referring expression like "John's office", more complex resolution processes are required to handle perceptually or spatially resolved entities. Thus, here we employ a semantics contextualization mechanism based on situation-sensitive contextualization functions that are applied to individual parameters within partial move specifications. Furthermore, given the phenomena of both anaphoric and elliptical underspecification, the contextualization model makes use of two distinct function sets: one function set for the *resolution* of specified content, and a second function set for the *augmentation* of move specifications in the case of elided information.

In this section we explore the details of the contextualization process independent of a specific domain model – it should be noted however that the contextualization process has been developed for task-oriented dialogue where the discussion of actions to be performed, and constructions of those actions, is of particular importance in interaction. The section begins with a discussion of the relationship between contextualization functions and the construction of so-called *solution sets* for particular dialogue moves. Then, in Section 6.4.2 we examine the contextualization process for individual moves. This is followed in Section 6.4.3 with an expansion of the contextualization process for the case of move complexes.

6.4.1 Solutions & State Descriptions

As detailed earlier in the chapter, any given move to be contextualized, μ, contains a signature derived template description, and the set of user supplied values. The template description, $\mu.attr$, consists of a set of role and filler restrictions derived from ontologically organized action signature specifications. User supplied values, $\mu.val$, are the unresolved semantic fillers for these slots – each instance of which is denoted ψ.

We informally define a solution, σ, as a set of values that parameterize a move specification. A solution is *partial* if one or more values for a minimal parameterization is not present. Conversely, a solution is *complete* if the solution set provides a minimal parameterization for the given move. For our purposes here, a solution parameter, ψ', is an entity defined in terms of the agent's domain ontology. In general, any given contextualization function, either an augmentation function, $\delta \in \Delta$, or a resolution function, $\gamma \in \Gamma$, may return multiple possible values for a given parameter. For example, in resolving a discourse reference such as "the bathroom", multiple possible *bathroom* referents may be available to a resolution function. Since these values will typically not all be equally likely, a rating, $r(\psi')$, is assumed assigned to each interpretation returned by a contextualization function. Specifically, we assume assigned ratings for each parameter to be in the interval [0,1].

The interpretations provided by any given contextualization function are dependent on both the shallow semantic categories provided through surface language (in the case of resolution) as well as the state of the interpreting agent. Such a state, s, is assumed to be application dependent. However, since we are developing this model for situated domains, we can assume here that such a state includes a description of the agent's physical as well as mental condition.

In addition to returning a resolved parameter entry, ψ', contextualization functions also return an updated state corresponding to the state of the agent if the agent was to act on that resolved parameter value. The importance of tracking resultant states during move contextualization can be seen in that in interpreting a move complex with a sequential relationship between two constituent moves, the second move may only be interpretable in the context of the post-condition state of the first move. Moreover, even within the contextualization of a single move, the state applicable to a contextualization function can become disjoint from the real-world state of the interpreting agent. For example, in contextualizing a Motion with two sequential path constraints, the second constraint can only be contextualized after the first constraint has been contextualized to derive an input state for the second constraint.

Rather than assuming that the application of a move, or the contextualization of move constituents, results in a single definite state, we assume contextualization functions result in a distribution of rated possible states even for a single symbolic contextualized interpretation. This is particularly necessary for the case of contextualizing spatial actions where a definite final state may not be known even for an unambiguous action description, e.g., "move forward". While each contextualization function returns such a state distribution, we also assume a given solution to have one specific state distribution associated with it – this distribution being the result of the last contextualization function applied for the given move. As with the ratings for individual parameters, we assume the ratings of resultant states to be in the interval [0,1].

While contextualization functions provide both the ratings for individual solution parameters – as well as the ratings for resultant states associated with a given solution parameter – it is necessary to calculate aggregate likelihoods for complete solutions. Given that the input state supplied to contextualization functions is uncertain, we will evaluate the likelihood of a given solution not only in terms of the ratings assigned to the parameters of a solution, i.e., $r(\psi')$, but also in terms of the ratings associated with the input state of each of these solution parameters. We return to the actual calculation of this likelihood in the next section; here we simply note that in addition to having a final state distribution associated with a given solution, each interpreted parameter for a given solution also includes the input state with associated rating against which the interpretation was produced.

To summarize, a given solution, σ, consists of a list of parameter interpretations and a final state distribution for the complete solution, i.e.:

$$\sigma = \{\Psi, S\} \tag{6.11}$$

where Ψ is a list of parameter interpretations, and S is the final state distribution associated with the solution, i.e., the value of $Set(\{s', r(s')\})$ for the application of the final contextualization function.

The list of parameter interpretations is in turn defined as:

$$\Psi = List(\{\psi', r(\psi'), s, r(s)\}) \tag{6.12}$$

where ψ' is a parameter value, $r(\psi')$ is a rating assigned to that parameter by a contextualization function, s is a state against which ψ' was obtained, and $r(s)$ is the rating for that state.

Since any given resolution or augmentation function can introduce more than one parameter interpretation, multiple solutions may be maintained at any given time during interpretation. We thus introduce the solution set, Σ, as the collection of solutions maintained at any given time for a given move. At each application of a contextualization function, Σ may be reduced or expanded. While the manipulations of Σ may be viewed as the development of a tree structure, it is not necessary to maintain previous steps in the manipulation of Σ since only non-applicable solutions are removed by contextualization functions. We thus define Σ straightforwardly as follows:

$$\Sigma = Set(\sigma) \tag{6.13}$$

where the size of the solution set can be retrieved directly when necessary from $\Sigma.size$.

With the notion of solutions and the solution set in hand, we can now also define the general form of a parameter resolution function as:

$$Set(\{\psi', r(\psi'), Set(\{s', r(s')\})\}) = \gamma_\psi(\psi, s) \tag{6.14}$$

where the augmentation function takes some unresolved parameter description ψ and input state s, and returns a set of possible interpretations of that value, ψ', along with associated ratings for that value, $r(\psi')$, as well as a resultant state distribution associated with that value.

Similarly, if a user omitted content for a given parameter, an augmentation function must be applied to derive content from the situational context. The general form of the augmentation function is similar to that for the resolution function:

$$Set(\{\psi', r(\psi'), Set(\{s', r(s')\})\}) = \delta_\psi(s) \tag{6.15}$$

with the exception of an omitted semantic content parameter.

Although contextualization functions do not manipulate solutions or the solution set directly, it should be noted that the application of contextualization functions, i.e., augmentation and resolution functions, may cause the solution set to grow in either or both of two ways. First, a given contextualization function can result in multiple possible interpretations for a given parameter – thus resulting in the expansion of the solution set itself. Second, even for a parameter that is contextualized with a single solution, the state distribution associated with that single solution can be anywhere from single valued in the ideal case to having a relatively large distribution of states in, for example, the case of capabilities that result in continuous behaviours. Each subsequent parameter interpretation is made with respect to each of these possible states. This management of solutions and their link to contextualization functions is the responsibility of the Generalized Contextualization Process to which we turn in the next section.

6.4.2 The Generalized Contextualization Process

The generalized contextualization process is the operation by which individual moves are contextualized against the situational state. In this and the next section we describe this generalized process – beginning here with the contextualization of individual moves.

Broadly speaking, the contextualization of individual moves can be viewed as a two step process:

- **Contextualization Function Application:** For each of the move parameters, an associated augmentation or resolution function is selected and applied within the context of a specific state. One or more possible interpretations of that parameter along with associated state distributions are then used to update the solution set before the next parameter is similarly processed.
- **Solution Set Analysis:** Following application of contextualization functions to all move parameters, the resultant solution sets are pruned and analysed to determine the overall likelihood of solutions.

Figure 6.12 sketches the outline of the contextualization function application process for individual dialogue moves. The input to the move contextualization process consists of: (a) the move to be contextualized, μ; (b) the set of augmentation functions, Δ; (c) the set of resolution functions, Γ; and (d) the initial solution set, Σ. Depending on whether a move is contextualized in the context of a move complex or not, the input initial solution set may be initialised to either the output solutions of the last processed move, or to a new empty solution which consists only of an associated state that has been initialised to the agent's current real-world state.

Referring to Figure 6.12, during contextualization the solution set Σ is updated for each contextualized parameter in μ. This solution set is first initialized to the input solution set, and then used as a basis for creating a revised solution set during the processing of each parameter. The contextualization process subsequently iterates through each of the move parameters and selects an augmentation or resolution function depending on whether the user supplied content for a given move parameter or not. This selection mechanism operates through the association of exactly one resolution and augmentation function with each parameter in a given move specification provided as input to the generalized contextualization process. Once selected, the augmentation or resolution function is then applied in the context of those states establishes by the previously applied contextualization function. The set of possible contextualizations for a given parameter are then used not only to update the set

```
Σ contextualize(μ,Δ,Γ,Σin){
    Σ = Σin;
    foreach(ψ : μ.val){
        Σnew = {};
        if(ψ is partially specified){
            foreach(σ : Σ){
                foreach(s : S){
                    Set({ψ',r(ψ'),Set({s',r(s')})}) = γψ(ψ,s);
                    foreach(element in returned set){
                        σnew.Ψ = σ.Ψ ⊕ {ψ',r(ψ'),s,r(s)};
                        σnew.S = Set({s',r(s')});
                        add(σnew,Σnew);
        }}}}
        else if(ψ' is unspecified) {
            foreach(σ : Σ){
                foreach(s : S){
                    Set({ψ',r(ψ'),Set({s',r(s')})}) = δψ(s);
                    foreach(element in returned set){
                        σnew.Ψ = σ.Ψ ⊕ {ψ',r(ψ'),s,r(s)};
                        σnew.S = Set({s',r(s')});
                        add(σnew,Σnew);
        }}}}
        Σ = Σnew;
    }
    return Σ;
}
```

Fig. 6.12 Contextualization Function Application Algorithm where: μ is a move, ψ is a move parameter, Σ is a Solution Set, σ is a solution, Δ is the augmentation function set, δ is an augmentation function, Γ is the resolution function set, and γ is a resolution function.

of solutions for the given dialogue move, but also to determine the associated final state distribution for that move.

To illustrate dialogue move solution development, Figure 6.13 abstractly depicts the relationship between the contextualization process itself and individual solutions for a simple move type with only two parameters – only one of which was explicitly filled with a value by the user. Referring to the left hand side of the diagram, the contextualization process proceeds by first applying the relevant augmentation function for the first slot of the move type to an initial state s_0. The application of this function results in two possible values to fill this slot type – one of which has an associated state distribution with three values and the other a state distribution with two values. Assuming the user provided some content for the second slot, the appropriate resolution function is then applied to that content and the resultant states of the first parameter to produce 5 possible interpretations of the second slot. As can be see from the application of the resolution function to input state s_{A3}, not all contextualization function applications necessarily produce interpretations for a given state-value pair.

A trace from the initial state of a contextualized move down through to a leaf

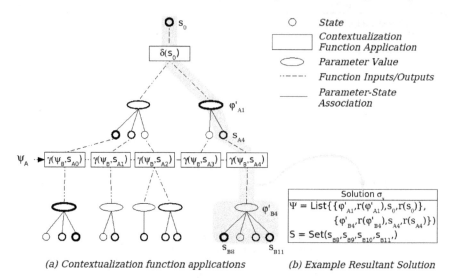

(a) Contextualization function applications   (b) Example Resultant Solution

Fig. 6.13 Illustration of the solution construction process. The left hand side of the diagram depicts the application of contextualization functions where the thickness of state and parameter value nodes is proportional to the ratings assigned to these values by contextualization functions. The right hand side of the diagram depicts one of the five possible resultant solutions generated by the contextualization process. A solution corresponds to a root to leaf trace of contextualization function application.

parameter at the maximum depth of the constructed tree corresponds to a single solution for a given dialogue move. This relationship is depicted on the right hand side of Figure 6.13 where a single solution for the dialogue move is related to the values constructed by contextualization function application.

Following the application of contextualization functions, equivalent solutions within the solution set are pruned before aggregate likelihoods are assigned to the remaining solutions in the solution set. To rate complete solutions we make use of the ratings assigned by contextualization functions to individual dialogue move parameters. As indicated in the last section, we not only make use of the actual rating for a given parameter possibility, but also the rating of the input state which was provided to the contextualization function. Ratings for both individual parameters and associated states are depicted abstractly in Figure 6.13 by way of line thickness for parameter value and state nodes respectively. Given that both parameter and state likelihoods are required to be real numbers in the interval [0,1], we straightforwardly make use of the product of these ratings as the aggregate rating for a given solution. Specifically, the total parameter rating for a given solution, $\Pi(\sigma)$, is calculated as the product of parameter and input solution state ratings for each slot, i.e.:

$$\Pi(\sigma_i) = \prod_{w=0}^{\sigma_i.\Psi.size} \sigma_i.\Psi_w.r(\psi') \cdot \sigma_i.\Psi_w.r(s) \tag{6.16}$$

where $\sigma_i.\Psi_w.r(\psi')$ denotes the rating for interpretation of slot Ψ_w in solution σ_i, and $\sigma_i.\Psi_w.r(s)$ denotes the rating for the input state which factored into that interpretation.

In performing contextualization, multiple input states may give rise to the same – or very similar – solutions. To illustrate, consider the case of interpreting an instruction such as "go along the pathway to the market square". Here, though one pathway may only be salient, the contextualization of a movement constraint such as "along the pathway" may result in multiple possible states for the interpreting agent. For each of these possible states, the generalized contextualization process generates interpretations for the second constraint "to the market square". While such an exhaustive strategy is useful in finding interpretations where the second constraint is less unique, in this case the interpretation of the constraint "to the market square" is likely to produce the same interpretation for many different input states. Thus, following solution construction and rating, we apply a pruning step to eliminate equivalent solutions.

Two solutions are considered equivalent if the values of their final state distributions are equivalent as determined by a domain-specific equivalence function. For each equivalent solution group, all but the solution with the highest total parameter rating are removed from Σ. It should be noted that since equivalence is based on resultant state equivalence rather than the equivalence of actual solution parameters, the pruning process prevents the unnecessary raising of clarification questions when the details of the interpretation of a single parameter is irrelevant to the final outcome of interpreting a dialogue move. For example, within the , which is developed in the next chapter, consider the case of a speaker having instructed an instructee to turn to the "left" in the case that both an *egocentric* and *allocentric* interpretation of left are possible and result in the same final state. In this case, despite the fact that the actual interpretation may be relatively uncertain, the contextualization process will keep only the marginally more likely interpretation (or first if interpretations are rated exactly equivalent and likely), and thus avoid raising an explicit clarification question which may be judged unnatural.

Following solution set reduction, a probability metric for a given solution, $P(\sigma_i)$, is then calculated based on $\Pi(\sigma_i)$ for each remaining potential solution, i.e.,

$$P(\sigma_i) = \frac{\Pi(\sigma_i)}{\displaystyle\sum_{j=0}^{\Sigma.size} \Pi(\sigma_i)} \tag{6.17}$$

It should be noted that for a solution with n parameter slots, i.e., where $\sigma_i.\Psi.size = n$, the ratings of resultant states for the nth parameter are not factored into either $\Pi(\sigma_i)$ or $P(\sigma_i)$. Thus, referring to Figure 6.13, no ratings for states $s_{B0}..s_{B11}$ are factored into solution probabilities for any of the five solutions produced in the contextualization of the assumed dialogue move. As already discussed, values for final solution distributions are however key to determining effective solution equivalence.

The calculated solution likelihood measure contributes to the determination of the task progression strategy in subsequent dialogue planning processes. For example, while multiple final solutions typically require explicit clarification, a solution with a likelihood measure notably greater than the next most likely solution's allows us to prune less likely solutions without the need for direct clarification. Such issues of dialogue planning will be addressed in Section 6.5. Before we do that, we first turn to the contextualization of move complexes.

6.4.3 Move Complex Contextualization

The process described in the last section related to the contextualization of individual moves. However, when processing move complexes, a solution's resultant states effect the contextualization process for subsequent moves. We must therefore extend the contextualization process to establish the necessary link between the contextualization of individual moves.

The move complex contextualization process can be viewed as a three-stage operation consisting of: (a) *Forward Propagation*, (b) *Branch Evaluation*, and (c) *Back Propagation*. During forward propagation, the move complex's constituent moves are contextualized in sequence as per the individual move contextualization process. However, the initial states applied to a given dialogue move may be based on the resultant states of a prior move. Specifically, the logical relation between moves – or rather the implied relation between resulting intentions – effects how the contextualization process operates over move complexes. In the current model we deal with two such relation types: the *sequential* relationship, and the *disjunction* relationship. If a sequence of intentions is implied, then each solution of move i is used to instantiate a solution for move i+1 – thus ensuring that the interpretation of a given move is made in the context established by a prior move. Whereas if the relationship between intentions is a disjunction, then no propagation of solutions is applied since a given intention will not be achieved within the post-condition state of another move. The handling of other relationship types can be more complex. For example, if the implied inter-intention relationship is a form of conjunction, then it is not at all clear how states should be propagated from one move to the next without establishing possible causality or interference relationships between capabilities – we leave the exploration of such issues for future work.

The result of forward-propagation is a tree structure whose trunk node is an initialization node and all other nodes are solutions. For each solution available for the last contextualized move, i.e., at the tree's maximum depth, we can establish a *solution-trace*, $\sigma_0..\sigma_n$, as the list of solutions from the trunk node down to a given final solution. For each solution trace the cumulative solution trace likelihood, $\bar{P}(\sigma_0..\sigma_n)$, is then calculated as the mean of solution probabilities in that trace, i.e.,:

$$\bar{P}(\sigma_0..\sigma_n) = \frac{\sum_{i=0}^{n} P(\sigma_i)}{n} \qquad (6.18)$$

To illustrate the relationship between moves, move complexes, and the solution

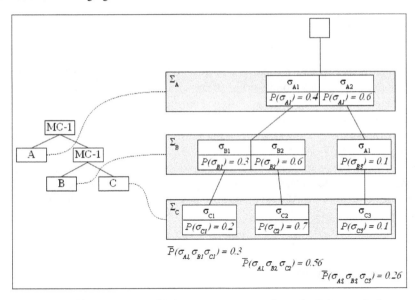

Fig. 6.14 Illustration of the relationship between a move complex and solution sets in the contextualization of a move complex. The left hand side of the figure abstractly depicts a move complex to be contextualized, while the right hand side depicts the progressive states of the solution set as each constituent dialogue move is contextualized.

tree constructed during contextualization, Figure 6.14 abstractly depicts a simplified example of move complex contextualization. In this example, let us assume that a move complex consisting of three atomic moves bound together by sequential move complexes are to be contextualized. This move structure is depicted on the left of Figure 6.14, with a view of the composed solution tree on the right hand side. Each level of the solution tree depicts the solution set for a given atomic move, i.e., Σ_A is the solution set for move A. Cumulative probabilities for solution-trace hypotheses are shown below solution set Σ_C.

Based on the calculated cumulative likelihood, solution sets are subsequently pruned of unlikely move solutions. This pruning operates through back-propagation up through low-likelihood solution traces. Since branching in the solution tree will have been due to ambiguous interpretations of particular moves or move constituents, this pruning back-propagation step has the effect of removing ambiguous move interpretations where a subsequent move had the effect of clarifying which of the ambiguous interpretations was most likely. If following the pruning of unlikely solutions there remains more than one possible interpretation solution at the solution tree's maximum depth, then no further solution tree pruning is possible without explicit clarification as per the case with individual move clarification.

6.4.3.1 Discussion

In this section we outlined the generalized contextualization process for dialogue moves and move complexes. The purpose of this process is to account for the application of situational context to individual dialogue moves and move complexes in order to augment omitted content and resolve content when provided. The result of contextualization is the updating of a set of solutions associated with a given move, while the process itself is *generalized* in the sense that it is domain independent within the range of task-oriented dialogue. In Section 7.4 in the next chapter we will expand on the contextualization process by introducing specific contextualization functions for a concrete situated domain. Before we do so however, we must first complete description of the AODM model. We proceed by introducing the dialogue planning processes.

6.5 Dialogue Planning

Dialogue planning encapsulates those processes responsible for composing what moves and acts, if any, should be pursued by the agent at a given point in the dialogue. Figure 6.15 depicts the components of the planning process applied in the AODM model. Here, rounded rectangles denote selections from the dialogue state structure, while regular rectangles denote processes. At this high level of abstraction, dialogue planning follows a three step mechanism consisting of *Move Planning*, *Act Planning*, and *Act Selection*, where a positive result from act selection is passed on to the language interface to finalize production. In the rest of this section, the three main stages of dialogue planning in the AODM model are detailed.

6.5.1 Dialogue Move Planning

In the AODM model, dialogue move planning determines which task-level communicative goals should be adopted by the agent at a given point in the dialogue. Thus, unlike dialogue act planning, which is a relatively domain independent process, dialogue move planning is more dependent on desired application logic. However, useful dialogue management components should, where possible, supply reasonable default conversational behaviours that can be extended or customized for particular domains. Thus, as depicted in Figure 6.15, the AODM move planning model is split into two functionality groups, where one of the two groups provide domain independent dialogue move planning behaviours. In the following, the dialogue planning process is examined in terms of these two functionality groups.

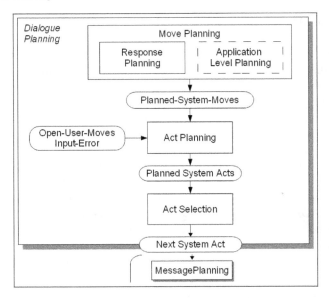

Fig. 6.15 Overview of the Dialogue Planning Process in the AODM Model.

6.5.1.1 Application Specific Planning

The first move planning functionality group is true *Application-Specific Planning*. As argued in earlier chapters, the nature of situated applications means that changes in the agent itself, or the agent's environment, often warrant the adoption of system-initiated dialogue goals. The *Application-Specific Planning* component accounts for the generation of these system initiated goals.

Application-Specific Planning can be realized through any number of decision making processes. Thus, the AODM model, and in particular the implemented Daisie framework described in Chapter 8, does not directly provide a solution for Application-Specific Planning. Instead, an interface is defined by which application logic can add dialogue goals to the Dialogue State Structure. Specifically, dialogue goals, modelled as dialogue move objects, may be placed into the Planned-System-Moves slot of the Dialogue State Structure through an application interface. Dialogue management processes, specifically *Dialogue Act Planning* and *Language Integration*, then pursue these goals in the normal way. Once achieved, application-specific dialogue goals are moved to the Closed-System-Moves slot of the Dialogue State Structure – which can also be monitored by application logic. Chapter 8 provides more information on the implemented Daisie framework, and the domain interface it provides for the adoption and monitoring of application-specific dialogue goals.

6.5.1.2 Response Planning

Having achieved an unambiguous interpretation of a user-initiated dialogue move, a cooperative dialogue partner should respond to that move. Although appropriate responses vary according to initializing move content and agent state, an appropriate response typically has two aspects. First, the receiving agent may alter their belief or intentional state in partial fulfilment of the performing agent's intended illocutionary effect. Second, the receiving agent may compose a backward-looking move to convey its response to the initial move. Although application-specific logic can be applied to both mental state update and response move composition, dialogue management should, wherever possible, develop generalizable solutions to common conversational intelligence tasks such as deciding upon appropriate responses. For this reason, the AODM Dialogue Planning process includes the *Response Planning* module that handles basic communicative protocols for addressing contextualized forward looking user moves.

The approach taken here to response planning is similar in concept to that of illocutionary logics described in Section 2.3.3, but has also been rationalized for cognitive modularity and the elimination of redundant dialogue contributions. Moreover, unlike in illocutionary logic based dialogue management models and systems, the response production protocol applied here constitutes only a small portion of the dialogue management model. Before describing this response planning protocol, we first examine the constraints imposed on the solution by modularization and redundancy concerns.

Modularization & Response Planning

Given the goal of modularization in dialogue management model design, a question arises as to how move response production should be distributed between dialogue and domain components. Two modelling options present themselves. The first option treats mental state update and response generation as distinct processes where the former process is an application-specific concern. The second option handles both response types within a single process. These two opposing options are best illustrated in terms of behaviour rules. Considering first the single process option, Equation 6.19 typifies an idealised response generation rule for a user request move:

$$Request(\alpha, \omega) \wedge CanDo(\beta, \omega) \rightarrow Intend(\beta, \omega) \wedge Accept(\beta, Request(\alpha, \omega))$$
$$(6.19)$$

where α is the agent making the request, ω is the action requested, β is the recipient of the request, and *CanDo* is a function that determines whether a given agent is capable of performing an action in the current state.

The second design option, more in keeping with the goal of modularity, is exemplified through the use of two distinct rules:

$$Request(\alpha, \omega) \wedge CanDo(\beta, \omega) \rightarrow Intend(\beta, \omega) \qquad (6.20)$$

$$Request(\alpha,\omega) \wedge Intend(\beta,\omega) \rightarrow Accept(\beta, Request(\alpha,\omega)) \qquad (6.21)$$

where the first rule is an element of a rational behavioural component, and the second rule is an element of the dialogue move planner.

The principle of modularity dictates that the latter distributed model is to be preferred to the former. Beneficial aspects of the distributed design include domain components retaining control over intention adoption, and the dialogue planning component remaining isolated from intention adoption strategies. However, a significant problem with the distributed design is that non-preferred reactions to user moves do not necessarily result in changes to intentional state that can be picked up on by a dialogue planner. For example, in the case that the domain logic chooses not to commit to the action requested by an interlocutor, no changes to mental state will be made by the application component that can result in the dialogue planner deciding to adopt a rejection move. In such a case it is necessary for the domain logic to adopt a dialogue goal which signals request rejection. This of course immediately mixes the two aspects of response planning that were to be separated.

Because of the problem just mentioned, and since synchronization of physical and verbal actions favours coherent communication, the AODM model adopts a single move response strategy. This response strategy is defined as part of the dialogue planning process rather than application-specific deliberation. Although the dialogue planning process thus takes control of mental state update in response to user dialogue moves, the domain logic of concrete applications retains final control over the decision on how the agent should react to individual dialogue moves. This control is exercised through precondition checks made by the dialogue planning component on the application state, i.e., by a mechanism similar to the `canDo` function used in Equation 6.19. Moreover, as mentioned earlier, concrete applications can augment dialogue induced deliberative processes with deliberative processes triggered by environmental or other stimuli.

Move Redundancy

The discussion above suggests that the dialogue agent should adopt dialogue goals in response to all user dialogue moves. Although convenient in idealised protocols, this strategy is inappropriate for many cases of natural dialogue even if the initiating move normally invites a response. For example, in the case of a user's request to perform a task, a verbal acknowledgement of the acceptance of that request is superfluous when the performance of that action itself is immediate and perceptible by the user – and so interpretable as an implicit acceptance of the initial request.

To reduce superfluous responses, the AODM dialogue planning process determines, on the basis of the behavioural response to a user move, whether it is appropriate to verbally respond to that move. However, a significant variable in developing such a strategy is whether the choice to explicitly acknowledge the user move is made at the level of dialogue move or dialogue act planning. Making this choice at the move planning level eliminates the needless generation of moves that are not realized, and thus simplifies the discourse state and reduces processing requirements. However, by always introducing the dialogue goal at the dialogue move planning level, and then making the realization choice at the act planning level, the dialogue

```
void plan_response(){
    foreach(Move_i : Set(Move)_OUM
                 ∧ (Move_i.State == Complete)
                 ∧ (Move_i.Σ_best.P > P_MIN)){
        remove(Set(Move)_OUM,Move_i);
        push(Set(Move)_CUM,Move_i);
        if(Move_i.type == Instruct){
            if(canDo(Move_i.content){
                add(Intentions, new Intention(Move_i.content));
                if((Move_i.content.earliestStartTime is later) ||
                   (Move_i.Σ_best.P < P_CLEAR)){
                    add(Set(Move)_PSM,AcceptInstruct(Move_i));
                }
            }
            else{
                add(Set(Move)_PSM,RejectInstruct(Move_i));
            }
        }
        else if(Move_i.type == Inform){
            if(integratable(Move_i.content){
                add(Beliefs, new Belief(Move_i.content));
                add(Set(Move)_PSM,Accept(Move_i));
            }
            else{
                add(Set(Move)_PSM,Reject(Move_i));
            }
        }
        else if(....

    }
}
```

Fig. 6.16 The Response Planning Strategy.

planner allows further constraints such as the modality of realization to factor into the final choice on realization.

Despite this concern, in the AODM model the choice on whether explicit dialogue moves should be adopted in response to forward-looking user moves is made at the dialogue move planning level. The chief justification for this is the cleaner discourse model achieved. The resulting response planning strategy is described in the next section.

The Response Planning Strategy

Figure 6.16 depicts pseudo-code for a portion of the AODM model's response planning strategy. It should be noted that in order to simplify discussion of the most salient points, we have generalized both atomic dialogue moves and complex dialogue moves simply to moves in this description. Also, as with other AODM process descriptions, the response planning strategy is described in a procedural manner.

However, it should also be clear that such a procedural account is operationally equivalent to the rule-based specifications applied for example in both formal descriptions of speech act semantics (see Section 2.3.3.1), and in the update-rule based strategies of the classical information state dialogue management (see Section 3.4.2 for an example).

Considering first the response planning strategy, each open user dialogue move that is both `Complete` and which has a solution likelihood that is greater than a minimal acceptable threshold (`P_MIN`) is first moved from the set of open moves to closed moves. The closed move is then used to perform an update of the agent's mental state, and generate a new dialogue move if appropriate. Both mental state updates and the generation of response moves are determined by the forward-looking move type, as well as a number of domain interface functions. To illustrate in the case of a user `Request` move, a domain function, `canDo`, is used to determine whether the requested capability can be achieved by the agent given its current state. This check is performed by evaluating the preconditions of the given capability. In the case of a negative result from pre-condition checks, a `RejectInstruct` dialogue move is instantiated and added to the `Planned-System-Moves` slot of the Dialogue State Structure. If capability pre-conditions are met, then an intention towards performing that capability is adopted. In the case of intention adoption, both the start time of the resultant intention and the likelihood of the given solution are considered to determine whether an explicit acceptance move should be made. If solution likelihood is below a clarity threshold, (`P_CLEAR`), or if the intended start time for the intention is not 'Now', i.e., a default start time where no start time was supplied by a user, then an `AcceptInstruct` move is instantiated from the initiating `Instruct` and added to the `Planned-System-Moves` slot. In practice values of 0.75 and 0.5 were applied for the thresholds `P_CLEAR` and `P_MIN` respectively, but further investigation of these thresholds is left for future work.

Similar logic is applied for handling each of the other initiating move types. To illustrate typical variation between the handling of different move types, Figure 6.16 also depicts pseudo-code for the handling of `Inform` types. Notable differences between the handling of `Instruct` and `Inform` moves concern the use of the `integratable` and `addBelief` domain interface functions rather than `canDo` and direct intention adoption. With respect to these functions, it should be noted that belief adoption mechanisms are typically considerably more complex than the addition of predicates to a belief state container as suggested by the simplicity of the `addBelief` function's signature. Another notable difference between the handling of `Instruct` and `Inform` types is that for `Inform` types, an explicit acknowledgement of information acceptance is always made in the current model. In practice this may not be necessary in multi-modal contexts where adopted information is denoted on a visual display, but for current purposes of situated interaction, where a display may not be present, explicit acceptance acknowledgement is more prudent.

6.5.2 Dialogue Act Planning

While the *Dialogue Move Planning* processes are responsible for generating system dialogue goals, the *Act Planning* component is responsible for analysing these goals to determine what, if any, dialogue acts should be performed by the agent. In the AODM model, *Act Planning* is also responsible for *communication management* aspects of dialogue management, including planning dialogue acts to address interpretation errors, or to fill missing information in open user dialogue moves. The result of *Act Planning* is a set of dialogue acts in the `Planned-System-Acts` slot of the Dialogue State Structure.

The AODM dialogue act planning process can be broken down into four distinct functionality groups based on the element of the Dialogue State Structure (DSS) that is processed by that group. These four groups are:

- **User Moves Processing:** Plan clarification acts to make progress towards single unique solutions for user dialogue moves that remain on the `Open-User-Moves` slot of the DSS.
- **System Goal Processing:** Plan dialogue acts to resolve adopted system goals that remain on the `Open-System-Goals` slot of the DSS.
- **Communication Error Processing:** Provide explicit statements of interpretation failure in response to errors remaining on the `Input-Error` slot of the DSS.
- **Conversational Act Processing:** Handle basic user-initiated conversational acts remaining on the `Non-Integrated-User-Acts` slot of the DSS.

In the following sections, these four components of the dialogue act planning process are described in more detail.

6.5.2.1 Addressing User Moves

If a user dialogue move has not yet reached a completed state through contextualization (user content resolution and augmentation), then additional information must be solicited from the user to progress the open move towards a closed state. This process of addressing open user moves has two stages: first, select the open move for which a clarification act should be composed; and second, compose that said clarification act as a selection of relevant information from the selected open move.

Figure 6.17 depicts a pseudo-code sketch for open user move handling. For the latest user move that is still open, the process first checks if the move is a simplex or complex move. While simplex moves can be processed directly, complex moves that are not yet complete require that a pivot, or focal point, move be selected from the move complex. This focal point move is the first move in a move complex which has not yet entered a complete state. This focal point move is selected through a straightforward search by the `selectFocalMove` function. Once selected, the process of question act composition is applied directly to this focal move rather

```
void handle_OpenUserMoves(){
    if(DSS.OPEN-USER-MOVES == null) return;
    Move_Top = pop(DSS.OPEN-USER-MOVES);

    if((Move_Top.State == OpenParameter || MultipleSolutions) &&
        (Move_Top is complex)){
      Move_Focus = selectFocalMove(Move_Top);
      if(Move_Focus.State == OpenParameter || MultipleSolutions
                                           || OpenQuestion){
          Move_Top = Move_Focus;
      }
      else return;
    }

    if((Move_Top.Σ_best.P <= P_MIN) ||
        (Move_Top.State == OpenParameter || MultipleSolutions)){
      Act_New = composeQuestionAct(Move_Top);
      Move_Top.State = OpenQuestion;
      Move_Top.child = Act_New;
      Act_New.parent = Move_Top;
      push(DSS.Planned-System-Acts,Act_New);
    }

    else if(Move_Top.State == OpenQuestion){
      if(Move_Top.child.timeStamp older than 30 seconds){
          push(DSS.Planned-System-Acts,Move_Top.child);
    }}}
```

Fig. 6.17 Addressing Open User Moves.

than its complex parent. Subsequently performed operations on the Dialogue State Structure are dependent on the state of the selected open move. If that move is in a OpenParameter or MultipleSolutions state, or if the probability of the best solution is below the minimum acceptance threshold, P_MIN, then a question act is composed for that move, mutually connected to that move, and added to the Planned-System-Acts slot of the dialogue state structure. If, on the other hand, the selected move already had a question act composed for it, but the question has not yet been addressed after a period of 30 seconds, then the agent will attempt to raise the question again.

As indicated, for a given open user move in the OpenParameter or Multiple-Solutions states, or with a low solution likelihood, an appropriate question act must be instantiated. Available AODM question act types have already been introduced in Section 5.2, i.e., elicitation, choice, and polar. For moves with low solution probability, a polar question is composed. For moves in the OpenParameters state, an elicitation question is composed which addresses the first open and mandatory element of the open moves template description. Specifically, the composed question object's content consists of the process type, the actor (if available), and the open slot to be addressed. For open moves in the `MultipleSolutions` state, then the composed question type is dependent on the number of solutions available. For open moves with two or three solutions, a choice question is composed, while for moves with more than three possible solutions, the system reverts to an elicitation question.

It should be noted that, in the case of composing choice questions, the act composition process relies on the domain model having returned a solution set where individual solution options have explicit distinguishing characteristics, i.e., the act composition process itself makes no attempt to determine distinguishing features between solutions.

6.5.2.2 Addressing System Moves

As described in Section 6.5.1, the dialogue agent may adopt both purely system initiated and responsive dialogue goals during an interaction. Unlike the handling of open user moves, which leads to the adoption of dialogue act questions to resolve ambiguities, the handling of system moves results in a wider variety of dialogue acts as the full breadth of dialogue move goals can be open to realization. Nevertheless, the initiation of dialogue acts for system-initiated dialogue moves is a straightforward process similar to the adoption of acts for open user moves.

The first step in addressing user moves is to select an open move to process. Similarly to the selection of open user moves, the act planning mechanism selects the topmost element from the DSS's `PLANNED-SYSTEM-MOVES` slot. The creation of a relevant dialogue act proceeds by copying all specified move content over into the constructed act. Although this is an extremely simplistic act creation method that leads to overly descriptive acts, it is sufficient for current purposes. A more complete and long term solution will require the extension of the act planning process with referring expression generation (see Dale [2007] for an introduction).

6.5.2.3 Handling Dialogue Processing Errors

The AODM model, and its realization in the Daisie framework presented in Chapter 8, can handle the signalling of a small number of dialogue processing error types. The range of error types handled, and the sophistication of the handling mechanisms, is basic – although it should be noted that the management of imperfect communication and processing has not been a priority in the development of the AODM model.

The range of processing error types accounted for in the model was introduced in Section 6.2.2. Occurrence of such errors is signalled in the Dialogue State Structure through the `Input-Error` slot. During act planning, this slot is analysed for such instances. If an error event has occurred, the system adopts a dialogue act of speech function type `SignalNonUnderstanding` to signal the occurrence of the processing error. The propositional content assigned to that act is that same input error event which led to the adoption of the dialogue act. The realization of an appropriate phrase to express the particular understanding failure is the sole responsibility of the Language Interface layer, and the transform-based semantics mapping process in particular.

6.5.2.4 Basic Conversational Acts

Finally, the most basic type of dialogue act planning is the handling of non-task related conversational acts. These acts, including greetings and leave-takings, are handled in the AODM model through an adjacency pair mechanism.

6.5.3 Dialogue Act Selection

The final component of the Dialogue Planning process is *Act Selection*. The *Act Selection* component is responsible at any given point in the dialogue for analysing the contents of the DSS's `Planned-System-Acts` slot, and choosing a consistent set of acts for production. The result of this selection is thus placed in the `Next-System-Act` slot of the DSS as the output of the complete Dialogue Planning process. The current model adopts a straightforward strategy of selecting all planned acts for realization in a current turn. Although a simple strategy, it is sufficient for current purposes. The act selection mechanism is also the place at which the operational states of the Production and Perception modules are taken into account. If either one of these modules are active, as signalled by their `isPerceptionActive` or `isProductionActive` functions respectively, then no act is moved to the DSS's `Next-System-Act` slot for realization in the present turn.

6.6 Discussion

Unlike information-seeking dialogues common in non-situated domains such as telephone based applications, the class of task-oriented dialogues in the situated domain characteristically involves descriptions of plans of action to be taken by dialogue participants. Such dialogues by definition require the description of complex action instructions, and also the discussion and description of an agent's intentional state during the course of an interaction. While the content of these action-oriented dialogues requires an appropriate synthesis of dialogue state with intentional state and system capabilities, the interpretation of action descriptions in the situated domain is complicated by the amount of content that is assumed to be retrievable from situational contexts by the speaker. In light of these issues, this chapter has aimed to draw together theories of rational agency and dialogue management in a practical way, while at the same time overcoming assumptions of perfect contextualization in situated dialogue processing.

Rather than starting from a state-based, frame-based, or probabilistic dialogue management modelling stance, the Agent-Oriented Dialogue Management (AODM) model follows in the style of the classical Information State Update (ISU) accounts exemplified by Larsson's [2002] IBiS. In particular, the model aimed to build upon the strengths of classical ISU theories, but to bring them into closer alignment with theories of rational agency. This was done principally by unifying intentional and information state in a single mental state model, and introducing the notion of a capability-oriented dialogue move as a more structured construct than dialogue plans as seen in classical ISU models. In this model, the dialogue move becomes a frame-like construct that is manipulated over the course of a dialogue, but unlike true frames is instead based on the range of capabilities available to the agent. The assumption of perfect contextualization was removed by introducing an explicit and well defined contextualization process for task-oriented dialogue which depends on the application of domain-specific contextualization functions to both resolve and augment content provided by speakers in given dialogue move instances.

From a core modelling perspective, the treatment of dialogue moves and dialogue acts as the central representation units in a discourse representation can be considered a partial realization of Poesio & Traum Theory (PTT) [Poesio and Traum, 1998]. However, whereas PTT focused on the basic tenets of the grounding process, the AODM model has been developed to explore the relationship between dialogue processes, agency and contextualization. Our consideration of the grounding process, the information state, and the problems of situated contextualization also distance the AODM model both from classical agent-based dialogue management models and also *neo-* agent-based dialogue management models such as Sadek et al. [1997]'s ARTIMIS system, or Egges et al. [2001]'s BDP dialogue agents.

Within the ISU school, the AODM approach and its implementation is probably closest in motivation to Gruenstein and Lemon's Conversational Intelligence Architecture [Gruenstein, 2002, Lemon and Gruenstein, 2004] introduced earlier in Section 3.4.4. Specifically, both models advocate a tight coupling between dialogue management and agency features – although the AODM model has attempted to push towards issues of representation and function-based language resolution and augmentation in an ontologically modular architecture. More specifically, while Gruenstein presented a case for the integration of the *activity tree* as a pseudo-reflection of the agent's intentional state into the information state, here we have teased apart Gruenstein's singular frame-like notion of activity into distinct concepts of capability, dialogue move and intention, and at the same time introduced the agent's true intentional state into the information state. More importantly however, whereas Gruenstein assumed perfect domain contextualization of commands – and relatively unnatural and non-elided instructions – the AODM model overcomes such limitations by introducing the generalized contextualization process as an explicit mechanism by which domain information can be applied to not only provide static default values for slots in an activity specification, but to provide domain-specific and rated possibilities based on dynamic context. Moreover, the current model allows the construction of complex plans based on the sequencing of individual activity elements.

6.7 Summary

In this chapter, we developed a model of *Agent-Oriented Dialogue Management*. This dialogue management model blends aspects of agency and intentionality with an information state analysis of dialogue processing which places emphasis on the dialogue move as a unit of dialogic exchange. The chapter first developed the model of agency assumed by the AODM model. This model, built on existing work in the area of agent-oriented programming, provides a description of action and intentionality used as the basis of a domain application interface. The dialogue state model was then developed. As indicated, this model put emphasis on the notion of a frame-like dialogue move as the central unit of task-oriented meaning in dialogic processing. An account of language integration was then developed which accounts for how dialogue acts perceived at the language interface are integrated into the dialogue state structure. Thereafter, a functional model of move content resolution and augmentation was introduced. The final element of the *Agent-Oriented Dialogue Management* model was the dialogue planning mechanism – including accounts of move planning, act planning, and act selection.

Though the generalized contextualization process was introduced, it was not instantiated for any specific domains. Thus it is not yet entirely clear how the contextualization process, or indeed the dialogue management model in general, relates to specific situational models. We address this issue in the next chapter by introducing a specific spatial language processing domain, instantiating the contextualization process for that domain, and thus providing a concrete worked example of the AODM model's application.

Chapter 7
Putting Routes in Context

Abstract This chapter moves to the instantiation of the dialogue architecture's contextualization process for a concrete situated domain. The chosen situated domain is the control of semi-autonomous robots, and we will look in particular at the case of verbal route int erpretation in this domain. The chapter begins in Section 7.1 by reviewing approaches to the representation and interpretation of verbal route descriptions, and shows why these may benefit from an agent-oriented dialogue interpretation model. Subsequent sections develop the necessary components for the situated interpretation process. In particular, Sections 7.2 and 7.3 develop models of spatial representation and capability for a route interpreting agent that are suited to both communicative and non-communicative processes. Section 7.4 then provides the essential link between route interpretation and the dialogue management model by instantiating the AODM model's generalized contextualization process for the presented domain. This instantiation involves the development of concrete contextualization functions that aid the augmentation and resolution of dialogue move parameters. The chapter is then brought towards a close in Section 7.6 with a comparison of the developed model with existing theories of situated contextualization.

7.1 Interpreting Verbal Routes

Verbal route instructions are a type of discourse where one speaker or *interlocutor*, the *Route Giver*, provides a second interlocutor, the *Route Receiver*, with information on how to reach a target location. Verbal route interpretation has a long-standing history in spatial language research and is particularly interesting to us here for two reasons: First, these route instructions contain a high quotient of spatial language, and are hence very prone to issues of situational contextualization. Second, route instructions can be considered a special case of plan interpretation – thus, any model of contextualization and interpretation for route instructions is applicable to a variety of task-oriented dialogue. Route instructions have been studied extensively in the linguistics and cognitive science communities [Denis, 1997, Prévot, 2001, Tversky

and Lee, 1999, Werner et al., 2000] as well as in the computational community for
language production [Richter, 2008, Dale et al., 2005, Stocky, 2002] and interpretation tasks [Levit and Roy, 2007, Bugmann et al., 2004, MacMahon, 2007]. In the
following, we review the properties of verbal route descriptions and the approaches
taken to their interpretation in computational models.

7.1.1 Verbal Route Descriptions

The properties of route descriptions as a patterning of language have been a source
of much research in the spatial cognition community over the past 15 years. One
notable source of work in this area is Michel Denis who has studied the structure
of verbal routes and attempted to postulate psychologically motivated models for
the production process. Having collected and analysed corpora of route instructions
in large scale unknown environments, Denis postulated that route instructions obey
particular schematized structures that consist of action descriptions and landmark
references [Denis, 1997, Daniel and Denis, 1998]. According to Denis' model, a
prototypical instruction begins with the initial orientation of the hearer in their environment so as to indicate initial movement direction. This is then followed by
one or more route segments that progress the interpreter towards their destination.
A route segment itself identifies some landmark, reorients the hearer, and instructs
the hearer on a particular action to be taken (a move or reorientation). Finally, the
hearer is oriented towards their destination.

This regular patterning observed by Denis indicates a correspondence between
the verbalization of routes and what is assumed as the speaker's internalization of
the space under discussion. Tversky and Lee [1999] built on this general premise
by expressing Denis' characterization in more strict terms. Moreover, Tversky argued that the regularities of verbal routes correspond to the regularities of visual
depictions of the same route information, and that both visual depictions and verbal directions were expressing the same underlying information. While the basic
assumption of a common underlying cognitive spatial representation is reasonable,
Tversky further argued that the similarities between verbal and visual route information was so close that it should be possible to directly map back and forward
between verbal and graphical models.

Such claims, and Tversky's strong characterization of Denis' schematization
model, led many to treat route descriptions as formalizable in the same terms as
the underlying cognitive spatial model, e.g., Werner et al. [2000]. Such a claim,
like Tversky's argument for direct correspondence between verbal directions and
visual depictions is however questionable. Tversky herself pointed out that while
verbal descriptions were sufficient for getting a hearer from source to destination,
the verbal instructions were missing much necessary information. It is only through
the application of context information and basic heuristics of route direction that
the hearer can determine this missing information. For example, following analysis of collected route instruction data, Tversky suggests two inference rules were

minimally required to extrapolate complete route instruction sets from actual surface data. The first rule, *the rule of continuity* assumes that segment start points are
equal to previous segment end points, while the *rule of forward progression* states
that directly subsequent reorientations require an interjected intermediary forward
movement.

Regardless of how precisely verbal route instructions obey structural patterning,
analysis of route structure descriptions suggests that their interpretation is a complex process that calls on a range of information sources. Prévot [2001], for example, took a detailed semantic approach to the analysis of route interpretation, and
concluded that verbalised routes are highly underspecified structurally when introduced through dialogue and that hearers can only produce the intended interpretation through the application of many layers of context – including both discourse
and situational information. This view is backed up by MacMahon [2007] who, in
an analysis of a verbal route instruction corpus, identified syntactic, semantic, pragmatic, and procedural inference heuristics that were required to produce intended
interpretations from a verbal route description.

Despite these difficulties, route descriptions do have properties that make establishing a correspondence between verbal instructions and underlying representations
not as complicated as for other types of spatial language. First, unlike the description of visual scenes, the description of route information is not prone to what Levelt
[1981] characterizes as the linearization problem. Namely, the nature of a route as
a series of actions to be performed in moving from source to destination imposes a
natural ordering on expressed spatial constraints; this in turn results in an assumed
tight correspondence between spoken and internalized form. Second, despite questions of overall structure, route instructions themselves make use of a relatively
small set of primitives. Kuipers [2000] for example characterizes route instructions
as being minimally behaviourally expressed in terms of just four primitives: "turnings" which change the agent's reorientation, "travels" which move the agent along
a path, "verifies" which describe checks of landmark presence, and "declare-goals"
which signals the completion of the route.

Of course, even the range of expressions for the four behavioural primitives identified by Kuipers can be extremely complex in themselves. As spatial action descriptions, "travels" and "turnings" rely heavily on models of spatial adverbs and prepositions. Prepositions in particular express both the location at which actions should
be performed, i.e., static prepositions, and also the basic trajectories of movement,
i.e., dynamic prepositions. And even though the study of spatial preposition meaning
has probably been the single most active area of research in spatial cognition and
language research (see e.g., Levinson [2003a] for a brief introduction to modern
thoughts on the issues, or Talmy [2000] for an analysis of spatial language within
a cognitive semantics framework), the actual meaning of prepositions and spatial
language as used in particular contexts is still subject to much debate. For example, although Talmy attempts characterizations of individual prepositions in terms
of constraints they place on the subject and ground, i.e., the locatum and relatum
in our chosen linguistic semantics model (see Section 5.3.1.1), these models remain
descriptive and some distance from representing the actual flexibility observed in

language.

Two features of spatial language that are of particular relevance to route instruction interpretation are the notions of perspective and spatial reference frame. When describing a particular relationship between two entities, it is often possible to describe the relationship from different points of view or perspectives. For example, given a reference object, or *ground*, and a *figure* that we are relating to that ground, a speaker could potentially express this relationship from the speaker's own perspective, the hearer's perspective, the *perspective* of the object, or even from an imaginary floating perspective above the scene. Thus, we talk about a particular scene or movement being defined with respect to a perspective. But, it is not the case that all perspectives can be used to express an equivalent set of prepositions and underlying relationships. Rather, they are constrained in part by a reference frame associated with that perspective.

Levinson [1996] describes three basic reference frames. The *intrinsic* reference frame is a reference frame inherently established by an object by virtue of it having a definite predominant surface, i.e., a front, top, etc. For example, considering Figure 7.1(a), the expression "the box left of the chair" defines the box (figure/locatum) as being left of the chair (ground/relatum) with respect to the chair's own intrinsic reference frame. On the other hand, the relationship between locatum and relatum can be expressed not with respect to the reference frame provided by the relatum but with respect to a third party object. This is the use of a *relative* reference frame, and is exemplified by Figure 7.1(b) where the expression "the box right of the chair" describes the same scene as 7.1(a), but with respect to an observer's reference frame. Levinson's third reference frame type is the *absolute* reference frame which defines a spatial relationships with respect to an intrinsic property of the environment rather than the figure itself. In English, the classic example of absolute reference frame is the cardinal reference system used to define the cardinal directions of north, south, east, and west.

For the dynamic preposition case, two other reference frames, or rather perspectives, are commonly made use of [Klatzky, 1998]. The *egocentric* perspective, tied to the intrinsic reference frame, is defined by a trajectory created by the direction of movement of the moving object. Allocentric perspectives on the other hand are related to absolute reference frames in that they are defined by virtue of global rather then mover properties. For the case of route directions given with respect to a map, a cardinal reference frame can often be applied. However, the nature of the map as an entity in the real world also allows the intrinsic properties of the map to be used as a form of absolute reference system for a moving object, e.g., "drive to the left" where left is defined with respect to the map's left side.

The diversity of perspectives and reference systems choices for a given situation introduces significant complication in mapping between verbal route descriptions and space. However, it is not even the case that speakers remain consistent in these choices within a single route description task. In a series of publications, Taylor and Tversky [1996] and Tversky et al. [1999] investigated perspective use in route descriptions and quantified perspective shift in such instructions. Despite the perceived wisdom that coherence maxims would favour the retention of a single perspective,

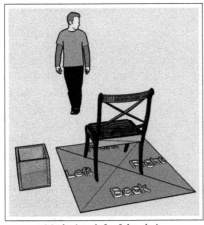

(a) the box left of the chair   (b) the box right of the chair

Fig. 7.1 Illustration of intrinsic (7.1(a)) and relative (7.1(b)) reference frame use for the same figure and ground.

Taylor & Tversky found that speakers frequently switched between so-called survey (allocentric) and route (ego-centric) perspectives. By way of explanation, Taylor & Tversky argue that though perspective shifting requires cognitive effort, that effort is not substantive in comparison to the effort already required in making focus shifts within the environment under discussion.

Taylor & Tversky's experiments, like most cognitive and linguistic experiments on verbal route instructions, focused on the case of monologic instructions provided by route giver to route follower in advance of the route follower's movement. Recently however, more work has become available on the study of route description properties in true dialogue. Shi and Tenbrink [2005] report on experiments wherein users were requested to give route directions to a robotic wheelchair. While the domain for this research was that of a robot negotiating route instructions with a user, the focus of the analysis presented was on the dialogic structures that might be required for an operating system, and as such relatively little is said about the actual spatial properties of the language used. Lawson et al. [2008] meanwhile reports on the use of computer-mediated communication between humans for spatial tasks including route navigation. Their findings, while preliminary, suggest considerable flexibility in perspective choice in dialogue. Also recently, Goschler et al. [2008] collected and analysed a corpus of dialogue route instructions and found that, while mixing of survey and route perspective was common, route followers did tend toward the use of monologic survey style indications of destination in later trials.

7.1.2 *Computational Models of Route Instruction Interpretation*

A number of approaches to the computational interpretation of routes from a language understanding perspective have been considered in recent years. These approaches can be split broadly into three categories based on the level of spatial representation afforded to the interpreting agent. *Perception-driven models* assume that the interpreting agent has no global representation of its environment and instead place emphasis on behavioural models and the agent's perceived environment (e.g., Bugmann et al. [2004], MacMahon [2007]). *Qualitative models* assume that the agent does in fact have a representation of its global environment, but that this global representation is expressed in qualitative rather than quantitative terms (e.g., Krieg-Brückner and Shi [2006], Bateman et al. [2007]). Finally, *quantitative models* assume that the agent does have access to a quantitative representation of its environment (e.g., Mandel et al. [2006], Levit and Roy [2007], Tellex and Roy [2006]). In the following, we review a number of the more prominent works in these three areas.

7.1.2.1 Instruction Based Learning

In the context of a project on Instruction Based Learning (IBL), Bugmann et al. [2004] developed a model of route instruction interpretation based on the direct translation of natural language instructions to primitive behaviours executable by a mobile robot. During a typical interaction, a user issued a series of spoken route instructions to the mobile robot to direct that robot towards a target destination in a 'toy town'. Figure 7.2(a) depicts the experimental setup with a user instructing the robot from a location proximal to the toy town, while Figure 7.2(b) shows an allocentric view on the toy town. Users were free to use whatever constructions they saw fit, thus resulting in spoken instructions such as *"go over a crossroads to a roundabout"* and *"okay if you're in the parking space go out"*. While the robot itself had no global map of the environment, it could perceive objects in its locality. Instructions from the user were then interpreted as commands and partial plans that could be executed by the mobile robot directly. The robot was capable of learning, however, in that once it had been instructed on how to reach a target, this procedural knowledge was maintained.

IBL's instruction processing model was organized into two main stages [Lauria et al., 2002]. During the first stage, a user's instruction was parsed into a DRT-based discourse semantics model for the resolution of anaphoric references.[1] During the second stage the partially augmented discourse semantics was translated into an application-specific procedural description language to be directly executed by the mobile robot. Specifically, the target of such translations was a process description language defining high-level control programs. This language includes both

---

[1] However, in the light of Knees [2002]'s work on developing anaphora resolution algorithms for IBL, it is not clear whether the discourse semantics was used for anaphora resolution in practice.

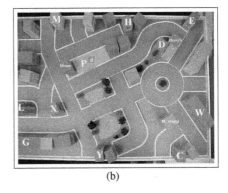

(a)   (b)

Fig. 7.2 IBL experiment set-up. Figure 7.2(a): user directing robot around toy town. Figure 7.2(b): allocentric view of map used.

primitive action descriptions comprising the action type, the objects used to parameterise the action, and a small number of logical connectives. Moreover, rather than having these basic actions defined arbitrarily by designers without reference to linguistic data, the IBL project applied Wizard-of-Oz studies to systematically collect information on which behavioural primitives are most frequently indicated through spoken language [Bugmann et al., 2004]. In total 15 behavioural primitives were identified from this analysis – although, on close inspection, it appears that the set of primitives may be reducible to a much smaller set given appropriate parameterisation, e.g., `go_until` versus `go`.

The IBL model was evaluated in part by transcribing corpus data directly to procedural instructions and having the robot follow the instructions. Despite the fact that these instructions were derived by hand, and thus likely avoided many of the difficulties of speech recognition, syntactic, semantics, and conceptual analysis, the system still only managed to find a target destination for 63% of input instructions [Bugmann et al., 2004]. Many of the problems encountered were attributed to the fact that the robot did not have a representation of the environment outside of its perceptual range. Moreover, behavioural primitives were mapped extremly closely to language form, and though Bugmann recognized the existence of instruction underspecification, no computational models were developed to handle such underspecifications.

7.1.2.2 MARCO

From a methodological perspective similar to the IBL project, MacMahon [2007] recently reported on a corpus-driven route instruction interpretation system for mobile robots endowed with an accurate perceptual and behavioural system, but lacking an explicit global spatial representation. As with the IBL project, MacMahon's usage scenario envisaged a user directing a mobile agent from a start position through an unknown maze-like environment. However, unlike the IBL project, the route follower is a software agent situated in a virtual environment, and, rather than being given an allocentric view on that environment, the route giver first learns the environment via a first-person ego-centric exploration of the virtual world, and must subsequently provide route instructions from memory of the environment only. Figure 7.3(a) depicts the view of the environment given to the user at a computer console, while Figure 7.3(b) depicts an allocentric view on a typical environment. It should be noted however that experimental participants were never provided with such a map.

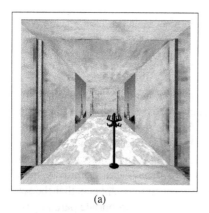
(a)
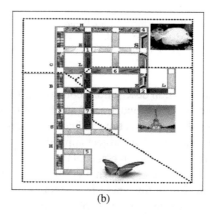
(b)

Fig. 7.3 Marco experiment set-up. Figure 7.3(a): view of virtual environment as perceived by user. Figure 7.3(b): allocentric view on map used.

MacMahon's interpretation model, named MARCO, is broadly similar to that used by the IBL project, but MacMahon placed greater emphasis on the layer of behaviour execution than was the case for IBL. More specifically, MARCO is composed of six processing modules; three modules for linguistic analysis, and three modules for task execution. Linguistic analysis modules included a *syntactic parser*, a *content framer* to perform semantic analysis, and an *instruction modeller* which transformed surface semantics to the reactive plan language used by MARCO's task executor. The task execution module on the other hand consisted of: (a) the *executor* which, when supplied with a reactive plan specification, interleaved execution of that plan with perceptual primitives; (b) the *robot controller* which provided a concrete robot implementation below the layer of reactive plans; and (c) the *view-description-matcher*, which compared plan-specified scene descriptions to the perceptual field in order to determine when plan steps were being achieved successfully.

According to MacMahon [2007], route instruction underspecification was principally addressed within two of MARCO's six processing modules. In transforming surface semantics to task plans, MARCO's instruction modeller reduced a wide variety of surface forms to functionally equivalent instructions through the application of hand-crafted rules. Such rules addressed the transformation of non-imperative statements and idioms to action primitives, and decomposed high-level commands into lower-level procedures. For instance, MacMahon indicates that the instruction "Take the third right to the end of the hall," would be reformulated as three distinct actions with meanings similar to "Go down to the third place with a path to the right. Turn right there. Go down to the end of that hall.". The executor module on the other hand was capable of applying different levels of task execution intelligence to achieving the stated actions within a reactive plan. Namely, when encountering an action for which the pre-conditions do not hold, the executor was capable of adopting plans towards achieving the pre-conditions of the goal. In many cases, such intelligence was straightforward in that the agent would straighten itself before proceeding down a corridor. In other cases, heuristics such as the need for a movement between two turn instructions were interjected through the executor module.

As indicated, MARCO did not have access to a complete global representation of its environment. Instead, MARCO's view-description-matcher had at all times only a single ego-centric perception of its environment. Moreover, rather than being a true perception of the environment which had to then undergo perceptual processing, MARCO's perception was a pre-computed symbolic abstraction over the environment from a given intersection. While this allowed MacMahon to ignore issues of symbolic abstraction at runtime, MARCO was thus limited to operating in a highly unrealistic discrete environmental model where only certain views of the environment were possible. Furthermore, the symbolic abstraction did not model the location or orientation of objects within the small-scale space of an intersection.

The MARCO architecture was evaluated by comparing the system's performance in interpreting a set of route instructions collected during earlier human-human interaction trials with the MARCO environment. However, despite the fact that route instructions were manually segmented and cleaned up by MacMahon for the parser, MARCO only achieved 30% successful interpretation in the third and most naturalistic experimental scenario. Given that MARCO was designed to systematically ignore instructions that it did not understand, and since MARCO's executor allowed searching of the environment for stated terminating landmarks, it is not clear how much of this 30% was due solely to the identification of a small number of landmarks in a search pattern.

7.1.2.3 Mandel's Fuzzy Interpretation Functions

A shared motivation for both IBL and MARCO was the interpretation of route instructions by robots which relied on local perception rather than having access to a global environment map. Thus, in both cases the robot could only interpret route instructions by performing the instructed actions within its environment (be that environment real (IBL) or virtual (MARCO)). This minimalist representation approach is appealing both because of the emphasis placed on behaviour-oriented robotics, and the natural correspondence of the approach to route interpretation by humans unfamiliar with an environment. However, in practice it is more common for robots to operate with highly detailed, though partial, maps of their environment. Thus, in practical tasks, it will often be the case that a robot has knowledge of its environment and can interpret a route instruction or other procedural instruction sets directly against its world knowledge before initiating action.

Adapting a research perspective where explicit global quantitative representations are available, Mandel et al. [2006] developed a route interpretation model for semi-autonomous wheelchairs operating in a partially known environment. Rather than considering the linguistic properties of routes and route interpretations, Mandel's focus was on developing a model of route interpretation that was compatible with a spatial representation suited to conventional robot navigation and localization tasks. To that end, Mandel made use of a global quantitative model comprising two layers. The first layer was a metric spatial model consisting of a 2D occupancy grid referenced by a Cartesian coordinate system. Such a layer is very common in the robotics community, and facilitates robot localization and local navigation [Frese, 2006, Magid et al., 2006]. The second layer was a Voronoi graph, a reduction of free space which describes maximal clearance to surrounding obstacles, and thus defines allowed paths through an environment. As with the occupancy grid, the Voronoi graph and other detailed graph-based representations are used extensively in the robotics community, in this case to perform global navigation tasks [Wallgrün, 2004, Mandel et al., 2005]. Figure 7.4 depicts the Voronoi-based representation for an office environment.

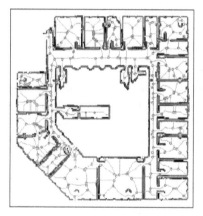

Fig. 7.4 Voronoi graph based spatial representation used by Mandel et al. [2006].

Rather than focusing on the linguistic elements of the route instruction process, Mandel assumed a route description model closely bound to the idealized schematized model proposed by Tversky. Namely, Mandel's route interpretation assumed as input a structural description of a route composed of a number of route segments where each segment in turn consisted of a well specified movement and turning description. Individual movements could make reference to landmarks, and to support this, the Voronoi graph representation was extended to allow annotations of graph nodes with string-based identifying labels, e.g., "kitchen", or "john's office". The interpretation process itself was a breadth-first search through the Voronoi graph space. Each step of the search involved the application of one of a set of fuzzy functions which defined likely interpretations of individual prepositional phrases and turnings. Thus, for a phrase such as "through the doorway", a function was available which provided likely end positions, modelled as Voronoi graph nodes, for a given start pose and environmental configuration.

The main advantage of Mandel's approach was the use of quantitatively well defined definitions of prepositional phrases, or rather trajectors, over a graph representation model that could be generated automatically from free space representations. Unfortunately however, the Voronoi graph model remains very fine-grained in comparison to the level of detail used by speakers in communication. Thus, the Voronoi graph often incorporated considerably more graph nodes than a user would typically assign as decision points in a route instruction description. This led to Mandel's approach favouring unnaturally long route instructions to aid the pruning of the search space, which while being well configured for robot navigation and localization, was unsuited to the requirements of natural language processing. Moreover, Mandel's approach presupposes a specification of the user's route instruction as a structural formalism which already includes a schematized view of routes, and so is already considerably abstracted from the realities of surface language.

7.1.2.4 Levit & Roy's Dynamic Programming Model

Levit and Roy [2007] recently presented a computational model of route interpretation geared to global map representations with geometric data, but with minimal explicit graph and topological information. The experimental scenario adopted by Levit & Roy, as well as the basic corpus of data used in their models, was derived directly from the HCRC Map Task corpus of human-human route instruction negotiation dialogues [Anderson et al., 1992]. In the Map Task, two participants were provided with highly schematized maps of a common environment. One participant, the *route giver* was then instructed to verbally direct the second participant, the *route follower* from an agreed starting position to a destination on the map. Route descriptions were not however monologic and straightforward; discrepancies between the maps used by route giver and route follower were purposefully introduced, thus introducing an element of negotiation in the route description process. Figure 7.5(a) depicts an extract from the Map Task based plan used by Levit and Roy in their route interpretation model.

As with other computational approaches to the interpretation of verbal route instructions, Levit & Roy's model was highly dependent on the notion of a lexicon of primitives used in the descriptions of routes. Using an extract of the Map Task corpus, the authors identified a total of 5 distinct *Navigational Information Units* (NIUs): *Positions, Orientations, Moves, Turns,* and *Compound References.* Con-

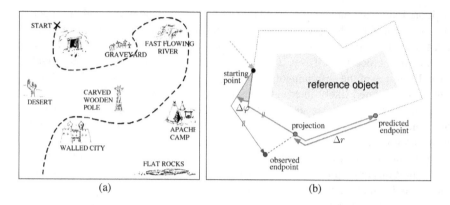

Fig. 7.5 Environment map and interpretation functions [Levit and Roy, 2007].

stituent components of NIUs were in turn analysed to identify the functional contri-
butions of individual moves. Constituents identified for the *Move* NIU type by Levit
& Roy included referenced landmark type, extent, and coordinate system (reference
frame). A notably important aspect of these constituents is the set of path descriptors
which describe trajectories as marked in language by dynamic prepositions. Levit &
Roy identified a total of 10 such path descriptors, i.e, TO, FROM, TOWARD, AWAY_-
FROM, PAST, PAST_DIRECTED, THROUGH, BETWEEN, AROUND and FOLLOW_-
BOUNDARY. The inventory of NIUs identified along with their constituents can thus
be seen as a type of route instruction ontology similar to aspects of the Generalized
Upper Model identified in Section 5.3.1.1, or the conceptual ontology introduced
later in this chapter.

For each NIU, an interpretation function was developed based on an initial manually designed prototype which was subsequently refined with respect to actual
sketched interpretations of the same NIU. For action NIUs, i.e., turns and moves,
parameterized prototypes were compared to observed human interpretations to calculate both radial and angular deviations from the manual prototype. Figure 7.5(b)
depicts comparison of the prototype and observed values for a *move* action parameterized with a FOLLOW_BOUNDARY path descriptor. Deviations from prototype
predicted end position were then used to construct target destination probability distributions for a given NIU. Since the original Map Task maps were purely geometric
and highly underspecified, a grid of rectangular regions was overlayed on the map
to discretize possible target locations. A given parameterized NIU then returned a
2D probability distribution over accessible locations within the map.

To perform complete route interpretation, the results of individual NIU application were sequenced into a complete path description. Given that the probability
distribution provided by any given NIU was comparatively broad (see Figure 7.6
for an example distribution where cell colour depth indicated likelihood of that cell
being the target of the current NIU), it was not generally appropriate to arbitrarily
select the most likely outcome of the nth NIU as an input for the $n+1^{th}$ NIU. Rather,

it was appropriate to calculate optimum interpretations for the nth NIU from its over-all contribution to the cost of a completely interpreted route. For this reason, a linear programming approach was adopted by Levit & Roy for calculating the complete route. The approach adopted is therefore broadly similar to the breadth-first graph based search adopted by Mandel in that all possible routes to a destination were searched through and the final path chosen was the most effective path to reach that destination.

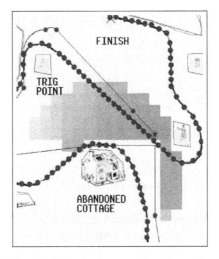

Fig. 7.6 State of the spatial model following interpretation of a number of route instruction segments. [Levit and Roy, 2007]

While the inventory of NIUs, their interpretation templates, and the general process of complete route interpretation presents a substantial contribution towards generalizable models of route direction interpretation, Levit & Roy did not consider the linguistic elements of the interpretation process and instead relied on the manual coding of NIUs from the corpus data provided by the Map Task. It is unclear how automatable this process might be since no clear indications of the annotation procedure are given by the authors. That said, the novelty of Levit & Roy's route interpretation model lies in the relatively unconstrained nature of the spatial model against which they operate, and how they deal with that unconstrained structure. More specifically, unlike the schematized structures considered by Bugmann, MacMahon, Mandel and others, the Map Task maps offer little in the way of inherent route structure. Landmarks are randomly rather than linearly distributed, thus resulting in the possibility for any given landmark to be passed on any side. Such properties required that the NIU functions developed by Levit & Roy be hard coded with respect to the given environments. This pre-computing of possible routes through the map, based on maximal distance potential fields, may have allowed the considerable simplification of NIU templates, but this was not considered by the authors.

7.1.2.5 Tellex & Roy's Spatial Routines

Although not concerned with route descriptions in the complete sense, Tellex and Roy [2006, 2007] recently reported on a metric model based approach to spatial action interpretation that has direct relevance to the interpretation of route instructions. Specifically, Tellex has investigated the definition of spatial routines as direct meaning giving functions for words used in the description of spatial actions typically used in routes. The approach, based on that outlined by Gorniak and Roy [2007], requires that concrete situation-specific contextualization functions be defined for individual words, and that those meaning functions be compositional to provide the complete and contextualized meaning of a given utterance. While an interesting premise, Tellex has focused on design of the low-level primitives based on perception routines, and as such it remains to be seen how effective and generalizable her spatial routines are.

7.1.2.6 Krieg-Brückner & Shi's Qualitative Route Graph

The reliance on quantitative representation and reasoning for route instruction interpretation presents a number of challenges. First, quantitative theories of language contextualization are still clearly in their infancy, and are highly dependent on assumptions about the nature of the interpreting embodied agent. Second, though quantitative reasoning in itself works well for individual cases, combination of quantitative reasoning models is often problematic. Third, quantitative models of space are extremely verbose and detailed to the point that they frequently become invalidated by minor changes in the real world. Fourth, and perhaps most interestingly for our purposes here, quantitative models do not necessarily map well to human conceptualizations of space – particularly where the space is large scale.

An alternative approach to spatial representation and reasoning is to forego explicit quantitative models in favour of qualitative accounts. Qualitative spatial reasoning through logical calculi (see Cohn and Hazarika [2001] for an introduction) investigates the intrinsic spatial properties of, and relationships between, entities such as regions [Randell et al., 1992], line segments [Moratz et al., 2000], and other spatial constructs. By operating at a high level of abstraction, qualitative spatial reasoning techniques focus on the combinatorial properties of spatial relationships, and can operate without the need for sometimes unreliable quantitative information.

Coming from the perspective of qualitative spatial reasoning, Krieg-Brückner has, over a series of works, investigated the usefulness of qualitative graph representations in the robotics domain, and for the route interpretation process [Krieg-Brückner et al., 2005, Werner et al., 2000, Krieg-Brückner et al., 1998]. At the heart of Krieg-Brückner's model is the Route Graph, an abstract graph-like representation of navigation space which may be instantiated to different kinds, layers, and levels of granularity. Attempting to emulate a cognitively plausible view of spatial knowledge, the Route Graph permits global representations of the environment, but eschews quantitative global knowledge. Thus, as an underlying spatial representation, the Route Graph lies between the representation-less models of IBL and MARCO, and the highly detailed models of Mandel, Levit, and Tellex. Instantiations of the Route Graph have been made both for large-scale navigation space such as tram-networks [Lüttich et al., 2004], as well as for robotic applications in medium-scale space such as office environments [Röfer and Lankenau, 2002].

The Route Graph is, at its core, a formalised graph-based spatial representation inspired by the schematized multi-modal view of route knowledge argued for by Tversky and Lee [1998]. As defined by Krieg-Brückner et al. [2005], the principle concepts within the abstract Route Graph specification are *Places*, *Segments*, and *Routes*. *Places* are anywhere that an agent can 'be', and are defined as having their own reference system related to an origin (position and orientation) associated with the place. In turn, local reference systems may or may not be rooted in a global reference system depending on the cognitive characteristics of the instantiated application. A *Place*'s origin is used to define the orientation of connecting *Segments* which are directed connections from one *Place* to another. A *Segment* is said to have a *Course*, an *Entry*, and an *Exit*, the latter two of which are defined with respect to the connecting place's origin as described earlier, while the *Course* is some description of the actual route taken between the *Entry* and *Exit*. A *Route* is then intuitively defined as a sequence of *Segments* without repetition (i.e., cycles are not permitted within route definitions). *Places* and *Segments* may be specialized with respect to particular applications. For example, in a Voronoi-styled instantiation of the Route Graph, *Places* may have a width denoting free space, while a *Segment*'s *Course* may be instantiated with quantitative information such as distance or width of the segment – or alternatively with qualitative information such as the action to be performed to traverse the route segment. In later formulations of the Route Graph, spatial relationships themselves were expressed through a variant of Freksa's [1992] Double Cross Calculus [Krieg-Brückner and Shi, 2006].

Whereas low-level robotics-oriented approaches to spatial representation have taken an ad hoc approach to the relationship between the spatial model and the ontological types of elements referenced by users in discourse, one of the useful features of the Route Graph model is that inter-ontological relationships have been considered explicitly. Specifically, Krieg-Brückner et al. [2005] set out some of the relations required between the world of route graphs (places, route segments, paths, etc.) and other modelling domains. Two such domains are explicitly identified: a spatial ontology that is expected to provide *spatial regions*, and a 'common-sense ontology', providing everyday objects such as rooms, offices, corridors, and so on. The relationships provided are intended to allow inferences back and forth between places in a route graph, the spatial regions that such places occupy, and the everyday objects that those regions 'cover'. These relationships were further refined by Bateman and Farrar [2005] in the context of a more general ontological framework.

One of the motivating factors for the development of the Route Graph is that it may be used for the interpretation of verbal route instruction. The assumptions of this interpretation approach are that: (a) a route graph may be used for representing the agent's spatial knowledge; (b) a partial route graph may be used for representing the route instructions given by a person; and (c) route interpretation can be performed by proving consistency between the route instruction specification and the spatial representation model through theorem proving and / or explicit qualitative spatial reasoning techniques. Such an approach essentially conflates verbalised route *instructions* with the spatial model which they are to guide a user through. This view of the route interpretation process owes much to Tversky's suggestions of correspondence between verbal, visual and internal representations of route information, but it is far from clear whether such an idealised model of route interpretation can work in practice. Not considering questions such as the validity of excluding quantitative descriptions of space and action, the most significant problem with the proposed route interpretation method is that it relies on a pre-formalisation of the verbalised route which is assumed to have removed many of the ambiguities present in surface language. However, achieving such a well specified input is itself far from trivial in practical language systems. For one thing, we know from linguistic analysis, including that of Tversky herself, that verbalised routes are underspecified and require inferences of segment information that is simply omitted in verbal descriptions. Moreover, even for explicitly mentioned segments, the contextualization of identified landmarks is a far from straightforward process. Thus, while the assumption of a route graph as a representation of space, or a representation of a route instruction, is in itself plausible, the interpretation process requires significantly more elaborate mechanisms than theorem proving and qualitative spatial reasoning techniques alone.

7.1.2.7 Discussion

While the various models just considered all include at their core the goal of interpreting linguistic descriptions of routes, the approaches taken vary widely in their cognitive plausibility, representational requirements, and computational costs. Formal structural models such as the Route Graph have the advantage of operating directly at a qualitative level, and thus do not require ad hoc abstractions from quantitative data during the language interpretation process. Formal approaches also offer greater potential to integrate spatial reasoning models with other symbolic reasoning necessary for language interpretation, e.g., ontological reasoning. However, the truly formalized approaches often have limited tractability, do not consider the details and difficulties of generating route descriptions from real natural language data, and are not necessarily easily integrated with real-world situated systems. Quantitative representation based approaches (e.g., Mandel, Levit, and Tellex's models) on the other hand offer improved tractability and tighter integration with other aspects of an embodied situated system, but have the distinct disadvantage that their knowledge models are often more suited to low-level robot navigation and self-localization processes than to communicative processes. Perception-oriented models on the other hand (e.g., IBL and MARCO) offer a route interpretation model that is most compatible with the notion of route descriptions as investigated by Denis, but by ignoring the possibility of internal representation they become divorced from the realities of modern service robotics.

Moreover, with few exceptions, the route interpretation models described above ignore issues of linguistic underspecification and ambiguity, and often rely on preformalised semantic structures. While such simplifications seem reasonable in the light of the assumed complexity of language processing and understanding, by ignoring these issues it is all too easy to assume that the route interpretation process is straightforward. Moreover, all of these route interpretation models have focused on route interpretation outside of a dialogic context, but yet, the dialogic process can be invaluable in clarifying referent choice, reference frame selection, and extent of motion in route descriptions [Goschler et al., 2008]. While it is self-evident that processing routes in a dialogic context would provide a means to handle the worst instances of ambiguity, this requires that the interpretation process must be organized so that the interpretations assigned to individual elements of a route instruction can be made explicit in such a way that dialogue processes can construct appropriate confirmation and clarification dialogue moves.

7.1.3 *Towards Action Oriented Route Interpretation*

In the remainder of this chapter, we develop a model of verbal route instruction interpretation based on the dialogue processing architecture developed in the previous chapter. The aims of this model are threefold. First, this model aims to illustrate and further explain the language contextualization process sketched in the last chapter. Second, this model aims to show the relationship between dialogue processes and the types of representation and reasoning necessary for meaningful communication in the situated domain. And third, this model aims to provide a solution to the route interpretation problem itself – a solution which removes the simplifying contextualization assumptions made by many of the route interpretation approaches introduced in the last section.

To investigate the route interpretation problem we will focus on one particular route interpretation scenario. In this scenario, which we will refer to as the *Navspace* scenario, a user gives verbal instructions to direct an autonomous robot around a partially known office environment. Here, the user and system play the roles of Route Giver and Route Follower respectively. Reflecting the real-world capabilities of robotic systems, both the Route Follower and Route Giver are assumed to have shared knowledge of their environment. However, while the Route Follower is capable of moving in the environment and following the Route Giver's instructions, it is the Route Giver who has knowledge of their target destination. Moreover, during a typical interaction with the robot, a user views the shared simulated environment from a plan perspective which includes corridors, various rooms, and the robot's position in the environment at any given time. A user, with a particular known target destination, is then free to direct the robot towards that destination in whatever way that user sees fit.

To instantiate a route interpretation process based on this scenario and the agent-oriented dialogue processing model developed in earlier chapters, we essentially require three components:

1. A physical context model.
2. An action inventory.
3. A resolution & augmentation function inventory.

The *physical context model* is the representation of space known to the interpreting agent. Rather than assuming a robot with no explicit representation of its environment, the approach taken here assumes that the agent has quantitative knowledge of that environment. This assumption is justified by the wide use of detailed quantitative models of space for robot localization and navigation. However, rather than assuming quantitative spatial models that are optimized for low-level robotic needs, the representation assumed here includes both qualitative route knowledge and an ontologically rich conceptual encoding of the environment. This physical context model will be outlined in Section 7.2.

The *action inventory* is that collection of elements that are used to create dialogue move objects operated over by the dialogue management process. In our case these actions are those that the agent can perform in navigating through its environment. Following from the characterization of routes by Denis and the many approaches to route interpretation outlined in the last section, we assume that the fundamental units of route descriptions are in fact these actions. While mobile robots can operate in terms of low-level actions, the interpretation of verbal routes requires that a sufficiently abstract level of action is available to the interpreting agent. While the actual inventory of actions can be very simple, e.g., potentially as few as just *moves* and *turns*, appropriate parameterizations of these actions are essential. The basic components of the action model will be outlined in Section 7.3.

The *resolution and augmentation functions* are used to instantiate the generalized contextualization process (see Section 6.4) for the specific case of route instruction interpretation. These functions provide the mapping between linguistic descriptions and the parameterization of actions used in describing routes. These mappings include not only the exophoric resolution of situational referents, but also the identification of elided information in the situational context. In Section 7.4 we detail a number of contextualization functions that provide situated meaning for the actions types outlined in Section 7.3 – and thus provide a working model of situated route interpretation.

It should be noted that the developed route interpretation model is not intended to be as complete an account as those offered by some of those models considered earlier. Indeed, the range of interpreted functions is in some way considerably less than that accounted for in some of the considered proposals. However, the aim here is not to provide an interpretation process that stands alone and operates on pre-formatted inputs, but instead to provide an interpretation model that fits into a complete dialogue processing architecture in a well understood way. Such a model thus becomes not only highly extensible, but can also be integrated with other aspects of conversational and situated intelligence.

7.2 The Physical Context Model

Depending on theoretical perspective and application needs, numerous spatial modelling solutions are available to situated applications. General purpose solutions include standards for the representation of geographic information [Cristani and Cohn, 2002], modelling environments for virtual reality and gaming applications [X3D, 2007], cognitive theories of spatial knowledge [Winkelholz and Schlick, 2006], and models of spatial representation for robot localization and navigation (see Murphy [2000] for an introduction). Moreover, even within our review of route interpretation models we saw variance between abstract graph based models [Krieg-Brückner et al., 2005], Voronoi-based models [Mandel et al., 2006], and strictly metric models [Levit and Roy, 2007]. This variance reflects the often contrasting representational needs of communicative and non-communicative tasks such as perception, action, localisation, and navigation.

Rather than assuming a single homogeneous model of space for both non-communicative as well as communicative processes, a multi-tiered representation of space will be adopted here. The adopted model follows from the graph based organization of space proposed by Krieg-Brückner et al. [2005], the subsequent ontological perspective on that model presented by Bateman and Farrar [2005], and the need for a metric space model to support localization, local navigation, and exophoric reference resolution. The Navigation Space (Navspace) Model ($\mathcal{NS}$) is therefore organized into three components:

$$\mathcal{NS} = \{\mathcal{RS}, \mathcal{CS}, \mathcal{GS}\} \tag{7.1}$$

where: (a) $\mathcal{RS}$, or *Region Space*, is a metric model of the agent's environment; (b) $\mathcal{CS}$, or *Concept Space*, is a knowledge-base-like representation of entities in the environment; and (c) $\mathcal{GS}$, or *Graph Space*, is a structural abstraction of navigable space that sets out possible navigable accessibilities between environmental entities. The remainder of this section motivates and describes each of these layers in more detail.

7.2.1 Region Space

Within the robotics community, *metric models* are representations of space that provide quantitative information on the size, shape and location of objects within an environment. Both flattened and hierarchical accounts of metric space are frequently applied to robot localization processes where the metric model is augmented with probabilistic methods to provide a reliable account of robot mapping and localisation in noisy environments [Frese, 2006]. Metric accounts of space are also highly useful in planning local navigation. Here, detailed knowledge of object shapes and positions is used to parameterize spline construction techniques that deliver smooth motion trajectories [Magid et al., 2006].

Metric spatial models have also been found useful for human robot communication. As indicated in Chapter 1, the resolution of exophoric references in language is partially governed by proximity and perceptual salience. Quantitative metric models of space that establish perceptual and proximity fields between environmental objects and an agent have thus been shown to facilitate reliable exophoric reference resolution [Kelleher, 2007]. Moreover, though metric accounts of space do not code static or dynamic spatial relationships, quantitative representations, via appropriate abstractions, can be used to resolve spatial relationships in computational models of language understanding and production [Kray et al., 2001, Kelleher et al., 2006]. It should be noted that metric models applied in human robot interaction tasks have one distinguishing feature over classical metric accounts used purely for robot localization and navigation; namely, whereas classical metric accounts of spatial representation need only note if a particular spatial parcel is occupied, metric space for interaction additionally requires information on the specific objects that occupy a given spatial parcel.

Within the Navspace model, $\mathcal{RS}$ is a quantitative metric representation of the agent's environment composed of a 2D Cartesian coordinate based grid system within which each grid element can be either unoccupied or occupied by one or more *Environment Things*. These *Environment Things* can be either physical in nature, e.g, walls, doors, pillars, or abstract, e.g., rooms, corridors, regions. The range of *Environment Things*, their features, and the possible relationships between them will be addressed under the discussion of *Concept Space*; here it is sufficient to know that more than one physical thing cannot occupy the same portion of *Region Space*, while abstract things can overlap with each other or with physical objects. Unoccupied *Region Space* grid points are uncommon since large abstract conceptual spaces such as *corridors* will occupy space which is free to move within. Thus, it should be noted that $\mathcal{RS}$ is not the same as an *Occupancy Grid* as used in localization and navigation routines, and that to be used for this, the $\mathcal{RS}$ should first be filtered of occupying abstract entities.

7.2.2 Concept Space

While *Region Space* provides a view of size and shape information for solid and non-solid entities, the entities that occupy region space can have additional conceptual attributes. While these *conceptual* attributes are not necessary for low-level robot navigation or localization, they are integral to communicative processing.

A symbolic representation of the agent's situational knowledge excluding metric and route navigation information is defined in terms of the $\mathcal{CS}$ (*Conceptual Space*)

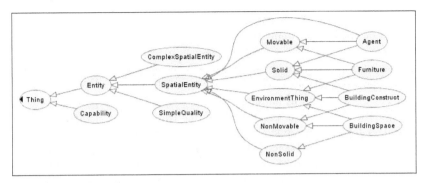

Fig. 7.7 $\mathcal{CS}$ layer's upper concept hierarchy.

Layer. In use, the *Conceptual Space* is analogous to a database which can be queried for entities that match particular conceptual descriptions, i.e., all known entities that are of type `Kitchen`, or all known entities that are `owned-by John`. We define the range of allowed entities, their possible features, and the relationships which can hold between them in terms of a domain ontology. The range of categories and relationships required are typical of what is termed General Ontology [Farrar and Bateman, 2004]. Given the tractability requirements for human-robot interaction, a simplified Description Logic based ontology was developed for this work. That notwithstanding, the particular choice of ontology is intended to be illustrative rather than complete, and would be well served through extension by, or eventual replacement with, an existing and more detailed conceptual ontology.

Figure 7.7 depicts the highest levels of concept definitions used by the $\mathcal{CS}$ layer as defined in the domain ontology. All $\mathcal{CS}$ layer objects subclass the `SpatialEntity` class. This class in turn includes any object types that have well defined physical extent, and are hence also captured explicitly within the $\mathcal{RS}$ layer. A number of classes directly subsume the `SpatialEntity` class. These classes are in turn partially cross-classified to result in four common-sense categories relevant to the definition of entities within the current target domain of indoor route interpretation. Specifically, the class `Agent` is the set of solid, moveable entities that can perform an action in the environment; the class `BuildingSpace` is the set of non-solid, environment things that are non-moveable; the class `BuildingConstruct` is the set of non-moveable, solid environment things; while the class `Furniture` is the set of moveable, solid, environment things. Further classification of concrete entities beyond those four classes is made, but will not be considered further here beyond noting such classes include notions such as `Wall`, `Corridor`, `Table` and so forth.

A number of *non-spatial* relations are also defined as a relation-hierarchy within the $\mathcal{CS}$ layer's backing ontology. These relations, e.g., `hasOwner` and `hasColor`, have ranges defined either in terms of concepts drawn from the `SpatialEntity` class itself, or alternatively from the `ComplexSpatialEntity` or `SimpleQuality` classes. The `SimpleQuality` class includes abstract entities which do not have a well-defined physical extent and hence are not captured explicitly within the $\mathcal{RS}$ layer. Subclasses of this class include concepts such as size, colour, spatial relationships, extent and time. The `ComplexSpatialEntity` class on the other hand includes complex constructs that pick out spatial regions and trajectories. These classes, elaborated later in this chapter, are the `Place` and `PathConstraint` which correspond to the static and dynamic cases of the linguistic semantic category `GeneralizedLocation` as introduced in Chapter 5. Such complex spatial entities play an important role in parameterizing the capabilities that the agent can perform, but do not have an explicit static presence within the $\mathcal{RS}$ layer.

The distinction between $\mathcal{CS}$ and $\mathcal{RS}$ layers is motivated both by pragmatic and ontological concerns. Namely, by separating out the geo-spatial properties of individuals from the non-spatial we can make use of the most appropriate reasoning system for the given information type, e.g., ontological reasoners for non-spatial category information, and explicit spatial reasoning techniques for the spatial properties of the model – such a distinction also follows from a more principled organization of spatial content for robotic systems as argued for by Bateman and Farrar [2005].

7.2.3 Graph Space

Unlike $\mathcal{RS}$ and $\mathcal{CS}$, $\mathcal{GS}$ is principally oriented towards the needs of robotic navigation and localization rather than communication. From an object perspective, $\mathcal{GS}$ is defined as a 2-tuple:

$$\mathcal{GS} = \{\mathcal{DP}, \mathcal{S}\} \tag{7.2}$$

where $\mathcal{DP}$ is the set of *Decision Points*, and $\mathcal{S}$ is the set of *Segments* between Decision Points.

Following Denis [1997] and others, Decision Points do not correspond to space occupying abstract or physical entities, i.e., `SpatialEntity` in the ontology presented in the last section, but are instead points of significance in route navigation space which may have a projection onto entities in $\mathcal{RS}$. Thus Decision Points can be co-located with one or more `SpatialEntities` and there is no one-to-one relationship between `SpatialEntities` and Decision Points. The projection of Decision Points onto physical space is made manifest through each Decision Point being assigned a location on the *Region Space's* Cartesian plane. Decision Points may be connected in the usual way via *Segments* which express navigability between Decision Points. Segments are defined in terms of their two connected Decision Points, and have a length corresponding to the direct distance between the two Decision Points.

Following Denis [1997] and Krieg-Brückner et al. [2005], an earlier version of $\mathcal{GS}$ also included $\mathcal{L}$, a set of landmarks, and $\mathcal{VR}$, a set of visibility relations that hold between Decision Points and Landmarks. Landmarks were defined as being visually or functionally salient at particular Decision Points in an environment. The set of Landmarks was thus a subset of the set of `SpatialEntity` instances known to the agent. Such a model has been found useful in the generation of naturalistic route instructions, see e.g., Richter [2008]. However, in these cases, unlike in the model presented here, actual visual salience information is not available due to the lack of a metric space model. The graph-based assertion of visibility is thus required as a substitute, but is typically hard-coded, established only for a number of toy landmarks, and difficult to scale over large spatial models because of the need to assert a prohibitively large number of relationships between decision points and landmarks. The approach taken here assumes that visibility relations should not be hard-coded in advance, but determined on the fly like any other salience phenomena.

7.2.4 Illustrated Example

To illustrate the Navspace model, Figure 7.8 depicts a simplified office environment, along with the three spatial representation layers for that environment. The office environment, which is used in simulation studies described later in Chapter 8, is illustrated with annotated rooms and landmarks in Figure 7.8(a); note here that the dark area in the centre of the environment is free space. Figure 7.8(b) shows the *Graph Space* layer which indicates navigable paths through the environment. Figure 7.8(c) in turn depicts an extract of the *Concept Space* model for the same environment. Finally, Figure 7.8(d) depicts the *Region Space* layer for the same environment; note here the overlapping of non-solid entities with both each other or with solid entities.

Figure 7.8 also indicates the nature of the relationships that can hold across elements of the different layers. Namely, we assume an individual in the $\mathcal{CS}$ layer occupies a region described by the $\mathcal{RS}$ layer. Similarly, a decision point in the $\mathcal{GS}$ layer can be thought of as having an access relationship with elements of the $\mathcal{CS}$ layer by virtue of the co-location of that Decision Point with a particular region in the $\mathcal{RS}$ layer.

7.2.5 Discussion

As a hybrid representation of space, the Navspace model has similarities to a number of other layered spatial representations. In particular, the $\mathcal{GS}$ layer has similarities to other graph-based accounts of spatial representation and navigation, e.g., Bos and Oka [2007], and in particular draws on Krieg-Brückner et al.'s [2005] abstract Route Graph model along with Bateman and Farrar's [2005] ontological reorganization of that model. One notable difference between the $\mathcal{GS}$ model and Krieg-Brückner's Route Graph formalism is that following Denis [1997] and Kuipers [2000], and general principles of ontological organization of space [Bateman and Farrar, 2005], the notion of Route Graph *place* or *Decision Point* is strictly zero-dimensional and does not import notions such as reference system from associated regions. Moreover, in

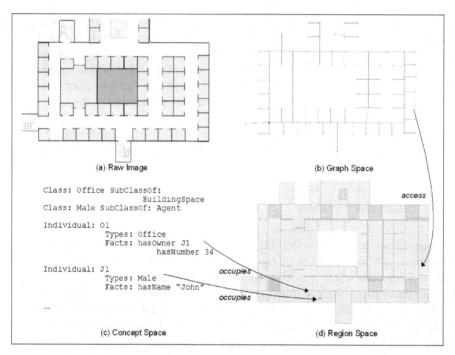

Fig. 7.8 Illustration of spatial representation layers.

the current model, graph nodes are not in themselves assigned common-sense types which correspond to entities in the $\mathcal{CS}$ layer. Rather graph nodes are simply decision points due to the particular navigation possibilities afforded by spatial arrangements.

With its assumption of ontologically distinct representation layers, the Navspace model also shares some methodological approach and structure with Kuipers's [2000] Spatial Semantic Hierarchy (SSH). Kuipers' SSH makes use of five representation levels: (a) the *metrical* level which provides a 2-D geometric map of the environment; (b) the *causal* level which uses the situation calculus to capture causal relations among the robot's views (symbolic abstractions over perception data) brought about by actions; (c) the *topological* level which defines places, paths and regions and the connectivity and containment relations that hold between them; and (d) and (e) the *sensory* and *control* levels which are sub-symbolic interfaces to sensory and behavioural capabilities. The greatest similarities between the SSH and Navspace models lies in the rough correspondence between Navspace and the SSH's metrical and topological layers. However, there are also many differences between the models. For example, unlike in the Navspace model, the SSH, motivated by human cognitive plausibility, eschews global quantitative information in favour of a qualitative topological organization of global space. Moreover, unlike the Navspace model, the SSH, through its causal, control, and sensory layers, provides a complete robot control architecture rather than purely a spatial representation. In itself the Navpsce representation model has fewer ambitions, and a comparison of the SSH model to the work presented in this book would necessarily also include the robot control architecture elements briefly discussed in the context of the Navspace application introduced in the next chapter.

While the Navspace model is modest in its architectural ambitions in comparison to the SSH, the Navspace model was expressly conceived to provide a suitable representation for communication tasks. However, before we introduce the interpretation process itself, and explain how this interpretation process relates to the agent's spatial knowledge, we must first introduce an inventory of available parameterizable capabilities.

7.3 The Action Inventory

Verbal route descriptions chiefly define those actions that an agent must perform in moving from a source to destination. A route interpretation process thus requires an account of those actions. The notions of action as used in the dialogue agent have already been introduced in Section 6.1.1. Here, we return to the topic of the agent's capability inventory, and outline the range of available actions, the constituents of those actions, and their likely effects.

Although the need for an action inventory is indisputable for any agent-oriented interpretation model, there is considerable variance not only in the range of actions available to an agent, but also in the particular effects that these actions are judged to have. Here we will make use of an action inventory derived from the linguistically motivated organization of situated action presented in the Generalized Upper Model [Bateman et al., 2008]. The derived model is however rationalized for the practicalities of behaviour modelling and execution in a limited domain.

The Generalized Upper Model (GUM) (See Section 5.3.1.1) provides a detailed ontological account of action and spatial meaning organization as indicated by the lexical and grammatical organization of language. Thus, it can be said that GUM provides a view on speakers' construals of spatial and action types including their constituents. Such a theoretical perspective provides an ideal basis for deciding upon actions that an agent should be capable of performing, even if that performance itself is implemented by way of lower-level actions and behaviours that have no natural correlate in language. In its description of spatial language, GUM offers several categories which are particularly relevant to the semantic description of verbal routes. Specifically these include: (a) the `NonAffectingDirectedMotion` configuration which describes spatial motion that makes use of path elements; and (b) the `OrientationChange` configuration that describes spatial motion that requires a change in the agent's orientation.

More importantly though, GUM goes beyond enumerating configuration types by describing these configurations in terms of their linguistically motivated constituents. Crucially, this methodology leads to the use of the `GeneralizedLocat-`

ion class as a concrete semantic reflex of the prepositional phrase. In other words, the GeneralizedLocation, consisting of a referenced object or *relatum*, and a spatial relationship descriptor or *spatial modality*, provides a semantic reflex for a prepositional phrase such as "at the junction", "along the hallway" and so forth. A GeneralizedLocation in turn then plays roles such as *placement* or *route segment* as appropriate in many different configuration types. This approach contrasts with traditional views on spatial term modelling where the preposition is a reflex of an explicit spatial relationship between two entities (in locating expressions) or between an entity or a process (in motion expressions). The advantage of this alternative approach is that the notion of giving meaning to a prepositional phrase in isolation as opposed to its involvement with other entities is given a natural foundation. That is, we can talk about the meaning of prepositional phrases such as "at the junction" or "along the hallway" in and of themselves. While there may of course be multiple interpretations of a given prepositional phrase which can only be resolved to a single meaning following composition with a process, the semantic organization offered by GUM offers a naturally compositional approach to contextualization.

However, as indicated at the beginning of Chapter 4, GUM is optimized for use in semantic interfaces to wide-coverage grammars rather than behaviour modelling languages. As such, GUM is not necessarily ideal for describing the contents of the action inventory directly. Of GUM's organizational features, one choice particularly influences this view. Namely, rather than making use of a process as the central organizational unit, GUM's central unit is the frame-like *configuration*. Like a process, the configuration picks out semantic constituents, but one of these constituents is itself typically a process. Thus, the representation of even a simple motion in GUM includes both a configuration and process type.[2] Working with both configurations and processes within the action description and execution model introduces an unnecessary complication that we will avoid.

In light of the above, a small inventory of actions broadly in keeping with GUM's organization – particularly at the level of prepositional phrase semantics – will be used for our current purposes. Figure 7.9 gives a taxonomic view of this inventory for a simple service robot capable of performing a number of movement operations. Though this inventory shares motivation with some computational route instruction primitives introduced in Section 7.1.2, we will see that the particular range of primitives - and crucially how they are modelled - is somewhat different to these works. However, while this inventory is intended to be useful for a range of basic tasks, including interpreting verbal routes, no claims of exhaustiveness or sufficiency are made here beyond what is required of the inventory by this chapter. In the remainder of this section, a number of these actions will be expanded upon in terms of their function and ontological constituents.

---

[2] The reasoning behind this approach is intended to support situations where processes have constituents which differ in some way from the constituents of the configuration as a whole [Halliday and Matthiessen, 1999, pp. 54-57].

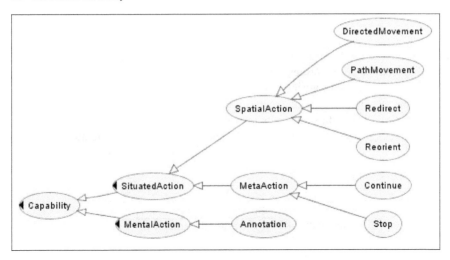

Fig. 7.9 Taxonomical view of the capability inventory

7.3.1 Spatial Actions

By far the most prominent action type in verbal route descriptions are those actions which move the agent through its environment. In the current model all such actions are subsumed under the `SpatialAction` type which defines a spatial action as beginning at a specific time and place, and being performed by a particular agent.

7.3.1.1 Path Movements

The most important of the `SpatialAction` types, the `PathMovement`, changes the agent's placement within its environment via a path following behaviour.

Within the context of route descriptions, movements are most often described in terms of a path trajectory constraint that expresses motion with respect to a reference entity. In English this is typically marked by prepositional phrases. Examples of instructed movement descriptions making use of path constraints include "drive along the corridor", "go to the church", and "go up the screen". Trajectory constrained movements can also be marked explicitly by a small number of verbs that describe paths of motion, e.g., "enter the corridor" or "leave the office". Broadly following the methodology adopted for the Generalized Upper Model, we model individual trajectory constraints in terms of a `PathConstraint` entity which captures both the referenced object, a path modality which describes the path constraint with respect to the referenced entity, and an extent, i.e.,

Definition 9. `Class: PathConstraint`
                 `SubClassOf: ComplexSpatialEntity`
                     `and (relatum exactly 1 EnvironmentThing)`

```
and (modality exactly 1 PathModality)
and (extent some PhysicalExtent)
```

where the class of concrete entities is any object that can be referenced in the world
– including both physical entities and socially constructed objects.

A path modality is the conceptual semantic reflex of a path preposition that is assigned a concrete situated meaning in the current model. Table 7.1 presents the set of
path modalities identified for the route interpretation task. Although the set is comparatively small, it is sufficient to demonstrate the principles of the contextualization
process. Path-like constructions that are not directly accounted for mostly consist of
projective relations used in a dynamic context, e.g., "walk over the bridge".

Modality	Example Realizations
Destination	Go to the second roundabout
	Enter that room
Source	Drive from that room
	Leave the car park
Approach	Head towards the university
	Approach the main junction
Pass	Walk past the bank
	Pass the entrance
Follow	Go along the corridor
	Follow the road
	Go up the corridor
	Drive down that road

Table 7.1 Available dynamic modalities with example realizations.

Path constraints are frequently modified by extent information. The purpose of
the extent information is to clarify the structure of the path where the use of modality and relatum alone is not sufficient. In the case of all path relations, the addition of extent information serves to modify the distance to be travelled while the
modality itself is used to indicate the overall trajectory structure. We define the class
`QuantitativeExtent` to model possible extent modifications for physical actions.

As a subclass of the `SpatialAction` type, the `Motion` type inherits properties common to all spatial actions. In particular, we assume that a `Motion` is performed with some speed, by some agent, beginning at some point of time. More importantly for our model here, we also assume that any motion is performed starting
at some place. This place is captured by the `SpatialAction` definition through
the `placement` role which is filled by an instance of `Place`. Moreover, following
the Generalized Upper Model, we capture a place, like a path constraint, in terms
of a referenced object and some spatial modality that identifies a generalised region
with respect to that referenced object. Ontologically, we define the `Place` class as
follows:

Definition 10. `Class: Place`
`                    SubClassOf: ComplexSpatialEntity`
`                        and (relatum exactly 1 SpatialEntity)`
`                        and (modality exactly 1 StaticModality)`

The range of static spatial prepositions that can be used to define a place is quite diverse, and far beyond the scope of this book to define and give situated meanings to. Thus, we limit the model to a set of three static modalities that can be used to define places. These three modalities are listed in Table 7.2 along with sample realization prepositional phrases. It should be noted that though prepositions such as "before" and "after" are most correctly viewed as temporal prepositions which express a process or event with respect to some other process or event, e.g., in the sentence "after you see the church turn left", these prepositions also have an equivalent static effect where the spatial area activated by these prepositions is determined by the area covered by a process up until the reference point, e.g., consider the spatial area carved out by the prepositional phrase in the utterance "after the kitchen turn right".

Modality	Example Realizations
Containment	in the kitchen
	at the junction
Precedence	before the church
Succession	after the bank bank

Table 7.2 Available static modalities with example realizations.

Folding in inheritance from the `SpatialAction` class, we can summarize the signature of a `PathMovement` as follows:

Definition 11. `Class: PathMovement`
`                    (actor exactly 1 Agent)`
`                and (placement exactly 1 Place)`
`                and (trajectory exactly 1 PathConstraint)`
`                and (speed exactly 1 AbstractSpeed)`

7.3.1.2 Directed Movements

In addition to path constraints, movements can also be characterised in terms of a generalized direction of movement. Directions of movement are typically realized by adverbs, e.g., "drive forward", and are frequently used when more specific path information is not available. In the current account, we model such directed movements with the distinct `DirectedMovement` type. The primary reason for choosing a distinct action type, rather than allowing a `PathMovement` to be parameterized in terms of a direction, is that this aids the development of straightforward low-level behavioural models. It should be noted that this same design criterion was applied in the definition of the `PathMovement` type – thus meaning that, unlike in the case of the linguistic semantic `NonAffectingDirectedMotion` described in Chapter 5, a complex path description is modelled at the conceptual level as a complex of simple `PathMovement` instances rather than a single instance with a complex internal structure.

The direction of movement is modelled in a `DirectedMovement` instance through the `moveDirection` property which is in turn filled by an instance of the `Direction` type. The `Direction` captures an actual projective or cardinal direction modality type. In summary, the class of cardinal modalities includes the conceptual semantics correlate of terms such as "north", or "south west", while the set of projective modalities includes the conceptual semantic correlate of terms such as "left", "forward", or "up". Rather than assuming that all projective terms are defined with respect to an egocentric perspective, we must account for the possibility of allocentric reference frame use. Thus, within the current model, static spatial relations including cardinal and projective relations are defined with respect to a given perspective. While the perspective for cardinal relations is fixed on an allocentric perspective, assigning perspective for projective relations is a more complex process that we shall return to later in this chapter.

The operational model of the `DirectedMovement` requires that the agent first turn in the stated, but contextualized direction, and then begin moving straight in that direction. The forward movement is modelled internally as a complex behaviour which will continue a path-following behaviour until a unique suitable forward path segment cannot be identified.

We can summarize the signature of the `DirectedMovement` action type as follows:

Definition 12. `Class: DirectedMovement`
>   `(actor exactly 1 Agent)`
>   `and (earliestStartTime exactly 1 Time)`
>   `and (placement exactly 1 Place)`
>   `and (moveDirection exactly 1 Direction)`
>   `and (speed exactly 1 AbstractSpeed)`

7.3.1.3 Reorienting

A `Reorient` action turns the agent about its vertical rotation axis resulting in a new orientation for the performing agent. Although a reorient behaviour is not typified by a change in location, in practice a reorient can result in a location change either because of the drive controls of the moving agent, or because the underlying reorient behaviour automatically aligns with salient pathways. The ontological signature of the `Reorient` action has already been introduced in Section 6.1.1.1.

Reorients are most frequently characterized in terms of a direction towards which the agent should turn, e.g., "turn right", or "turn northwards". We model directions in the `Reorient` class in terms of a `direction` property which is filled by a `GeneralDirection`. The `GeneralDirection` defines the direction of movement which is one of either a cardinal or projective modality. A turning motion, and hence the GeneralDirection, can also be modified through the inclusion of quantitative extent information, e.g., "turn 90 degrees to the left".

7.3.1.4 Redirections

As noted in Section 6.1.1.1, the Reorient action may only be invoked when the agent is stationary. For the case of an agent that is already moving, we introduce a fourth action: Redirect. The distinction between Redirect and Reorient actions is motivated partly by the operational differences between turning on the spot versus effecting a turn while following a path. The distinction is also motivated in part by linguistic differences in how turns while stationary versus turns while moving are construed by language. Namely, whereas the construction "turn right" can be used to indicate either a reorientation (if the interpreting agent is stationary) or a redirection (if the interpreting agent is in motion), path oriented turning constructions such as "take the second right" can never be used to indicate a turning of the agent on the spot having been in a stationary state.

At the level of ontological signature, the Redirect and Reorient share the same features. It should be noted though that following performance of the orientation change, a Redirect spawns a forward directed DirectedMovement instance. We summarise the signature of the Redirect action as follows:

Definition 13. Class: Redirection
                SubClassOf: SpatialAction
                    and (actor exactly 1 Agent)
                    and (earliestStartTime exactly 1 Time)
                    and (placement exactly 1 Place)
                    and (speed exactly 1 AbstractSpeed)
                    and (direction exactly 1 GeneralDirection)

7.3.2 Mental Actions

In addition to physical actions, route descriptions frequently entail the performance of non-physical actions. These non-physical actions are manipulations of the agent's mental state or checks of the agent's spatial model or perceptual field. We model such actions in terms of the MentalAction type which simply requires that the mental action be performed by some agent.

7.3.2.1 Placement Assertion

One mental action type considered in the current model is the assertion of a placement that signals the end of a monologic route instruction, e.g., as might be realized by "the university will be to your right". We introduce the AssertPlacement action to capture this phenomenon.

The purpose of a placement assertion is to inform the agent of the location of the originally targeted destination. Following the modelling approach used earlier in the linguistic semantics, the `AssertPlacement` type captures this final destination through the use of a `locatum` role filled by a `SpatialEntity`. Unlike the case of concrete entity use in motion instructions, it is assumed that the interpreting agent is not aware of the locatum's location in advance. Thus, it is not necessary that a signalled relatum be resolved in advance of the actual location assertion.

As in the case of motion descriptions, the actual location is asserted through the use of a `placement` role filled by an instance of `Place`. The signature of an `AssertPlacement` can thus be summarized as follows:

Definition 14. `Class: AssertPlacement`
  `        SubClassOf: MentalAction`
  `        and (locatum exactly 1 SpatialEntity)`
  `        and (placement exactly 1 Place)`

7.3.2.2 Placement Verification

A second non-physical act seen frequently in route instructions is the verification of location as typically realized by phrases such as "after the junction you will see a postoffice". As noted by Denis and Tversky, these constructions are used to provide confirmation information rather than to specifically alter the path to be taken by the agent. We introduce the `VerifyPlacement` action to capture this phenomenon, and, like the `AssertPlacement` action, the signature of this class is defined in terms of a `locatum` which signals the object which should be observable from a place that fills the `placement` role. The signature of a `VerifyPlacement` can be summarized as follows:

Definition 15. `Class: VerifyPlacement`
  `        SubClassOf: MentalAction`
  `        and (locatum exactly 1 SpatialEntity)`
  `        and (placement exactly 1 Place)`

7.3.3 Meta Actions

A number of meta-actions are also frequently applied in human-robot or human-vehicle interaction where motion control is shared between user and system. Although this class of action is not frequently used in *in-advance* route instruction descriptions, they are frequent in *online* route instructions.

`Stop` is a meta-action which can be used to suspend an active process. The most important parameter, and hence signature constituent, of the `Stop` type is the process which is to be affected. While this process may be indicated explicitly in language, e.g., "stop turning", the process is often omitted under the assumption that the process to be stopped is obvious, i.e., as in the single word utterance "stop!". Moreover, as a subtype of `SituatedAction`, a `Stop` typically has a specific spatio-temporal location at which the suspension of the referenced action is to be effected. In the case of route type instructions, this is often signalled explicitly in language, e.g., "stop after the church".

Following the explicit suspension of an action, it is typical for a speaker to explicitly invoke a new action to replace the suspended action. However, in some cases speakers may suspend an action with the intention of having that action continued. We thus introduce a second meta-action, `Continue`, which, like the `Stop` action, is parameterized principally by a process that has been suspended. We can summarize the signatures of the `Stop` and `Continue` actions as follows:

Definition 16. `Class: Stop`
        `SubClassOf: SpatialAction`
        `and (controlledProcess exactly 1 SpatialAction)`

Definition 17. `Class: Continue`
        `SubClassOf: SpatialAction`
        `and (controlledProcess exactly 1 SpatialAction)`

7.3.4 Discussion

This section has described the inventory of actions that the route interpretation model depends upon. From the perspective of route instruction interpretation, these actions are the basic constituents of route instructions, and as such describe the actions an agent must perform in following a typical route instruction. From the perspective of the AODM dialogue management model presented in the last chapter, these actions constitute the interface to the domain model and the content of contextualizable dialogue moves.

Although the inventory presented has similarities with the action modelling assumptions made by MacMahon [2007], Levit and Roy [2007], and Bugmann et al. [2004], there exist subtle but important differences between those action sets and the inventory presented here. Most significantly, while both the current action inventory as well as those presented by MacMahon [2007], Levit and Roy [2007], and Bugmann et al. [2004], were broadly linguistically motivated, the model developed here attempts to hold more tightly to spatial action description without becoming too specific with respect to action types. Namely, in comparison to the model suggested by Bugmann et al. [2004], a more abstract and consistent action inventory has been adopted here. Thus, rather than including highly domain-specific actions such as, for example, a specific capability for entering a roundabout, the inventory here is more generic, and thus applicable to a wider range of domains. In contrast, whereas Levit and Roy [2007] require that linguistic constructions other than simple instructions be broken down into a sequence of very fine-grained NIUs, the current model adheres more tightly to the surface form, and thus allows a complex verbal action description to be encapsulated within an individual capability instance. Similarly, whereas MacMahon [2007] extrapolates circumstantial placement into distinct actions during the interpretation process, here such placement information remains within the original action description.

By modelling actions close to the surface form, but in a generic way, the approach taken here binds the level of capability modelling relatively tightly to the semantics-pragmatics interface. While this partially breaks a tenet of modularity, the principle motivation behind this stance is that those capabilities, and resultant intentions, can thus be more easily integrated with a dialogue capable language contextualization process. Moreover, since the development of actions at this level of granularity is typically in the service of a user interaction layer, and since these behaviours already abstract considerably above low-level robotics behaviours, we argue that there is little reason not to model these actions relatively tightly to the level of linguistic semantics. These coarse grained actions can, and do, become expanded out into constructions of finer-grained robotics-oriented actions below the level of capability – but, to the most part, this highly domain-specific level of representation is not necessary for verbal human-machine interaction. That said, the capability inventory included here is of course illustrative, and further investigation of capability abstraction for communicative behaviour remains open for future investigation.

7.4 Dialogue Move Contextualization

The model of route interpretation pursued in this chapter assumes that simple spatial actions are the principle units of route descriptions. In the last section we introduced a number of spatial actions which we argue can be used in following routes. While these actions were described in terms of required parameter types and the basic operation of these actions, it was assumed that the required parameters for a specific action instance have unambiguous meaning within the situational context when the action is invoked. This condition in turn requires that the dialogue move that requests a particular action be specific with respect to the parameters taken by that action. In the case of situated language interpretation, such specificity requires that the values for dialogue move parameters be collected from either linguistic content or the situational context as appropriate.

In Section 6.4 we introduced a generalized contextualization process for the resolution and augmentation of parameters in dialogue move instances. The general idea of that contextualization process was that for a given user move, domain-specific contextualization functions are applied to dialogue move parameter slots in order to provide concrete values for these parameter types. Specifically, if content was provided by a user for a given move parameter, ψ, a resolution function, γ_ψ, must be applied against the user provided content and situation state to provide a concrete parameter ψ'. Similarly, if no content was provided by the user for a given slot, an augmentation function, δ_ψ, must be applied against the situational state alone to generate possible values for that parameter. The general form of these contextualization functions were provided by Equations 6.14 and 6.15.

While the model presented in Section 6.4 does provide a general framework for contextualization, no specific augmentation or resolution functions were provided to instantiate the model for a given domain. Therefore, in this section we will instantiate the generalized contextualization process for the types of action introduced in the last section. This instantiation thus allows sequences of route describing dialogue moves to be given context specific meaning. The aim of this section is not however to provide a complete inventory of all contextualization functions for each dialogue move type, but rather to provide an illustration of the approach, and sufficient detail to provide worked examples of the route interpretation process.

Instantiation of the generalized process requires the development of appropriate resolution functions to resolve content provided by speakers, and augmentation functions to model that information omitted by the user. Considering for a moment the types of actions and parameters described in the last section, the instantiation of resolution functions aims to answer questions like 'what are the meanings of the semantic correlates of "right", "at the kitchen" or "along the corridor" in a given situational context?'. Similarly, the instantiation of augmentation functions aims to answer questions like 'given the current situation of the interpreting agent, in which direction should it turn if the user simply asked it to "turn" without specifying a specific direction?'.

We proceed by describing contextualization functions for the augmentation and resolution of a number of the intrinsically spatial roles introduced for the capability types listed in the previous section. The spatial placement of a physical action is a vital role that sets out where the action is to be performed, and hence constrains the interpretation of subsequent parameters. Therefore, the first set of contextualization functions that we will consider are for the `placement` role in Section 7.4.2. Then, in section 7.4.3 we address the `direction` role which is key to the parameterization of the `Reorient` and `Redirect` capabilities. Similarly, in Section 7.4.4 we will look at the specifics of the `trajectory` role as used by the `PathMovement` capability type. Though due to space limitations, we limit ourselves to discussing the contextualization functions for only these three role types, the principles of the developed models extend to the other spatial role types used in the last section. We begin by introducing the state variable type to be used in contextualizing dialogue moves in the Navspace domain.

7.4.1 The Navspace State Variable Type

As described in Section 6.4, the AODM model's generalized contextualization process assumes the existence of a state variable type that describes the state of the agent before, during, and following the performance of particular actions. Before we introduce the specifics of contextualization functions, we must first instantiate the state type, s, for the route interpretation model.

Given the importance of the agent's physical state to the interpretation of spatial language, we include both the agent's physical and mental state in the state description for the contextualization process. Specifically, we define the interpreting agent's state type as follows:

$$s = \{x, y, \theta, sp, Bel\} \tag{7.3}$$

where x and y are the position of the agent against the spatial model's metric *RS*
layer; θ is the agent's angular displacement measured counter-clockwise with respect to the positive x axis of that *RS* layer; sp is the agent's speed; and *Bel* denotes
the agent's non-dialogic mental state including the *NS* model itself (see Section 7.2)
and a more conventional belief state container. Instances of this state variable type
are both used to parameterize contextualization functions, and are used to populate
the resultant state distributions associated with resultant contextualizations.

7.4.2 Contextualizing Places

Having introduced the state variable type for the Navspace domain, in this section
we begin the illustration of the contextualization process for route instruction interpretation by considering the contextualization of `placement` roles as used in
spatial actions such as the `Reorient` and `PathMovement` types. Contextualization of placements involves either the assignment of default placements in the case
that the user did not explicitly indicate a specific placement, or the resolution of the
semantic `Place` category if a speaker did in fact provide a description of where an
action is to take place. We will begin by looking at this case of resolution before
moving on to briefly address the issue of placement augmentation.

7.4.2.1 Resolving Places

In Section 7.3.1.1 we defined a `Place` as a generalized region described in terms of
a referenced object and some spatial modality. In the context of the route interpretation process, the significance of place resolution is to identify the point at which
an action is to be performed. Following the general form established by Equation
6.14 for resolution functions, the signature of the placement resolution function is
simply:

$$Set(\{Place', r(Place'), Set(\{s', r(s')\})\}) = resolve_{placement}(Place, s) \tag{7.4}$$

where *Place* is the non-resolved filler of the `placement` role supplied by the
speaker, s is an input state description; *Place'* is a contextualized `Place` instance
including the concrete relatum and modality identified; and $Set(\{s', r(s')\})$ is the
resultant state distribution for a given place interpretation.

Given the function signature just introduced, we see that the resolution of a given
place in a given situation both picks out a resolved place instance, as well as a
set of states which describe the state of the agent were it to move to that location.
Moreover, given the constituents of a `Place` element, the process of resolving the
content of a `placement` role involves the resolution of a concrete relatum, the

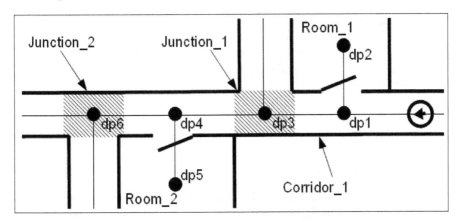

Fig. 7.10 Differing interpretations of "at the junction" for alternatives "turn left at the junction" and "turn right at the junction". Thick lines denote the walls in the environment, thin lines denote segments in the agent's *GS* spatial layer, circles denote decision points in the agent's *GS* spatial layer, and hatched areas correspond to the areas occupied by the junctions themselves in the agent's *RS* spatial layer.

identification of the place described with respect to that relatum along with the spatial modality, and the update of the agent's state to the region defined by the denoted place. In the remainder of this section, we concretely describe how these issues are dealt with by the placement resolution function.

Relatum Resolution

The resolution of relata used in place descriptions is a specific case of reference resolution. The resolution of referents is however a complex phenomenon that has been a staple of linguistics research across theories of grammar [Cowper, 1992, ch. 10], semantics [Saeed, 1996, ch. 7], and pragmatics [Yule and Widdowson, 1996, ch. 2] for decades. Although we will not attempt to provide a comprehensive model of reference resolution here, a basic model of resolution is key both to the situated dialogue domain and to the illustration of the dialogue move contextualization process.

Before outlining the specific approach to reference resolution applied here, let us first consider some factors that may influence a reference resolution process:

- **Intrinsic Features:**
  Most referring expressions encode information about the intrinsic features of a referent. In the ideal case, this information is a socially agreed upon proper name, e.g., "room A1024", but in most cases the information encoded may indicate only sortal features, e.g., "the cup". Moreover, non-semantic features may also be marked in the referring expression, and thus influence the resolution process, e.g., grammatical gender marking in pronouns.
- **Discourse Salience:**
  The reference resolution process is well known to be subject to models of dis-

course structure and salience. Numerous discourse based referring expression resolution mechanisms have been proposed ranging from syntax-oriented binding theory (see Cowper [1992] for an introduction), through the many discourse semantics oriented models introduced in Chapter 2.

- **Situational Salience:**
  So called *exophoric* references refer to objects in the environment rather than objects previously introduced through discourse. Thus, like discourse salience, proximal and visual saliency models also have a direct influence on the reference resolution process.
- **Affordances:**
  Not all referents of a type may be applicable to a task or description at hand. For example, referring to Figure 7.10, a turning to a particular direction may not be possible at a given *junction*. Thus, in Gibson's [1977] terminology, the *affordances* offered by an entity can serve as an important distinguishing characteristic in reference resolution.

An ideal reference resolution process should seamlessly integrate all such factors into a single model. However, given our focus here on aspects of situated language use, the reference resolution strategy which we have applied focuses on the interplay between models of situational salience and intrinsic features. The influence of affordances will also be considered, but later in the context of complete move contextualization where the contextualization of subsequent parameters can be applied as a filter over those places identified by the more fundamental place resolution strategy.

In the following we describe the simple model of exophoric reference resolution which we have applied to relatum resolution against the Navspace model.

Intrinsic Feature Matching: Before accounting for the influence of situational salience on object resolution, the applied resolution process must first acquire a situation independent resolution of references. As introduced earlier in the chapter, the Navspace model's *CS* spatial layer captures the intrinsic features of entities known to the interpreting agent. Thus, the reference resolution process makes use of this *CS* layer, backed up in implementation by a Description Logic reasoner, to directly resolve a base set of entities that match an existentially qualified referent description. Straightforwardly, we summarize this as follows:

$$Set(SpatialEntity) = query(CS, Descriptor) \qquad (7.5)$$

where $Set(SpatialEntity)$ is the set of `SpatialEntity` objects in the *CS* layer that have been judged to meet the *Description* supplied by the user as filler of a `Place`'s `relatum` role.

Salience: The static query function just defined will return all objects of a particular type known to the Navspace model. To refine the identification of referents that are not uniquely defined in terms of intrinsic features, a spatial salience measure is applied. The simple model of salience used here is based on four principles:

- **Perceptibility:** Objects which are directly perceptible from the situated agent's current location are more salient than objects not perceptible.

- **Focus:** Objects in the direct field of view, and in particular in the centre of that field of view, are more prominent than those on the periphery. The periphery in this case is assumed to include the entire rotational axis of the agent.
- **Proximity:** Objects closer to the agent are more salient than objects further from the agent.
- **Quantity:** Objects which take up a larger field of view are more prominent than those with a smaller field of view.

Given these four principles, we can build a model of object salience that operates against the concrete description of entities in the Navspace model's *RS* spatial layer. First, working under the assumption of the 2D spatial model used in the *RS* layer, we define a focal prominence measure, $f(\theta)$, as follows:

$$f(\theta) = e^{-\left(\frac{\theta^2}{A}\right)} \tag{7.6}$$

where θ is an angular displacement measured with respect to the agent's anterior sagittal plane, and A is a distribution constant selected experimentally to value 150 to provide a relatively wide acceptability range for angles. $f(\theta)$ thus assigns to a given angle a relative importance of objects located at that angle over the range $0 \leq f(\theta) \leq 1$. Figure 7.11(a) depicts the plot of this function over the domain $-180° \leq \theta \leq 180°$.

As indicated, we assume that objects which are directly perceivable by the agent are more salient than those that are obstructed by other solid objects in the environment. We thus introduce the function r, which for a given object, obj, and angle, θ, returns a scalar weighting value that varies based on whether a view of the object is obstructed at the given angle:

$$r(obj, \theta) = \begin{cases} 0.1 & \text{object view obstructed at } \theta \\ 1.0 & \text{object view not obstructed at } \theta \end{cases} \tag{7.7}$$

where the scalar values of 1.0 and 0.1 were empirically found to be sufficient for the salience model. While this is one possible set of values for these weights, we make no claims that these values, nor the salience model in general, have any empirical basis with observed human behaviour. Empirically investigating the factors which influence a salience measure where interlocutors share a survey view on an environment remains a question open for future work.

Given $f(\theta)$ and $r(obj, \theta)$, we define the salience S of an object, obj, as follows:

$$S(obj) = \sum_{\theta=\theta_a}^{\theta_b} (f(\theta) \cdot \delta(obj, \theta) \cdot r(obj, \theta)) \tag{7.8}$$

where θ_a is the angular displacement of the first perceivable part of an object from the agent's centre axis; θ_b is the angular displacement of the last perceivable part of an object from the agent's centre axis; and $\delta(obj, \theta)$ is a function which returns the distance between the agent and the surface observed at θ. Figure 7.11(b) illus-

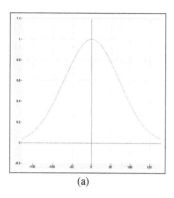
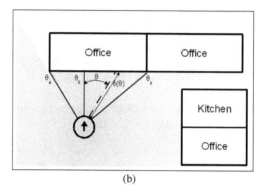

(a)   (b)

Fig. 7.11 Calculating the salience of objects. Figure 7.11(a) illustrates the focal prominence measure $f(\theta)$ over the domain $-180° \leq \theta \leq 180°$. Figure 7.11(b) provides an illustration of the values used in calculating the salience measure S for a given object.

trates the relationship between θ_a, θ_b, θ, $\delta(obj, \theta)$, and $r(obj, \theta)$ for a given object relative to the situated agent.

Given the salience measure, we can now define a function, *resolveEntity*, which returns the set of objects with associated salience probability that match a given semantic *Descriptor* for a given situation, s, i.e.,

$$Set(\{Entity, P(Entity)\}) = resolveEntity(Descriptor, s) \tag{7.9}$$

where

$$P(Entity_i) = \frac{S(Entity_i)}{\sum_{j=0}^{\#Matches} S(Entity_j)} \tag{7.10}$$

Referring to the example illustrated in Figure 7.10, and assuming no other junction instances were available in the agent's spatial model, the *resolveEntity* function straightforwardly returns a set of two resolved junctions, i.e., Juncation_1 and Junction_2, along with associated probabilities such that $P(Junction_1) > P(Junction_2)$.

Modality Application

As indicated, Place instances are defined not only in terms of a referenced relatum, but also in terms of a spatial modality. The placement resolution function thus not only makes use of a relatum resolution model to identify a specific entity in the environment, but also makes use of spatial modality definition and application models. The range of modalities supported for place description have been introduced earlier in the chapter, and were summarised in Table 7.2. For each of those three modalities, a concrete spatial mapping function is provided within the reference resolution process that establishes the relationship between the modality as a symbolic description and the spatial layers of the Navspace model. In this section we briefly summarize the concrete application functions for each of these three modalities.

The application function for the `Containment` modality is defined in terms of the spatial footprint carved out by a relatum against the *RS* representational layer. Specifically, the resolution function for a `Containment` modality is defined as construing a place identical to the bound area of a given relatum. As outlined earlier, the bounds of an entity in the Navspace model are explicitly defined by the *RS* layer. Thus, referring to Figure 7.12, the place defined as being "in R2" is depicted as the light grey area.

Two dynamic modalities are also supported by the Navspace model for the construal of places. These modalities, `Precedence` and `Succession`, define regions of space constructed in part by a path from a given start position to a specified relatum. The specific regions constructed are a section of the path just prior to the relatum in the case of `Precedence`, and a section of a path that extends that first path in the case of `Succession`. To provide an interpretation model for these modalities we make use of the *GS* representation layer, a path planning algorithm that operates on that layer, and an assumption that rather than intersecting with a given relatum, the path carved out by these modalities extends towards the relatum but remains on the main path.

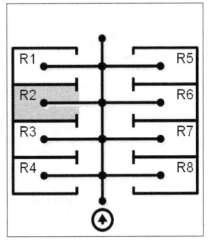

Fig. 7.12 Illustration of the application of the `Containment` modality to a particular environment entity.

Figures 7.13(a) and 7.13(b) illustrate the application of the interpretation functions for `Precedence` and `Succession` modalities applied against the relatum R2 from the position of the agent. As indicated, the modality interpretation functions first calculate a path from the input state variable to the Decision Point which lies closest to the relatum without being co-located with the relatum. In the case of the `Succession` modality, this path is then extended through a standard forward search behaviour by a distance proportional to the length of the relatum's average diameter. In both cases, a likelihood distribution function is then applied over those paths to provide an estimate of resultant positions.

The distribution functions, shown below, capture the core meanings of the `Precedence` and `Succession` modalities in the Navspace model:

$$f_{precedence}(d, B_A, s_0) = \begin{cases} \frac{d}{B_A} & s_0 \leq d < B_A \\ 0 & B_A \leq d \end{cases} \tag{7.11}$$

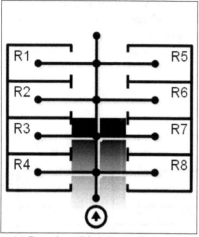 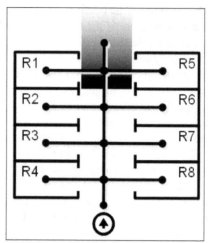

(a) Precedence Modality applied to R2   (b) Succession Modality applied to R2

Fig. 7.13 Illustration of the application of dynamic placement modalities Succession and Precedence to a particular environment entity. Specifically, the graded area indicates the area carved out between a given observer position and the referenced entity, R2. Area gradient figuratively reflects likelihood of positioning at that point.

$$f_{succession}(d, B_B, r) = \begin{cases} 0 & s_0 \leq d < B_B \\ \frac{-(d-B_B)}{3r} + 1 & B_B \leq d \leq (B_B + 3r) \end{cases} \tag{7.12}$$

where B_A is the projection of the relatum's initial boundary onto the path; B_B is the projection of the relatum's final boundary onto the path; s_0 is the path start; r is the average width of the relatum; and d is a distance along that path.

While Equations 7.11 and 7.12 provide a likelihood estimate along a line segment, the generalised interpretation of a Place in the Navspace model is however a region defined with respect to the relatum by the given modality. Thus, following identification of the path portion on the *GS* layer, the modality interpretation function maps this path back onto the *RS* layer through co-location of decision points on that path with regions in the *RS* layer. Thus, referring again to Figures 7.13(a), the conceptual semantics of the locative prepositional phrase "before R2" is captured by the placement resolution function as a potential field of possible locations which occupy the region between the agent and the given relatum.

Place Discretization

As mandated by Equation 7.4, the placement resolution function identifies concrete candidates for Place instances along with possible states of the agent which correspond to being at that place. As just described however, a resolved place may be thought of in terms of a potential field, and, thus, may result in a continuum of possible states for the agent – all of which can be classified as being conformant with a given placement. However, to limit resource use, instead of returning a state distribution which spans the entire definition of a given place, the final operation performed by the `Place` resolution function is to abstract over a place's footprint to the set of Decision Point locations that are co-present with the defined place. Specifically, for each Decision Point, the Cartesian coordinates of that Decision Point along with the most probable pose of the agent at that point (calculated through a path planning routine from the current value of s), are used to derive the resultant state distribution for each possible place interpretation.

To illustrate this final step, and the outcome of the `Place` resolution function, consider again the example of placement resolution presented in Figure 7.10. Table 7.3 summarizes typical results for the interpretations of two prepositional phrases against the situation depicted. Here, the resolution of the semantic correlate of "at the junction" with respect to the agent's pictured pose, results in two possible interpretations corresponding to `Junction_1` and `Junction_2` respectively. For each of these interpretations, a single resultant state value is associated. Such a state value includes the location of the decision point centred in that junction, along with the pose of the agent were it to move directly from its current state to that location. Similarly, the resolution of the conceptual semantic correlate of the locative prepositional phrase "before the junction" also results in two candidate `Place` interpretations. However, in such a case, while one possible state is associated with the `Junction_1` based interpretation, i.e., the state of being at `dp1`, three possible resultant states are associated with the `Junction_2` based interpretation, i.e., states of being at `dp1`, `dp3`, or `dp4`.

ψ	ψ'	$r(\psi')$	s'	$r(s')$

("at the junction")
Place
  relatum Junction
  modality Containment
	Place
	  relatum Junction_1
	  modality Containment
		0.84	x=300,y=100,θ=180,...	1.0

	Place
	  relatum Junction_2
	  modality Containment
		0.68	x=100,y=100,θ=180,...	1.0

("before the junction")
Place
  relatum Junction
  modality Precedence
	Place
	  relatum Junction_1
	  modality Precedence
		0.84	x=400,y=100,θ=180,...	0.89

	Place
	  relatum Junction_2
	  modality Precedence
		0.68	x=200,y=100,θ=180,...
			x=300,y=100,θ=180,...
			x=400,y=100,θ=180,...
				0.78
				0.51
				0.3

Table 7.3 Typical results of conceptual semantic resolution of the phrases "at the junction" and "before the junction" for the spatial situation depicted in Figure 7.10. Note that r values for both ψ' and s' refer to ratings and not probabilities over the set.

We thus see that placement resolution can do no better than identify two possible, but weighted, interpretations of "at the junction" for the situation in Figure 7.10. The return of these multiple candidates, along with their associated updated state values, by the placement resolution function results in the maintenance of two dialogue move solutions by the generalized contextualization process described in the last chapter. These two similarly likely possibilities can only be reduced by considering additional constraints in the interpretation of the complete dialogue move. In the case of the example presented in Figure 7.10, this cannot occur until the direction of turning is taken into account. We will return to this issue and the specific example in Section 7.4.3. First however, we briefly address the augmentation of placement information in the elided case.

7.4.2.2 Place Augmentation

If no explicit placement information is provided by a speaker for a given dialogue move, the generalized contextualization process invokes the placement augmentation function rather than the placement resolution function. Following from both the general form of augmentation functions as defined by Equation 6.15 in the last chapter, and the specific placement resolution function captured by Equation 7.4 in the last section, we define the signature of the placement augmentation function as follows:

$$Set(\{Place', r(Place'), Set(\{s', r(s')\})\}) = augment_{placement}(s) \qquad (7.13)$$

where s is an input state description; $Place'$ is a contextualized $Place$ instance including a relatum and spatial modality; and $Set(\{s', r(s')\})$ is the resultant state distribution for a given place augmentation.

Unlike the relative complexity of placement resolution, the placement augmentation function operates straightforwardly by assuming that the default place for performing an action is always derived directly from the current input state, s, of the agent, i.e., by default we perform an action where we already are. Given the initialization operation of the generalized contextualization process described in the last chapter, this input state can either be the actual physical state of the agent if the move under consideration is an atomic dialogue move contextualized in isolation; or, alternatively, the input state can be one of the resultant states of contextualization of the previous dialogue move if the current move is the non-initial element of a move complex.

While the placement augmentation function defaults to assuming that the input state captures the place at which the capability is to be performed, it is necessary to derive an appropriate $Place$ instance to provide an intentional level representation of the state. In an earlier version of the current model, presented in Ross and Bateman [2009], we made use of a special place instance $Here$ to denote this default at the symbolic level. While this is an appealingly simple way of capturing this default placement information, a more explicit and less ad-hoc approach to modelling the $Place$ instance allows a more descriptive representation of placement knowledge.

s	ψ'	$r(\psi')$	s'	$r(s')$

x=400,y=100,θ=180,.. (collocated with dp1)	Place
	  relatum Corridor_1
	  modality Containment
		1.0	x=400,y=100,θ=180,..	1.0

x=200,y=200,θ=270,.. (collocated with dp5)	Place
	  relatum Room_2
	  modality Containment
		1.0	x=200,y=200,θ=270,..	1.0

x=100,y=100,θ=180,.. (collocated with dp6)	Place
	  relatum Junction_1
	  modality Containment
		1.0	x=100,y=100,θ=180,..	1.0

Table 7.4 Typical results of conceptual semantic augmentation of the placement role for three spatial situations depicted in Figure 7.10.

Unfortunately however, in contrast to always defaulting to a special Here type, the generation of a more meaningful Place instance from a given situation can often result in a relatively large set of place descriptions. Consider for example how even with just the Navspace model's Containment modality, the state of being collocated with Junction_1 and facing to the left of the page in Figure 7.10 could be easily described by Place instances corresponding to either "in the corridor" or "at the junction". Indeed, if more static modalities were made available – including for example projective modalities corresponding to "left", "right", or "behind" – then a very large number of possible Place instances could be use to describe the same physical state.

A full model of automatic place description for a given state must thus overlap with many of the features of referring expression generation in the situated domain. Rather than pursuing such a complete model here, we have adopted a simplified model, which, for any given state, s, generates a Place instance consisting of the Containment modality along with the smallest EnvironmentThing entity which overlaps with the position entailed by s. To illustrate, Table 7.4 presents the results of place augmentation for three distinct input states against the situation depicted in Figure 7.10.

As with placement resolution, these results of applying the placement augmentation function against the input state, s, are used by the generalized contextualization process of Section 6.4.2 to update any maintained dialogue move solutions. The states associated with these solutions are in turn used as input states by the generalized contextualization process for the contextualization of subsequent parameters in the dialogue moves under consideration. In the next section we move on to consider such subsequent parameters by addressing the contextualization of direction information as used for example in Reorient and DirectedMovement instances.

7.4.3 Contextualizing Directions

Having described the placement contextualization functions in the last section, this section describes the set of contextualization functions for the `direction` dialogue move parameter. Within the inventory of activity types introduced in Section 7.3, the `direction` parameter plays an important spatial descriptive role in the `Reorient`, and `Redirect` activity types. As with the description of the contextualization functions for the `placement` role, the purpose of detailing the direction resolution and augmentation functions is to both illustrate how the AODM model's generalized contextualization process is instantiated for specific task-oriented dialogue types, and to illustrate how this contextualization process can in itself provide a straightforward approach to the interpretation of route descriptions.

7.4.3.1 Direction Resolution

The direction resolution function, associated with the `direction` role by the AODM model's generalized contextualization process, operates on the contents, if present, of a `GeneralDirection` semantics unit (see Section 6.1.1.1) to provide concrete values for direction modality and extent as well as the resultant agent state if the agent was to face that direction. Following the general form established by Equation 6.14 for resolution functions, the signature of the direction resolution function is simply:

$$Set(\{GeneralDirection', r(GeneralDirection'), Set(\{s', r(s')\})\}) =$$
$$resolve_{direction}(GeneralDirection, s) \quad (7.14)$$

where *GeneralDirection* is the non-resolved filler of the `direction` role supplied by the speaker, s is an input state description; *GeneralDirection'* is a contextualized `GeneralDirection` instance including a concrete direction and extent; and $Set(\{s', r(s')\})$ is the resultant state distribution for a given direction interpretation.

As indicated, a `GeneralDirection` consists both of a modality and an extent, and either or both of these pieces of information can be present in a surface form – and thus may be present in the derived partial dialogue move specification. The model presented in this section deals with either of those cases, whereas we handle the case of an entirely omitted `GeneralDirection` in the next section. We begin by considering the specifics of the direction model itself.

The Direction Model

The Navspace direction interpretation model supports sixteen projective and eight cardinal direction types; these types are summarized as three eight-point stars in Figure 7.14. We thus differentiate between a set of egocentric projective modalities and allocentric projective modalities. For the sake of brevity, the latter set shall be referred to as simply allocentric modalities, but it should be noted that both these and the cardinal modalities are defined with respect to an allocentric frame of reference.

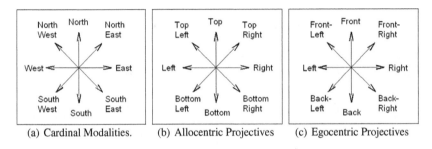

(a) Cardinal Modalities.   (b) Allocentric Projectives   (c) Egocentric Projectives

Fig. 7.14 Supported Direction Modalities.

We assign an idealised quantitative value to be used in the reasoning process. These values establish mappings between the symbolic and non-symbolic representation layers. Table 7.5 depicts these mappings. As noted, both cardinal and allocentric projective modalities are defined with respect to a single allocentric reference frame. This reference frame sees the angle 0° as pointing to the right of the page, with increments moving counter clockwise from that zero value. The quantitative mapping of egocentric modalities is dependent on the orientation of the interpreting agent, which we denote as $\theta°$.

The direction type mappings just provided are idealised definitions rather than definitions that can be used in practical circumstances. To provide more robustness in direction applicability, it is necessary to introduce acceptance functions which define whether a given orientation can be assigned to a specific direction type. One possible means of providing a more robust definition of specific direction types would be to assign these types to particular 45° segments of an eighth-plane (see for example Moratz et al. [2001] for a similar model). Such an approach can however be overly rigid. Instead, we will apply a graded half-plane acceptability function to describe possible mappings between direction types and actual spatial situations. Specifically, the graded acceptability function is based on three assumptions:

- **Idealized Projective Directions:** For each of the projective modalities, there exists an idealised quantitative value measured against the agent's *RS* layer as per the definitions provided in Table 7.5.
- **Angular Displacement Dependency:** The acceptability of a directional modality is maximal at the idealised angle and falls off proportionally towards 0 as angular displacement from the ideal increases towards an absolute 90°.
- **Radial Displacement Dependency:** The acceptability of a projective modality is maximal at the relatum and inversely proportional to the distance from the relatum.

Based on these three principles, we can define a generic projective modality acceptability measure as a radial potential field as follows:

$$f(\Delta\theta, d) = \frac{A}{d} e^{-\left(\frac{\Delta\theta^2}{B}\right)} \qquad (7.15)$$

Cardinal Modalities	Allocentric Projectives	Egocentric Projectives
North $= 90°$	Top $= 90°$	Front $= \theta°$
NorthWest $= 135°$	TopLeft $= 135°$	FrontLeft $= (\theta° + 45°)$ mod 360
West $= 180°$	Left$_{Allo}$ $= 180°$	Left$_{Ego}$ $= (\theta° + 90°)$ mod 360
SouthWest $= 225°$	BottomLeft $= 225°$	BackLeft $= (\theta° + 135°)$ mod 360
South $= 270°$	Bottom $= 270°$	Back $= (\theta° + 180°)$ mod 360
SouthEast $= 315°$	BottomRight $= 315°$	BackRight $= (\theta° + 225°)$ mod 360
East $= 0°$	Right$_{Allo}$ $= 0°$	Right$_{Ego}$ $= (\theta° + 270°)$ mod 360
NorthEast $= 45°$	TopRight $= 45°$	FrontRight $= (\theta° + 315°)$ mod 360

Table 7.5 Mapping modalities to idealised quantitative values

where $\Delta\theta$ is the angular displacement of a candidate angle, θ, with respect to the idealised angle for a given modality; d is displacement with respect to the relatum; A is a granularity constant; and B is an angular dispersion constant which defines how quickly acceptability decreases as $\Delta\theta$ increases. To illustrate, Figure 7.15 graphically depicts the application of the potential field based definition of the allocentric and egocentric left projection types against the agent's position between two objects.

Identifying Movement Candidates

While a user can in principle ask to turn or move in an arbitrary direction, here we assume that the route description task entails that any direction described by a user should point along a viable movement direction. Hence, rather than processing a direction without regard for the movement affordances offered by the environment, we make use of the agent's spatial model to naturally constrain possible direction interpretations. Within the Navspace model, allowed movement directions are captured by the *GS* representation layer. This layer thus provides a natural abstraction over available route navigation paths, and hence an evaluation of salient movement directions afforded by the environment.

However, to evaluate possible movement directions in the *GS* layer, we must assume that the agent has a pose at a particular decision point in that layer. Although the application of contextualization functions for various movement actions do typically result in the agent having a pose at a particular decision point, the Navspace model does not however inherently require the agent to be located directly at a decision point. Thus, before we can evaluate candidate graph segments with respect to a stated `GeneralDirection`, the direction resolution function first abstracts the agent's location to the decision point closest to the agent's actual location – whilst maintaining the same orientation. As indicated earlier, decision points have an associated catchment area, and we thus make use of this catchment area to identify the abstracted location.

Candidate Evaluation

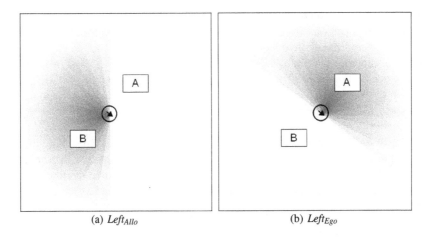

(a) *Left$_{Allo}$*   (b) *Left$_{Ego}$*

Fig. 7.15 Application of acceptability fields for two projective relations.

Given the agent's location at a particular decision point, the direction resolution function evaluates the applicability of segments connected to that decision point with respect to the `GeneralDirection` instance under consideration. This evaluation is principally made by way of the direction modality definitions listed in Table 7.5, along with the modality acceptability measure given by Equation 7.15. However, as indicated earlier, a `GeneralDirection` is defined in terms of both a direction and extent parameter, and either or both of these parameters can be present in a given instance. The contextualization process can thus be reduced to the evaluation of each connected segment for candidate direction modality and extent quantities. The evaluation process model for each of these three cases is briefly summarised below:

- **Modality but no extent:** The `GeneralDirection` contained a modality type but no extent information. Candidate concrete modalities are selected and used to evaluate each segment with respect to the ideal angle for that modality. Each segment is evaluated through the use of Equation 7.15 with $A = 1$ and $d = 1$, and $B = 3000$ to provide a high dispersion of acceptability. If more than one concrete modality is applicable to the incoming semantics, e.g., if both Left$_{Ego}$ and Left$_{Allo}$ may be applicable, then each segment is evaluated with respect to each potential interpretation. Since no extent information was available, extent is assigned to each direction interpretation based on the actual angular displacement of a given segment from the input state orientation.
- **Extent but no modality:** The `GeneralDirection` contained an extent value but no direction. The angular displacement of each candidate segment with respect to the state orientation is evaluated. Without direction, extent information is scalar. Thus, it is necessary to evaluate each segment both clockwise and anticlockwise with respect to the state orientation. To evaluate the applica-

bility of a given segment with respect to a stated displacement, we once again make use of Equation 7.15. However, we assume that when stated, angular information is more specific than directional. Thus, we set $B = 1000$ to provide a tighter acceptance distribution. For each segment it is possible to assign a specific egocentric modality. This is done by evaluating each segment with respect to each of the idealised egocentric projective modalities. The idealised modality which results in the highest acceptance value is used to augment modality information.

- **Modality and extent:** In the case of the `GeneralDirection` containing both extent and direction information, each segment is evaluated with respect to that information as per the individual cases just given. The total likelihood of a given segment is then provided as the average of modality and extent acceptability values.

To illustrate the direction contextualization process, Figure 7.16 provides three worked examples of `GeneralDirection` contextualization for the cases just outlined. For each of the three cases, contextualization is performed in the same situational context, i.e., with the agent at a decision point with four connecting segments while facing downwards (survey view). Specifically, Figure 7.16(a) graphically depicts the application of two modality functions for allocentric and egocentric interpretations of "left"; note that the best result is judged to be an allocentric interpretation of left corresponding to the direction of segment d. Similarly, Figure 7.16(b) on the other hand graphically depicts the application of the extent interpretation function in the resolution of a `GeneralDirection`; note that extent is evaluated with respect to both positive and negative deviations from the agent's forward heading. Finally, Figure 7.16(c) depicts the application of both modality and extent evaluation functions for a complete `GeneralDirection` instance. This analysis corresponds to a cross combination of the individual analyses of Figures 7.16(a) and 7.16(b). Table 7.6 summarizes the input values with resolved parameters and resultant states for a number of the cases depicted in Figure 7.16.

ψ	ψ'	$r(\psi')$	s'	$r(s')$
GeneralDirection modality Left	GeneralDirection modality Left-Allo extent 90	1.0	x=200,y=200,θ=180,..	1.0
GeneralDirection extent 45	GeneralDirection modality FrontLeft extent 45	1.0	x=200,y=200,θ=315,..	1.0
GeneralDirection modality Left extent 45	GeneralDirection modality Left-Ego extent 45	1.0	x=200,y=200,θ=315,..	1.0

Table 7.6 Typical results of conceptual semantic resolution of the direction role for the situations depicted in Figure 7.16.

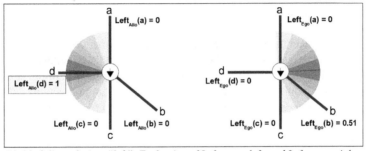

(a) *Modality only, i.e., "left". Evaluation of $Left_{Allo}$ on left, and $Left_{Ego}$ on right.*

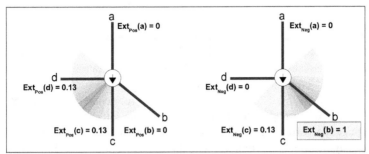

(b) *Extent only, i.e., "45 degrees". Evaluation of positive extent on left and negative extent on right.*

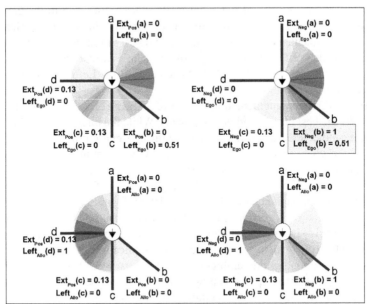

(c) *Modality and extent, i.e., "45 degrees to the left". The four figures depict combinations of the evaluation results from Figures 7.16(a) and 7.16(b).*

Fig. 7.16 Illustration of `GeneralDirection` resolution for three types against a situational context. Note the highlighting of the best result of resolution in all three cases.

7.4.3.2 Direction Augmentation

In the case that a speaker completely omitted direction information, a direction augmentation function is triggered by the AODM model's generalized contextualization process to determine the most likely orientation for the performance of the activity under consideration. Following the general form of augmentation functions defined by Equation 6.15, the signature of the direction augmentation function is as follows:

$$Set(\{GeneralDirection', r(GeneralDirection'), Set(\{s', r(s')\})\}) =$$
$$augment_{direction}(s) \quad (7.16)$$

where s is an input state description; *GeneralDirection'* is a contextualized General-Direction instance including a direction modality and extent; and $Set(\{s', r(s')\})$ is the resultant state distribution for a given direction augmentation.

In modelling direction augmentation, we assume that if a user utters a highly elliptical instruction such as "turn" to indicate that a reorientation is to be made, then the direction of turning to be made is either to the left or to the right of the interpreting agent's current orientation. This assumption is encoded directly into the direction augmentation function. As with direction resolution, the direction augmentation function makes use of the *GS* representational layer to abstract over reorientation possibilities afforded by the environment. Also as per direction resolution, the physical location and pose of the agent in *RS* is first abstracted to the closest decision point in *GS*. Segments leading from that decision point are then evaluated to determine their likelihood as reorientation candidates. Given our assumption of turnings being made to either the left or right of the agent's current orientation, these connecting segments are therefore both evaluated against the Left$_{Ego}$ and Right$_{Ego}$ modalities by way of Equation 7.15.

The result of direction augmentation is a set of GeneralDirection objects instantiated for the best fitting candidate segments. While the direction modality for each of these candidates is determined based on the egocentric projective modality which best fits a given segment, the extent property is determined directly as the angular displacement between the agent's initial orientation and the orientation of the candidate segment. To illustrate direction augmentation, Table 7.7 summarises the result of direction augmentation for the spatial configuration which was used in Figure 7.7. As can be seen from the multiple results, the outcome of direction augmentation in this situation is ambiguous. Thus, in the current model it would only be through explicit dialogue or through the application of subsequent dialogue move constraints that this ambiguity could be resolved. It should be clear however that in an alternate spatial configuration, where the segment (b) was not present, that augmentation would straightforwardly result in a single prominent GeneralDirection candidate.

s	ψ'	$r(\psi')$	s'	$r(s')$
x=200,y=200,θ=270,...	GeneralDirection modality Right-Ego extent 90	1.0	x=200,y=200,θ=180,...	1.0
x=200,y=200,θ=270,...	GeneralDirection modality Left-Ego extent 45	0.51	x=200,y=200,θ=315,...	1.0

Table 7.7 Typical results of conceptual semantic augmentation of the direction role for the spatial configuration depicted in Figure 7.16. Note that only those interpretations with $r(\psi') > 0$ are presented.

7.4.3.3 Resolving Ambiguous Placements

To conclude our description of the contextualization of direction roles, we return to the ambiguous placement resolution example presented in Figure 7.10, along with the results of placement resolution provided by Table 7.3.

From the earlier results it was clear that the resolution of the conceptual semantics for "at the junction" was ambiguous with respect to the spatial configuration depicted. However, having now also considered resolution functions for the contextualization of direction constraints, we can now see how the results of placement resolution can be clarified through the application of additional constraints. Specifically, as per the generalised contextualization process described in Section 6.4.2, the results of placement resolution are fed through as initial states to the resolution of a subsequent direction parameter. Considering the case of a request to turn right, the direction resolution function when applied to the situation corresponding to Junction_1 will fail to produce a high rated interpretation, while, when applied to the situation corresponding to Junction_2, will produce a high rated interpretation as expected. These results are in turn used by the generalized contextualization process to eliminate low rated solutions.

A more complete example – depicting not only the application of contextualization functions in sequence, but also the complete stages from surface forms through linguistic and conceptual semantics, as well as information states – will be presented shortly in Section 7.5. Before presenting this example, we complete our illustration of the general operation of dialogue move contextualization functions with an overview of contextualization for the trajectory role.

7.4.4 Contextualizing Trajectories

The final case of spatial parameter contextualization that we will consider here is that of the path trajectory. The path trajectory, modelled in terms of a PathConstraint instance, plays a key role in the parameterization of PathMovement activity instances. The contextualization of `PathConstraint` instances is similar to the contextualization of `Place` instances as dealt with in some detail in Section 7.4.2. Namely, features such as the identification of candidate relata and the discretization of resultant candidate states are comparable to that for `Place` contextualization. Thus, here we will keep our discussion of `PathConstraint` contextualization brief, and focus on the aspects of the trajectory resolution function which diverge from the case of `Place` resolution.

7.4.4.1 Resolving Path Trajectories

As indicated, the path trajectory role is filled by instances of the `PathConstraint` type, which in turn includes a relatum that the trajectory is made with respect to, and a spatial modality that defines the geometry of the trajectory. The `PathConstraint` may also contain an optional extent qualification, but we will not consider this feature further here.

If a speaker includes information in a dialogue move which partially fills a `trajectory` role, then the trajectory resolution function is invoked on that content by the generalized contextualization process for a specific initial state. Following the form of earlier resolution functions, the signature of the trajectory resolution function is simply:

$$Set(\{PathConstraint', r(PathConstraint'), Set(\{s', r(s')\})\}) =$$
$$resolve_{trajectory}(PathConstraint, s) \quad (7.17)$$

where *PathConstraint* is the non-resolved filler of the `trajectory` role supplied by the speaker, *s* is an input state description; *PathConstraint'* is a contextualized `PathConstraint` instance including a concrete relatum and path modality; and $Set(\{s', r(s')\})$ is the resultant state distribution for a given direction interpretation.

Path Construction

As with the resolution of `Place` instances, the resolution of path constraint instances first determines a set of candidate relata which match the features given for the relatum. This resolution process is based directly on that presented in Section 7.4.2.1 for the resolution of relata in place resolution. Relatum resolution thus makes use of both the intrinsic features of the described entity, as well as the spatial proximity of an input state, *s*, to candidate relata in order to compose a rated set of possible relatum interpretations.

Also as with `Place` resolution, the meaning of a resolved path constraint can be described in terms of the region of possible states carved out by the application of a modality to an identified region. But, unlike in the case of place resolution, the meaning of a `PathConstraint` is described chiefly in terms of a trajectory that relates the input state *s* and the identified candidate relatum. Hence, following the identification of candidate relata, the trajectory resolution function builds candidate paths between *s* and identified relata. It is in turn against these paths that the modality definition functions (described below) are applied. However, it should be noted that rather than being simply a connecting path between two poses, constructed paths extend beyond the relatum by a distance proportional to the relatum's average diameter as determined from the *RS* layer.

The result of path construction is a set of augmented paths which connect s with a given relatum, and which denote the following points along the path:

- s_0 - the starting point on a path
- B_A - the projection of the relatum's initially passed boundary onto the path
- B_B - the projection of the relatum's exit boundary onto the path

It should be noted however that not all constructed paths with respect to a relatum explicitly intersect the boundaries of the relatum. Rather, the path planning algorithm is weaker than a standard path planning algorithm in that it also generates candidate paths which are proximal to a relatum rather than directly intersecting with that relatum. This distinction is necessary to plan paths to capture the conceptual semantics of dynamic prepositional phrases such as "past the kitchen" where the kitchen itself is seen as a path indicating landmark rather than a specific decision point along the path. To demonstrate this point, and illustrate resultant augmented paths in general, Figure 7.17 on the right depicts one possible candidate path generated with respect to the agent's initial pose and the room marked R2. Note

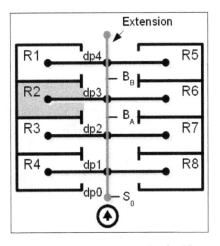

Fig. 7.17 Candidate augmented path with respect to R2.

the projection of the room's bounds onto R2, and that the path itself does not intersect R2. It should be noted though that this is only one of the candidate paths generated, and other paths, including one which does terminate with the decision point collocated with R2, are also generated.

Modality Definitions & Application

Following the construction of the basic path set with respect to a given relatum, the trajectory resolution function applies the PathConstraint's path modality to produce likelihoods of resultant location along those paths. As described in Section 7.3.1.1, five path modality types are currently supported by the Navspace model. We assume that these types, which were listed in Table 7.1, determine a final position distribution likelihood over a path constructed between the named relatum and input state. Following the approach used to define the dynamic placement modalities in Section 7.4.2.1, we therefore give meaning to these path modalities by defining them as functions which operate over constructed candidate paths as follows:

$$f_{destination}(d,s_0,B_A,B_B) = \begin{cases} 0 & s_0 \le d < B_A \\ 1 & B_A \le d \le B_B \\ 0 & B_B < d \end{cases} \tag{7.18}$$

$$f_{source}(d,s_0,B,r) = \begin{cases} 0 & s_0 \le d < B \\ \frac{-(d-B)}{2r}+1 & B \le d \le (B+2r) \\ 0 & (B+2r) < d \end{cases} \tag{7.19}$$

$$f_{approach}(d,s_0,B,r) = \begin{cases} 1 & s_0 \le d < (B-0.5r) \\ \frac{-(d-(B-0.5r))}{0.5r} & (B-0.5r) \le d \le B \\ 0 & B < d \end{cases} \tag{7.20}$$

$$f_{pass}(d,s_0,B,r) = \begin{cases} 0 & s_0 \le d < (B) \\ \frac{-(d-B)}{3r}+1 & B \le d \le (B+3r) \\ 0 & (B+3r) < d \end{cases} \tag{7.21}$$

$$f_{follow}(d,s_0,B_A,B_B) = \begin{cases} 0 & s_0 \le d < B_A \\ \frac{0.5(d-B_A)+0.25}{B_B-B_A} & B_A \le d \le B_B \\ 0 & B_B < d \end{cases} \tag{7.22}$$

where s_0 is the starting point on a path; B_A is projection of the relatum initial boundary onto the path; B_B is the projection of the relatum's exit boundary onto the path; d is a distance along the path; and r is an average diameter of the relatum. Graphically depicted in Figure 7.18, these modality models constitute naive quantitative descriptions of the path modality terms that can be applied directly to paths generated over the *GS* spatial representation layer.

Since the above functions are continuous over a candidate path's length, we must limit the size of the resultant state distribution produced by the trajectory resolution function. As was the case for the resolution of Place entities, this is achieved by considering only states which correspond to specific decision points along the candidate path. Thus, rather than evaluating the modalities for the continuous space along the path, we instead evaluate the modality for each decision point along the candidate path, where the distance, d, is the distance between that decision point and the start state, s_0. Thus, referring to the situation depicted in Figure 7.17, the resolution of the conceptual semantics for a dynamic prepositional phrase such as "towards R2" consists of a candidate solution ψ' along with two possible resultant states corresponding to the locations of decision point dp1 and dp2.

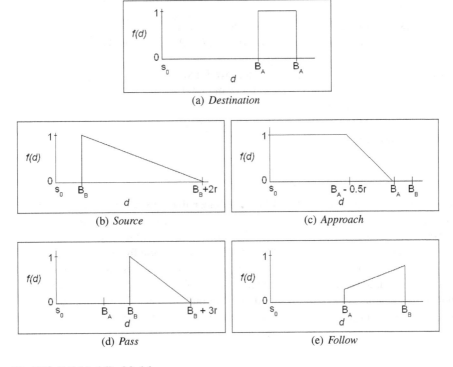

(a) *Destination*

(b) *Source*

(c) *Approach*

(d) *Pass*

(e) *Follow*

Fig. 7.18 Path Modality Models.

7.4.5 *Discussion*

In the preceding three sections, the operation of a number of contextualization functions has been outlined for handling the resolution and augmentation of overtly spatial roles seen in a number of activity types related to the interpretation of verbal route instructions. The purpose of describing these functions was threefold. First, to illustrate how a concrete instantiation of the AODM model's generalized contextualization process interacts with a domain model. Second, to demonstrate how the contextualization of action type parameters necessarily effect a model of route interpretation based on an action-oriented perspective of route instruction meaning. And third, to illustrate how multiple layers of spatial meaning and representation must be pulled together in the interpretation of spatial language.

However, due to length constraints for a relatively short book, I have not attempted to describe the complete set of contextualization functions used in the route interpretation model, nor have I even attempted to exhaustively describe the details of the contextualization functions which were considered. For example, the activity types which we make use of are all also parameterized by non-spatial features including the performing actor, the earliest start time of performance, and so forth.

Label	Speaker	Utterance
(a)	*User:*	turn left, then drive towards the kitchen
(b)	*System:*	OK, I will turn to my left, then I will move towards the kitchen
		agent starts moving
(c)	*User:*	turn at the junction then move forwards
(d)	*System:*	should I turn to my right or my left?
(e)	*User:*	your right
		agent turns at junction and proceeds

Table 7.8 Sample Dialogue

While not discussed here, it should be made clear that appropriate augmentation and resolution functions for such additional non-spatial roles are applied by the generalized contextualization process in the construction of solution sets for all relevant dialogue move types. Moreover, the trajectory resolution functions outlined in the last section also handle the inclusion of extent constraints to further refine the likelihood distribution following the interpretation of a path constraint – but I have omitted discussion of this feature here due to not only to space limitations, but also in order to not diverge too greatly from the core goals of the chapter.

7.5 A Worked Example

With a concrete domain model and a number of contextualization functions in hand, we will now work through a sample dialogue to not only illustrate the dialogue move contextualization process, but also to show how this process integrates into the AODM model as a whole. For the worked example, we take an illustrative dialogue from the Navspace scenario. As introduced earlier, in this scenario a user plays the role of a route giver in directing a mobile robot around an indoor office environment. The user is not co-located with the robot, but rather interacts via an interface that provides a survey view on the agent's location. This survey view introduces the possibility of misunderstandings in reference frame use, but provides an accurate indication of the robot's current state. The scenario focuses on the provision of so-called incremental route instructions rather than purely in-advance route instructions. However, a user can, if they so wish, provide a complete in-advance instruction.

Table 7.8 and Figure 7.19 respectively depict the sample dialogue and the corresponding spatial configuration. The depiction of the configuration includes the position and orientation of the interpreting agent at two different points in the dialogue, as well as an overlay of the *GS* spatial layer. Details on the concrete application in which this dialogue was executed will be discussed in the next chapter. In the rest of this section meanwhile, I break down the sample dialogue into two exchanges, and describe the AODM model's handling of these exchanges with a particular emphasis

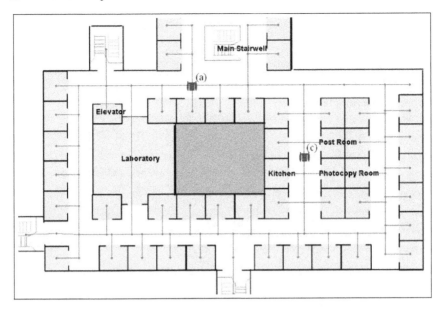

Fig. 7.19 Situation model for worked example.

on the contextualization process.

7.5.1 Exchange 1: An Unambiguous Move Complex

7.5.1.1 Initial Analysis

As dictated by the AODM model's control cycle (see Section 6.2.4), if a user contribution is perceived and no processing errors occur, then input processing steps including language analysis, act abstraction, act integration, and move contextualization are performed. The first three processing stages, i.e., perception, analysis, and abstraction, provide a mapping from surface language through to application-dependent conceptual semantics by way of an application-independent linguistic semantics. These mapping processes were described in terms of a *Language Interface* in Chapter 5. Referring to the user's contribution (a) in Figure 7.8, if this contribution is perceived in an initially empty state, then the contents of the Dialogue State Structure following perception, analysis and abstraction is as follows:

Example 29.

```
LUU: ``turn left, then drive towards the kitchen'', 0.8
LUM: (lm:SequenceMC#M1(
        gum:alphaParticipant
                (gs:NonAffectingOrientationChange#NAOC(
```

```
       uio:hasSurfaceFunction        (uio:CommandSSF#CSF1),
       gum:processInConfiguration (lm:Turning#T1),
       gs:orientationDirection       (gs:GeneralisedLocation#GL1(
        gs:hasSpatialModality         (gs:LeftProjection#LP1)))))
       gum:betaParticipant
                   (gs:NonAffectingDirectedMotion#NADM1(
       uio:hasSurfaceFunction        (uio:CommandSSF#CSF1),
       gum:processInConfiguration (lm:Driving#D1),
       gs:route                      (gs:GeneralizedRoute#GR1(
        gs:pathPlacement             (gum:GeneralizedLocation#GPP1(
         gs:hasSpatialModality
                   (gs:GeneralDirectionalNearing#T2)
           gs:relatum                (lm:Kitchen#K1)))))))))
LUA:  (uio:ActSequence#AS1(
       uio:alpha (uio:Act#A1(
       uio:function   (uio:Instruct),
       uio:performer  (uio:User),
       uio:content    (ns:Reorient#R3(
         ns:actor        (ns:System)
         ns:direction    (ns:GeneralDirection#GD1(
           ns:modality (ns:Left)))))))
       uio:beta (uio:Act#A2(
       uio:function   (uio:Instruct),
       uio:performer  (uio:User),
       uio:content    (ns:PathMovement#M3(
         ns:actor        (ns:System)
         ns:trajectory  (ns:PathConstraint#PC1(
           ns:relatum       (ns:Kitchen#j1),
           ns:modality      (ns:Approach#A1))))))))))
```

where LUU, LUM, and LUA respectively denote the Latest-User-Utterance, Latest-User-Message, and Latest-User-Act slots in the Dialogue State Structure.

7.5.1.2 Act Integration

Following production of the application-dependent conceptual semantics captured by the Latest-User-Act slot of the DSS, *act integration* is invoked to integrate the dialogue act complex (AS1) into the DSS. Since we assume that the DSS was initially empty, the integration of constituent acts into already open user or system dialogue moves fails, and the integration process thus looks to the capability library to determine if new user dialogue moves can be instantiated for input user acts. With respect to the Navspace model, corresponding dialogue moves are identified for each of the two input atomic acts (A1 and A2). These two dialogue moves are instantiated within the same dialogue move integration step with all relevant information transposed from their dialogue act counterparts.

The result of this integration step is a dialogue move complex consisting of two atomic moves as depicted schematically in Figure 7.20. In their filler slots, both atomic moves include content transposed from the initiating dialogue act, but remain in a *new* state with empty solution sets, since no contextualization has yet been performed.

7.5.1.3 Dialogue Move Contextualization

During the control cycle's next processing step, the presence of the new user dialogue move complex triggers execution of the generalized contextualization process. Figures 7.21 and 7.22 depict this contextualization process through the construction of specific solutions for each of the constituent dialogue moves. The contextualization process for the first atomic move (Figure 7.21) begins with initialization of a single solution with a state distribution based on the agent's real-world state. Augmentation and contextualization functions are then applied in sequence to each of the move's parameter types to develop the concrete solution

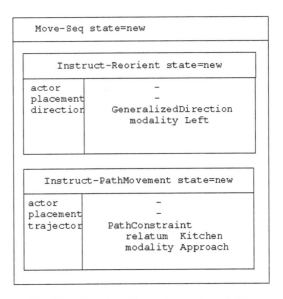

Fig. 7.20 Move Complex following integration of initial user acts.

set. Placement and actor information are for example filled by augmentation functions. The third contextualization function, i.e., $\gamma_{direction}$, is then applied to the user supplied filler for the `direction` role in the context established by solution σ_{A-0}. Since, in this context, two possible interpretations of the direction *left* are possible, the solution set diverges into two distinct solutions with differing, but equally rated, resultant states.

Following the application of contextualization functions for each move parameter, likelihood values associated with each of the two resultant solutions are then calculated by the generalized contextualization process. Given that both possible interpretations of the third parameter were judged equally likely by the direction resolution function, and given that there was no divergence in interpretation for any other parameter, both of the resultant solutions are rated equally with $\Pi(\sigma) = 1$ in this simplistic case. Subsequently, both solutions are judged equally likely in the given situation with $P(\sigma) = 0.5$ in each case.

The contextualization of the second atomic move (depicted in Figure 7.22) then begins on the basis of the results of the first move's contextualization. Specifically,

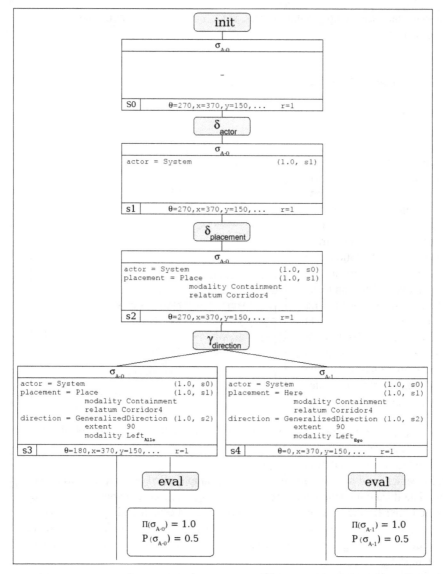

Fig. 7.21 Contextualizing "turn left, then go past the kitchen" pt. 1. Rounded grey boxes denote the application of functions in updating individual solutions (large white frames). The figure read from top to bottom shows the development of the solution set at four distinct stages of the contextualization process.

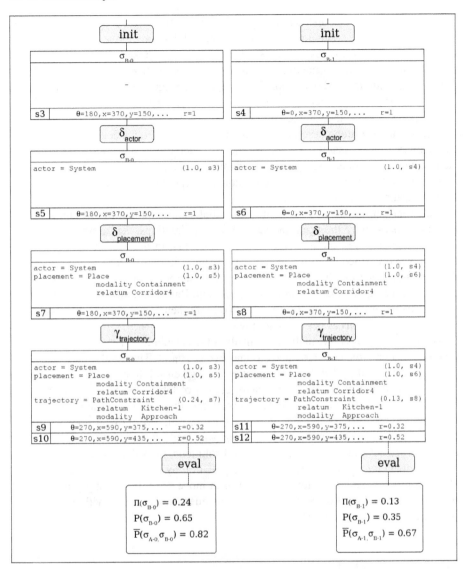

Fig. 7.22 Contextualizing "turn left, then go past the kitchen" pt. 2.

given the semantics of an instruction sequence assumed by the generalized contextualization process, the contextualization of the second dialogue move is instantiated
with two solutions whose associated state distributions are taken directly from the
result of the first move's contextualization. Then, as per the first move's contextualization, augmentation and resolution functions are subsequently applied to each of
the dialogue move's parameters. In the case of the trajectory role, a resolution function is applied against the supplied user content. Since only one kitchen is salient to
the agent in both solution cases, the number of possible solutions does not fork further. However, given that the *Approach* trajectory modality is uncertain in terms of
the resultant agent state, multiple state variables now become associated with each
of the two move solutions – each state variable corresponding to a decision point on
the path leading up to the kitchen.

Importantly however, given that the two initial states supplied to this contextualization process have the agent facing in different directions, the ratings assigned
to the interpretation of the trajectory role are not equal in the case of the two solutions. Namely, since for solution σ_{B-0} the agent is initially pointing away from
the referenced relatum, the rating assigned to the interpretation is less than for the
case where the agent is facing towards the relatum. Based on individual parameter
ratings, overall interpretation likelihoods, $P(\sigma)$, are then calculated for the second
move's set of solutions as per the first move.

The final stage of contextualization involves the calculation of cumulative solu-
tion trace likelihoods, i.e., $\bar{P}(\sigma_{A-0}, \sigma_{B-0})$ and $\bar{P}(\sigma_{A-1}, \sigma_{B-1})$. These values are calculated as the mean of probabilities for solutions in a given solution trace. In this case
the joint likelihood of an egocentric interpretation of left followed by the agent's
movement towards the kitchen is judged more likely than an allocentric interpretation of *left* – this difference being due to the lower rating of moving *towards the
kitchen* if an allocentric interpretation of *left* was made in the first move. Given
the difference between the two net likelihood solutions, the solution trace with the
higher probability is judged the correct interpretation by the generalized contextualization process, and the solution sets for each move are subsequently pruned of
solutions σ_{A-0} and σ_{B-0} respectively. Both constituent moves along with the move
complex are now in a *Complete* state.

7.5.1.4 Dialogue Planning

Following contextualization, the multi-stage dialogue planning process is invoked
to determine what actions should be taken by the agent on the basis of its current
mental state.

The move level *Response Planning* algorithm (described in Section 6.5.1.2) first
handles the completed complex user move by shifting it from the open to closed user
move set on the DSS. Then, given the act type for the constituent atomic moves, and
following the verification of capability pre-conditions, an intention towards the content of the move complex is adopted. Moreover, since the cumulative probability of
the move complex solution is less than the clarity threshold for move contextualiza-

tion, the response planning algorithm also results in the adoption of a new system
dialogue move that explicitly accepts the individual user instructions. Thus, the updated DSS following response planning includes the following complex move in the
`Planned-System-Moves` stack:

Example 30. PSM : (uio:Complex#MC2(
```
    uio:alpha (uio:Move#M3(
      uio:function  (uio:AcceptInstruct),
      uio:performer (uio:System),
      uio:content    (ns:Reorient#SR1(
        ns:actor        (ns:System)
        ns:placement   (ns:Place#PL1(
            ns:modality (ns:Containment),
            ns:relatum  (ns:Corridor#Corridor4)))
        ns:direction    (ns:GeneralDirection#GD1b(
            ns:modality (ns:Left-Ego)
            ns:extent      90))))))
    uio:beta (uio:Act#M4(
      uio:function  (uio:AcceptInstruct),
      uio:performer (uio:System),
      uio:content    (ns:PathMovement#PM2(
        ns:actor        (ns:System)
        ns:placement   (ns:Place#PL2(
            ns:modality (ns:Containment),
            ns:relatum  (ns:Corridor#Corridor4)))
        ns:trajectory (ns:PathConstraint#PC2(
            ns:relatum        (ns:Kitchen#k1),
            ns:modality       (ns:Approach))))))))
```

while the following primary intention is added to the agent's intention set for subsequent sequencing when the intention management process is called:

Example 31. Int : Intention(SEQ(
```
    operand((ns:Reorient#SR1(
      ns:actor        (ns:System)
      ns:placement   (ns:Place#PL1(
          ns:modality (ns:Containment),
          ns:relatum  (ns:Corridor#Corridor4)))
      ns:direction    (ns:GeneralDirection#GD1b(
          ns:modality (ns:LeftEgo)
          ns:extent      90))),
    operand(ns:PathMovement#PM2(
      ns:actor        (ns:System)
      ns:placement   (ns:Place#PL1(
          ns:modality (ns:Containment),
          ns:relatum  (ns:Corridor#Corridor4)))
      ns:trajectory (ns:PathConstraint#PC2(
          ns:relatum        (ns:Kitchen#k1),
          ns:modality       (ns:Approach)))))
```

The Dialogue Planning process's *Act Planning* and *Act Sequencing* steps are then
called to determine what concrete dialogue acts should be performed by the agent.
Specifically, a realization act for the newly planned system dialogue move is instantiated and added to the `Planned-System-Acts` slot of the Dialogue State
Structure. It should be noted, however, that the current model's planning of dialogue

acts in response to system dialogue moves is extremely simplistic. Namely, all information from a given dialogue move is transposed into a new dialogue act during the act planning process. Thus, the newly created complex dialogue act has effectively identical content to that of the underlying dialogue move.

With no other acts to plan, the *Act Sequencing* step is then invoked. In this final step of the dialogue planning process, the newly instantiated act complex in the `Planned-System-Acts` slot is copied to the `Next-System-Act` slot for subsequent processing by the AODM model's *Message Planning*, *Realization*, and *Production* modules. Although we will not detail these three production steps here, it should be noted that it is at the level of Message Planning that unnecessary content is filtered from the dialogue contribution before the system's response, (b), is realized. This filtering is achieved through the appropriate conditioning of functional transform rules as described in Section 5.5.

Finally, following execution of the intention management sequence and the language production steps, the dialogue state structure input and output abstraction slots are cleared.

7.5.2 Exchange 2: An Ambiguous Move Complex

7.5.2.1 Initial Analysis & Contextualization

Following intention adoption in response to the user's first instruction, the agent begins moving from the location marked (a) in Figure 7.19 towards the location marked (c) where the user initiates the second exchange. The second exchange in the sample dialogue, i.e. *(c) - (e)*, is similar to the first exchange in that the user's initiating utterance instructs a complex move consisting of two atomic moves. However this second exchange is less straightforward since the user's instruction is more ambiguous in the given context than was the case for the first exchange.

Since the first utterance of the second exchange is similar to that for the first exchange, we will not explicitly list the results of each of language input, analysis, abstraction, and integration processes here. The results of language integration are however, as with the first exchange, a newly instantiated dialogue move complex consisting of two atomic moves. This move complex, in its new state and without any solutions, is depicted in Figure 7.23.

Figures 7.24 and 7.25 depict the development of the contextualization process for this move complex. As can be seen, the contextualization of the first dialogue move is similar to that in the first exchange except that in this case placement information rather than turning direction were provided for resolution. The placement *at the junction* is ambiguous in that there is more than one junction salient to the interpreting agent at any time. Other junctions, notably the junction behind the agent, are also possible interpretations but we omit them in Figures 7.24 in order to avoid visual clutter; it should be noted however that the ratings for these alternatives is considerably less than that for the junction depicted.

Since it is possible for the agent to turn either to its left or right at the most salient junction, two equally likely solutions are introduced by the direction augmentation function. In the move complex's second atomic move, the user instructs the system to move forward after having made the initial reorientation. Since the `moveDirection` resolution function produces an interpretation of "forward" which is equally likely in both cases, the subsequently calculated solution likelihoods and cumulative solution likelihoods are rated equally for both solution traces. Further automatic reduction of the solution set by the generalized contextualization process is not possible, and further reduction is only possible through dialogic means.

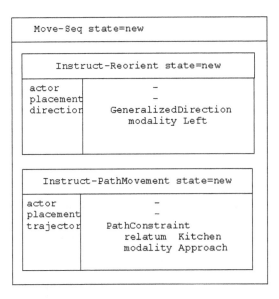

Fig. 7.23 Move Complex following integration of initial user contribution.

7.5.2.2 Dialogue Planning

Following the completion of contextualization, the newly created user move remains in the `Open-User-Moves` field of the Dialogue State Structure. However, unlike in the case of the first exchange, this newly created move is in a `MultipleSolutions` state rather than a `Complete` state. Thus, the first phase of dialogue planning, i.e., response planning, takes no action in response to this user move.

The second dialogue planning phase, i.e., act planning, does however take action to further the open user move. Specifically, as described in Section 6.5.2.1, the pivot move in the open user move complex is first identified so that an appropriate clarification question can be composed. This pivot move is the first dialogue move in which multiple solutions were introduced – in this case, the instruction to make a reorientation at the junction. Following identification of the pivot move, a choice question is subsequently composed to resolve the ambiguous move. This choice question, listed below, is subsequently added to the `Planned-System-Acts` slot of the DSS, which in turn eventually gives rise to the system's clarification question utterance being raised following message planning, realization and output.

Example 32. `uio:Act#A10(`

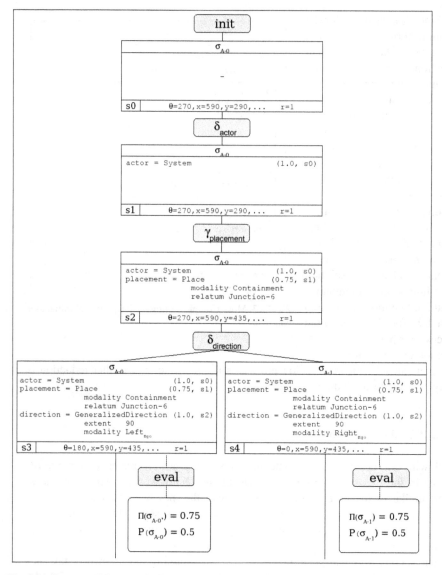

Fig. 7.24 Contextualizing "turn at the junction then move forwards" pt. 1.

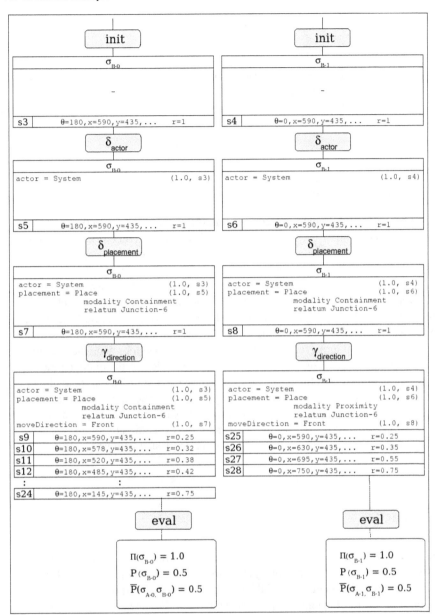

Fig. 7.25 Contextualizing "turn at the junction then move forwards" pt. 2.

```
uio:function   (uio:ChoiceQuestion),
uio:performer  (ns:System),
uio:content    (ns:Reorient#R5(
     ns:actor(ns:System),
     ns:placement((ns:Place#PL3(
          ns:modality (ns:Containment),
          ns:relatum   (ns:Junction#Junction6))))
     ns:direction(SetContainer#sc1(
          setElement(?ns:LeftEgo),
          setElement(?ns:RightEgo))))))
```

7.5.2.3 Response Integration

For the final stage of this contextualization example, the user has correctly perceived the system's clarification question and thus replies with utterance (e). This contribution is subsequently interpreted through analysis and abstraction to an *Answer* dialogue act, before this act is in turn used to reduce the solution set associated with the still open user move. Specifically, having been triggered, the dialogue integration process attempts to determine the relevance of the user's answer act to the information state. Following from the methodology outlined in Section 6.3.2.2, the user's contribution is unified with the system's question to produce a proposition. This proposition is in turn used to update the dialogue move for which the system's dialogue act was raised. This update involves the reduction of the dialogue moves solution set, and the cascading removal of all subsequent solutions in the dialogue move complex which were dependent on the pruned dialogue move. With the open user move now in a completed state, no further contextualization is required during the contextualization process. Dialogue planning thus acts directly on the reduced solution, and an intention towards performing the requested actions is adopted by the agent.

7.6 Discussion

From the outset of this book we have argued that the processing of dialogue in the situated domain requires a tight integration of spatial reasoning and spatial modelling resources with models of spatial language and general purpose dialogue management systems. In Chapter 6, we presented the Agent-Oriented Dialogue Management model, including its generalized language contextualization process, with the aim of addressing these requirements. However, to clearly demonstrate that model, it has been necessary in this chapter to instantiate the interpretation model for a specific domain. Having chosen verbal route instruction interpretation as that domain, it has been necessary to: (a) develop a layered spatial representation to meet the dual needs of communicative and non-communicate reasoning types required by the interpreting agent; (b) develop an inventory of capabilities that the agent applies in processing verbal route instructions; and (c) develop contextualization functions which provide the key mapping between the language interpretation processes required by dialogue and the agent's own spatial representation and action inventory.

While we recognize the schematization of long routes into a series of 'segments' as a general characteristic of route descriptions in unknown environments, and one which has particular significance to the production of clear unambiguous routes [Richter and Klippel, 2004], we argue, following Tversky and Denis's own comments, that the robust resolution of verbalised routes by artificial systems should not be overly dependent on such well-formed segmentation. Based on this argument, the approach to route interpretation taken here has eschewed a dependence on high-level, segment-oriented route interpretation in favour of a discretized action-oriented model. The particular model employs a search space technique similar to Mandel et al.'s [2006] methodology, but has been applied to spatial representations more suited to the processing of spatial language, so as to reduce the complexity of the search space and hence be practical even for short language inputs.

While the approach taken has attempted to build on some important qualitative aspects of representation and reasoning, i.e., the importance afforded to an ontologically-grounded conceptual description level, as well as the use of a relatively abstract graph-based representation of space, the approach taken to spatial reasoning has been a quantitative one. The justifications for taking this quantitative approach are twofold. Most significantly, the quantitative approach affords relatively straightforward development and implementation of models which are intuitive. Such a property is of considerable advantage when the research focus is not the investigation of qualitative reasoning models. The second justification comes from a more philosophical perspective in that the quantitative reasoning approach represents a just-in-time and least-effort approach to spatial relation categorization. That is, the approach taken to spatial relations applied here assumes that objects lie in configurations known to the agent, but that explicit symbolic spatial relationships between these objects need not be asserted until if, and when, necessary. Of course, this assumption relies on the existence of a quantitative spatial representation to begin with, and cannot account for transitivity in spatial relations, or, indeed, functional rather than geometric or path-based meanings. Thus, it remains open for future work to investigate the augmentation of Navspace's quantitative reasoning methods through qualitative techniques.

Given the quantitative approach to reasoning employed in Navspace, there are clear similarities between the models of spatial prepositions – or rather spatial modality – used here, and some existing preposition interpretation models. Within the route interpretation domain, the quantitative modality models used here are most similar to the models employed by Levit and Roy's [2007] and Mandel et al.'s [2006]. However, whereas Levit and Roy's [2007] models were hard-coded to the particular Map Task spatial configuration, the models outlined here are dependent only on the overall organization of space assumed by Navspace – and not any specific spatial configuration. Outside of the route interpretation domain, the Navspace modality models have broad similarities to a number of projective modality definitions. For example, Moratz et al. [2001] outlined a simple segment-based configuration of projective relations around an agent's ego. Similarly, Klippel et al. [2004] offers an empirically derived model of action descriptions for an agent making a turning. Quantitative models of so-called path prepositions are less frequently encountered. However, one model of quantitative path prepositions which holds some similarity to the work presented here is due to Kray et al. [2001], who in a series of experiments, empirically studied the properties of so-called path relations through the properties of the German *path prepositions* "entlang" (along) and "vorbei" (past).

The Navspace route interpretation model is also naturally comparable to some of the existing route interpretation approaches presented in Section 7.1.2. Of these, the Navspace model is most similar to the work of Mandel et al. [2006] and Levit and Roy [2007] in that explicit quantitative representations of space are assumed, and that the route interpretation problem reduces to a state space search. However, unlike both of these models, the Navspace model has, as a matter of design, provided a far tighter coupling between the spatial and linguistic processing components. Not only does this tight coupling remove many of the simplifying assumptions concerning language input made by Levit and Roy [2007] and Mandel et al. [2006], but also crucially allows the interpretation process to fall back to dialogic interaction in the case of ambiguity. That said, both Lauria et al.'s [2002] IBL and MacMahon's [2007] MARCO have attempted to couple the route interpretation process to linguistic resources. However, in the case of IBL, it is far from clear how successful this was in practice, and, even for the more recent MARCO, a considerable simplification of input data was assumed.

As noted, the approach to route interpretation pursued here – and to action interpretation in general – relies on the notion of context-sensitive resolution and augmentation. As such, the enhancement of user supplied content through context also has similarities to some existing content enhancement models in dialogue management. For example, as detailed in Section 3.4.3, the SmartKom project applied an *overlay* process to compose together search constraints assembled in discourse. However, this notion of augmentation and overlay is concerned more with the assembly of individual dialogue acts into a single meaningful *request,* and thus has more in common with the construction of complete plan sequences in Navspace rather than in the particulars of the contextualization process. Indeed, the approach to content enhancement applied here is in some ways more comparable to SmartKom's multi-modal input fusion processes. However, whereas their notion of information fusion focuses on combining language and specific gestural elements, the notion of non-linguistic context applied in the Navspace model is broader. Moreover, due to the relativistic nature of spatial models, the approach taken to enhancement here has relied on a functional model. Such a functional approach to spatial language interpretation has also recently been investigated by Tellex and Roy [2006, 2007], but, unlike Tellex, the model presented here has considered not only the enhancement of user supplied information, but has also considered the resolution of elided information through augmentation.

There are a number of ways in which the Navspace model could be extended to produce a more comprehensive account of route interpretation. Probably most significantly, the Navspace model as presented here has made no attempt to couple directly to a discourse semantics, or other surface-oriented dialogue history, to resolve true anaphora or to fill ellipses. Such an approach was attempted by Lauria et al.'s [2002] through a DRT framework which pre-processed linguistic input before a procedure specification was constructed. The inclusion of such an approach would seem beneficial, but performing such resolution prior to contextualization in the current framework would likely not give optimal results. Instead, direct fusion of the physical situation contextualization process with a discourse resolution process would seem more beneficial. Recent works that fuse discourse and visual salience influences in reference resolution, e.g., Kelleher [2007], offer useful insights on how such an extension might be achieved in the future.

Also notably, since the development of the Navspace model was motivated chiefly to provide a concrete instantiation of the generalized language contextualization process, the model has considerable room for extension in terms of its breadth of coverage in action and spatial modality types. Although broadly comparable, the breadth of such types presented in this chapter is smaller than some of the contemporary models considered earlier – in particular in comparison to Levit and Roy's [2007] inventory of *Navigational Information Units*. However, as noted, our primary motivation has been to couple the route interpretation process to a dialogue process in a principled way – and in so doing to provide an account of route interpretation which is in itself both plausible and easily scaled through the addition of new spatial modalities and actions.

7.7 Summary

This chapter developed a verbal route instruction interpretation model which applies an iterative contextualization process to action specifications over a hybrid spatial representation. The primary motivation for developing this model has been to provide a concrete instantiation of the generalised contextualization process introduced in Chapter 6. The secondary motivation meanwhile was to address the lack of language process coupling in existing route interpretation models.

The chapter began with an examination of the properties of route instructions as well as current computational interpretation models. We saw that most existing approaches to route interpretation make unrealistic assumptions about the nature of their inputs, and are cast too far from the realities of a natural language interface. In light of this, an action-oriented route interpretation model was then outlined and fleshed out over subsequent sections. This model consisted of three elements: (a) a hybrid spatial representation suited for communicative as well as non-communicative purposes; (b) an inventory of actions which map to the route instructions given by speakers; and (c) concrete resolution and augmentation functions which instantiate the contextualization process for the hybrid spatial model and action inventory.

The presentation of the Navspace scenario and modelling assumptions given in this chapter tended to focus on the underlying models rather than their implementation. In the next chapter we switch perspective, and instead discuss the instantiation of the models presented both in this and previous chapters through concrete dialogue systems.

Chapter 8
Route Instruction Dialogues with Daisie

Abstract In this penultimate chapter, the concrete implementation and application of the dialogue architecture and spatial language interpretation models presented in this book are briefly described. The chapter begins in Section 8.1 with an introduction to *Daisie*, a dialogue processing framework which implements the language architecture developed earlier. While Daisie itself is intended to be application independent, in Sections 8.2 we present the *Navspace* application as a concrete solution to the dialogic route interpretation problem built on top of Daisie.

8.1 The Daisie Dialogue Toolkit

The dialogue processing elements of the architecture presented in Chapters 4 through 7 have been implemented within *Daisie*, a Java based, open source, dialogue systems toolkit[1]. Specifically, Daisie (Diaspace's Adaptive Information State Interaction Executive) implements core dialogue processes including the language interface, and the Agent-Oriented Dialogue Management model.

Daisie's design has been influenced by the requirements of the agent-oriented dialogue architecture presented in this book, but also by two highly pragmatic concerns:

- **Modularity:** The experimental nature of natural language processing means that it is typically possible to substitute alternate solutions to particular processing tasks. This is seen for example in language analysis where grammars ranging from ad hoc keyword spotting to mildly context-sensitive formalisms such as Joshi's [1987] Tree Adjoining Grammars (TAGs) can be applied. As long as the applied grammar supports certain data import and rendering standards, i.e., obeys the properties of the language interface defined in Chapter 5, then a variety of grammars can be adapted. The same principle is seen not only in other low-level language processing areas, i.e., language recognition, generation, and

---

[1] See http://www.daisie.org

synthesis, but also to a limited degree in higher level language processes such as language integration and planning. The Daisie toolkit therefore supports a high degree of modularity in design.

- **Embeddability:** Early dialogue system implementations typically placed dialogue functionality at the centre of a complete application that included one or more data sources that instantiate the application for a particular domain. However, in real-world applications, spoken dialogue elements are often regarded as an interface or view on a complete application. Thus, rather than being implemented as a stand-alone application, the Daisie toolkit has been implemented as a library of functionalities that can be embedded within a third-party application.

With these requirements in mind, Daisie consists of three functionality groups that are integrated into an application alongside specific linguistic resources. These three Daisie functionality groups are: (a) *The Daisie Plugins*, which provides a range of language technology solutions that have been pre-defined to populate Daisie with a minimum inventory of language parsers, speech synthesizers, and so forth; (b) *The Daisie Backbone*, which integrates language technology plugins with dedicated Daisie dialogue management code into a coherent dialogue system; and (c) *The Daisie Application Framework*, which is used directly by an application to provide the minimum interface to Daisie functionality.

Figure 8.1 depicts Daisie's principle processing components. The Daisie Application Framework and Daisie Backbone are depicted explicitly in the centre of the figure. The Daisie Plugins on the other hand are depicted as trapezoids around the boundary of the Daisie Backbone. These plugins in turn interface with linguistic resources depicted as cylinders. In the following, the three Daisie functionality groups are described further.

8.1.1 The Daisie Backbone

Referring again to Figure 8.1, at the centre of the Daisie Backbone sits an implementation of the Dialogue State Structure that provides a shared memory for each of the primary processing components (regular rectangles). The default Dialogue State Structure is based on the specification given in Section 6.2.3. However, this default structure may be extended as necessary to support new dialogue strategies in conjunction with the language integration and dialogue planning components.

Following the stated modularity requirement, overall control of Daisie's backbone is not hard-coded, but is provided by a pluggable mechanism. In the basic case, this mechanism allows the use of simple information control flows such as that described in Section 6.2.4.2. But more importantly, the pluggable control mechanism also allows the development of control methodologies based on fundamentally different technological principles. For example, a formal methods based approach to the control of dialogue management and dialogue system flow outlined by Shi et al. [2005] has also been directly applied within the Daisie backbone. This

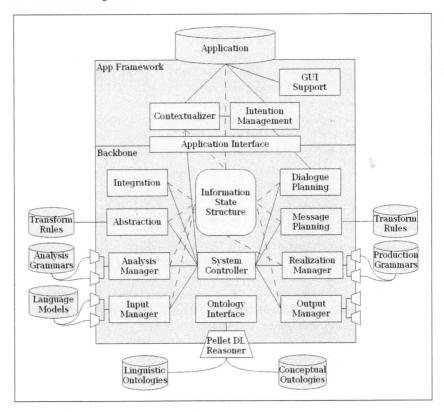

Fig. 8.1 Daisie Toolkit Components.

formal methods based approach applies Hoare's [1985] process algebra *Communicating Sequential Processes* for the specification and model based analysis of the dialogue system's primary control processes. The ability to make such an analysis can improve the safety and reliability of dialogue systems – qualities which will prove critically important as dialogue systems are applied to safety-critical situated applications.

Moving outwards from the Dialogue State Structure and System Controller, the Daisie Backbone includes a range of language technology manager components and dialogue processing modules. The main dialogue processing modules, i.e., *Abstraction, Integration, Dialogue Planning*, and *Message Planning* follow directly from the models developed in earlier chapters. The language technology manager components, i.e., *Input Manager, Analysis Manager, Realization Manager*, and *Output Manager* provide abstractions over language technology solutions. The language technology plugins, illustrated as trapezoids at the periphery of the Daisie Backbone, each implement a well-defined interface for that language technology component type. As well as maintaining a set of plugins, manager modules also include functionality to select the best result from managed plugins. For example, in the case of the *Analysis Manager*, sortal filtering constraints and chunking heuristics are applied to select the most complete parse from any included plugins.

Finally, the Daisie Backbone also includes Knowledge Management support which is made use of throughout the Daisie framework. This support includes basic representation and reasoning functionality as used for example by the Dialogue State Structure. Moreover, knowledge management also includes ontological reasoning support such as type checking as used throughout the dialogue system. This ontological reasoning is in turn provided by another Daisie plugin – in this case an instance of the open-source Pellet Description Logic reasoner [Sirin et al., 2007].

8.1.2 The Daisie Application Framework

The Daisie Application Framework provides functionality that is embedded within, and typically extended by, a given application. This potential for extension, or indeed replacement, is in contrast with the backbone modules which are stand-alone and not directly interacted with by applications. Referring again to Figure 8.1, the Daisie Application Framework modules include the *Application Interface*, *Intention Management*, *Contextualization*, and *GUI Support* modules.

The *Intention Management* module provides an implementation of the intention structure and management framework presented in Section 6.1. We shall see later that this system is used in the Navspace application to drive the operation of that application at a behaviour executive level. That said, the default Daisie Intention Management module is lightweight in comparison to complete theories of agent-oriented design and programming. In principle, Daisie's default intentionality model could be replaced straightforwardly with an alternate agent theory such as a Procedural Reasoning System (PRS) implementation (e.g., Konolige [1997]), a Golog interpreter (e.g., Soutchanski [2001]), or a fully formed agent-oriented programming solution, (e.g., Collier and O'Hare [2009]). For this reason, the *Intention Management* module is implemented as part of Daisie's Application Framework rather than the Daisie Backbone. This implementation is optional and may be replaced at development-time with alternate theories or implementations.

The Daisie *Contextualizer* module implements the iterative contextualization framework presented as part of the Agent-Oriented Dialogue Management model in Chapter 6. Individual applications extend this framework with definitions of resolution and augmentation functions. As with Intention Management, alternative solutions to the contextualization process are possible but still compatible with the agent-oriented dialogue management model captured by Daisie. Thus, the Contextualizer module is also developed as part of the Daisie Application Framework rather than Daisie Backbone.

In keeping with the embedded functionality approach taken to Daisie's design, a fully functional Graphical User Interface (GUI) is not included in Daisie. Instead, the Daisie Application Framework provides a GUI support module that allows an application to embed views of the dialogue system and dialogue state structure di-

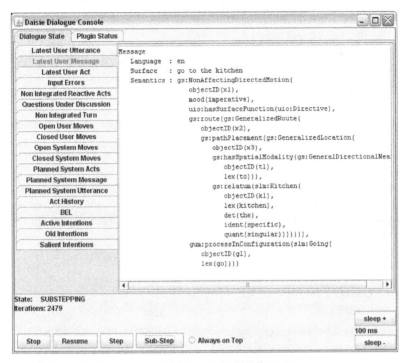

Fig. 8.2 The Daisie Dialogue System Console with Dialogue State View

rectly within a developed application. The primary Daisie Application Framework panel is the Dialogue State Structure Viewer (DSSV). As the name suggests, the DSSV provides a view of the dialogue state at any given time. This view is depicted in Figure 8.2. To save space, the DSSV provides a multi-tabbed pane where each tab provides a view of one slot in the Dialogue State Structure. Red highlighting on tab labels indicates if information under that tab has changed within the last dialogue system execution cycle. To aid debugging, the DSSV also allows the stopping, slowing down of, and stepping through the dialogue system's execution cycle. Two additional panels, not depicted, allow the viewing of the state of Daisie plugins, as well as the state of the system controller implementation.

The *Application Interface* provides a gateway between the Daisie Backbone and application functionality proper. As a true interface, this module must be instantiated by a given application to ensure functionality assumed by Daisie Backbone components is available. In particular, the Application Interface provides a number of interface function groups:

- **Move Definition Retrieval:** Functionality assumed by the AODM integration process for accessing an application's library of potential user initiated moves.
- **Contextualization:** Functionality assumed by the main dialogue processing architecture for contextualizing user moves following move integration.

- **Mental State Update:** Functionality assumed by the dialogue planner for adopting new beliefs and intentions in response to user moves.
- **Mental State Query:** Functionality assumed by Daisie components for checking whether parametrizing constraints hold in the current situation.

As indicated, contextualization and the update and query of intentional state can be provided directly by the Application Framework's included Contextualizer and Intention Management modules. Non-intentional state query and update must however be provided on an application-specific basis.

In addition to providing functionality to the Daisie backbone for querying and updating the application state, the Application Interface conversely provides functionality to the application for querying and updating the dialogue state structure. This is used primarily to allow an application to adopt system-initiated dialogue goals, but can be used by an application to access the entire Dialogue State Structure if required.

8.1.3 The Daisie Plugins

Daisie includes a number of language technology plugins that can be used to construct a minimal operational spoken dialogue system. These plugin types are grouped according to Daisie's language technology manager classes. That is, four plugin types are assumed by Daisie: input plugins, analysis plugins, realization plugins, and output plugins. For each plugin type, an interface is provided which specifies required functional and ontological constraints that must be met by an implemented plugin solution. Input plugins include a speech recognition plugin based on the proprietary Vocon system from Nuance, and a Java-based text input component that can be embedded within application GUIs. Analysis plugins include a custom Daisie keyterm spotting module, and a complete language analyser based on the OpenCCG system [Baldridge and Kruijff, 2002]. Realization plugins include a solution based on the KPML language generator [Bateman, 1997]. Finally, output plugin solutions include a speech synthesis module based on the MARY system [Schröder and Trouvain, 2001], and a text output system that can be embedded within application GUIs.

8.1.4 An Alternative Integration Approach

While the fitness of individual dialogue system components is crucial if we want to develop intelligent dialogue systems, McTear [2002] and others argue that how we integrate language processing components into a working system is at least as important as individual component design. Daisie represents one approach to the creation of spoken dialogue systems. The characteristics of the approach are tight-coupling of resources, a dependence on a plugin metaphor, and the view of Daisie as a library of functionalities to be embedded within an application. This is certainly not the best or only way to approach the dialogue systems integration problem. Before moving on to look at an applications of Daisie, the following briefly comments on an alternative approach to the dialogue systems development and integration problem.

In earlier work, I investigated the use of a middleware-oriented solution for integrating the elements of a spoken dialogue system [Ross et al., 2004d]. Starting from the set of macro-scale, loosely-coupled components, the first integration strategy was to wrap each individual language technology component within an intentional agent wrapper written in the ALPHA agent language [Ross et al., 2004b]. These agents were subsequently interpreted through the Agent Factory framework [Collier, 2001]. The resultant system was dynamic in that new agents could be started or killed at any time during the dialogue systems operation, and these agents automatically made or terminated connections with any agents providing relevant services registered with a central directory service agent. Each individual agent included a number of reactive style rules which triggered behaviours based on either communication from other agents within the architecture, or based on perceived input, for instance from the speech recognition system. The perceived advantage to such an approach is that it followed recent trends toward intelligent integration strategies based on service provision. Such approaches have been common in the dialogue system design literature; in particular, Martin et al.'s [1999] Open Agent Architecture has been used extensively to integrate language technology components within the context of the DARPA Communicator project and elsewhere [Bos et al., 2003].

While the multi-agent approach was useful in that: (a) it provided the capabilities to dynamically change the configuration of the agent's within the dialogue architecture; and (b) because it provided a logical language formalism that had potential to allow ready use of well-defined ontologically mediated communication between the individual agents, it did however also produce two significant drawbacks:

- First, though the intentional agent *shells* provided a unique communication interface between the various agents, these shells also added a significant overhead in communication time and synchronization difficulties. Although these overheads seem minimal in the idealized view of a dialogue system as for example typified by Figure 3.1, they cause considerable more complication in true dialogue systems development. In practice, dialogue systems require a considerable amount of communication between components even in processing a single system utterance. While such communication can be achieved through a highly decoupled mechanism, it is inefficient and complex in comparison to the use of a more tightly coupled approach.
- Second, the multi-agent integration strategy reflects trends in the loose-coupling of software systems, and that loose-coupling was found to be not necessarily the most appropriate strategy for dialogue system development. In particular, dialogue systems are not in general dynamic systems, and the configurations of components within these systems displays a particular architectural arrangement which must always be achieved. While it is interesting and sometimes necessarily to try different system configurations based on variants of a compo-

nent, e.g., different speech recognizer or synthesis products, the most important issue to be considered there is the existence of a commonly accepted communication language for such components. Such a common interchange language can be developed for particular component types, and this is not dependent on the use of an intelligent agent framework.

Daisie can thus be viewed as a revised, more tightly coupled framework for dialogue systems integration in the light of the above observations.

8.2 The Navspace Application

In this section, an implementation of the route interpretation model presented in Chapter 7 is presented. The implementation, *Navspace*, is a simulated robot control application that has been used to both study the application of Daisie, and to investigate models of spatial modelling and linguistic interaction in the robotics domain, and for the control of robotic wheelchairs in particular.

The Navspace application was originally built to investigate models of spatial language interaction for robotic wheelchair platforms. As an example of a (semi-)autonomous system in a situated environment, the wheelchair platform requires dialogic interaction with a relatively high quotient of spatial language that goes beyond locative expressions. Moreover, such platforms are particularly interesting from a situated applications perspective since they incorporates the state-of-the-art in many fields of robotics research, but yet are tempered by the realities of system development in the medium term. On the other hand, while the application domain has real-world relevance, the general principles of spatial modelling and interaction with the system should be generalizable to other areas of situated application development including more advanced robotic systems as these become available.

Figure 8.3 shows the Navspace application's main interface window. The centre of this window depicts the schematized environment and the location of an autonomous wheelchair robot within that environment (iconic figure located in the upper centre of the screen). The screen is a direct view on the system's own knowledge, and is also the information provided to the user; thus ensuring mutual knowledge of the shared environment. The only discrepancy between user and system knowledge is the user's knowledge of the target destination for an interaction. This is displayed to the user as a highlighted red area (upper right of the figure).

Interaction with the Navspace application is based on textual rather than spoken channels. This has been done to minimize the difficulties introduced by speech recognition and synthesis, and thus allow greater research focus on representation and modelling issues. However, to more accurately reflect qualities of the spoken interface, a history-less chat interface was applied. This interface, shown at the bottom of Figure 8.3, consisted of two text fields. The first text field is used by the user to type in free-form natural language, while the second was used to display messages from the system for a duration proportional to the length of the message.

Figure 8.4 depicts the Navspace application architecture. Dark rectangles indi-

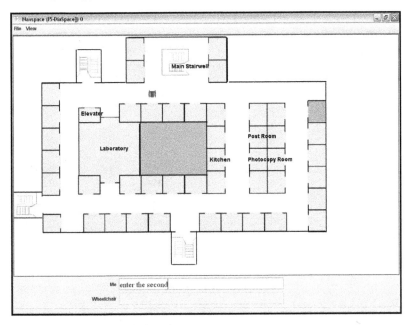

Fig. 8.3 The Navspace application's main screen

cate those modules that directly extend application support provided by Daisie, while other paler rectangles are Navspace specific processing components. It should be noted that while some inter-component relationships are depicted to illustrate notable information access relationships between modules, as well as general processing flow, Figure 8.4 does not provide an exhaustive description of inter-module relationships.

The Navspace architecture can be broken down into three coarse-grained components. The first coarse component, the *Agent Model*, provides an implementation of a cognitive robotic simulator. The model applies: (a) the intention management system provided by Daisie, (b) the spatial representation and reasoning model presented in chapter 7, and (c) a number of custom components to provide simple models of perception, actuation, and route planning. The second coarse-grained component, the *Contextualizer*, builds directly on Daisie's contextualization model with a number of augmentation and resolution functions. These functions, based on the models presented in Chapter 7, make extensive use of the spatial reasoning functionality included in the agent model. The final coarse-grained component, the *User Interface*, is the Graphical User Interface depicted in Figure 8.3.

Since most of the models employed by Navspace have already been introduced in earlier chapters, we need not describe Navspace component design further here. However, it should be noted that in order to complete integration with the Daisie framework, linguistic resource development was necessary. Moreover, requirements placed on the Navspace application by the limited supply of study participants, and

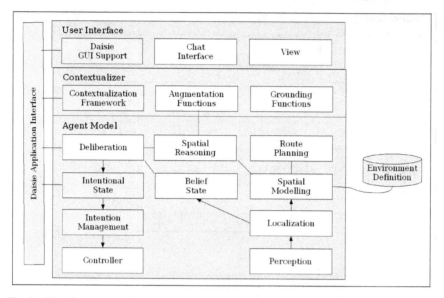

Fig. 8.4 The Navspace application architecture.

the need for an interactive demonstrator for visiting researchers, meant that it was necessary to instantiate Daisie for both German and English language coverage. This was achieved by developing grammar resources in the KPML and OpenCCG frameworks for both languages. However, bar localization of the Navspace user interface, no changes were needed to semantic abstraction rules or higher level Daisie or Navspace components. This simple customization of the Navspace application for multiple languages based on only grammar development is an important benefit of the highly modular design provided by the agent-oriented processing architecture and its implementation in Daisie.

8.3 Summary

In this chapter the implemented systems based on the models presented in earlier chapters were described. The first of the implemented systems described was Daisie, a tight-coupled dialogue architecture framework based on the agent-oriented architecture and dialogue management model developed in earlier chapters. Daisie has been developed as a library of functionality that can be embedded in domain applications. A spatial language interpretation system based on Daisie was then described. Navspace, a simulated robot environment for processing route instructions, makes use of the incremental language contextualization process described in earlier chapters. Both Daisie and Navspace are freely available open source tools which fully implement the models presented in earlier chapters.

Chapter 9
Summary & Outlook

Abstract Having explored the issues of language interpretation and dialogue processing in the situated domain, this book now comes to a close. This concluding chapter begins in Section 9.1 with a summary of the theoretical and applied contributions made in this work. This is followed in Section 9.2 with a comparison of these contributions to some related theories of language processing and situated intelligence. Then, in Section 9.3, suggestions for future research that extend this book's contributions are presented. Finally, the chapter, and the book, conclude in Section 9.4 with some remarks on the position of this work with respect to the goals of situated language development.

9.1 Summary

Over the past eight chapters, this book has made a number of contributions to our understanding of language interpretation and dialogue processes in the situated domain. These contributions can be collated into three groups:

- **Agent-Oriented Dialogue Processing:** The first major contribution was the development of an agent-oriented dialogue processing architecture. The agent-oriented architecture resulted from an analysis of the state of the art in dialogue theory which concluded that contemporary dialogue management models are poorly suited to the situated domain since they rely on simplified non-agentive domain models as well as idealized contextualization assumptions. The agent-oriented dialogue processing architecture was developed to overcome these limitations. Notable features, or sub-contributions, of this architecture include:

    - **The Language Interface:** The Language Interface accounted for the ontological diversity of grammatical and knowledge processing resources in practical, situated dialogue system construction. Amongst other processing and representation layers, this interface included two levels of semantics representation, i.e., the linguistic semantic and the conceptual semantic

levels. This two-level approach was adopted to account for the contrasting
requirements of reusable grammar interfaces and specific domain application design. To provide flexible runtime alignment between these layers, a
functional transform model was also contributed which provides an ontological and context-sensitive mapping mechanism.
- **The Agent-Oriented Dialogue Management Model:** The Agent-Oriented
Dialogue Management (AODM) model fused an information state based
account of dialogue processing with a declarative theory of agent capability. This capability model provides the link between dialogue processes and
a theory of agency used to drive situated domain applications. The specifics
of the AODM model included a discourse state structure, as well as solutions to the problems of dialogue move selection, content integration, and
dialogue planning.
- **The Situated Contextualization Model:** The situated contextualization
model provided an integration of the AODM model with the realities of
situation specific language meaning through an iterative function application process. This model thus provided a solution that allows removal of
idealised contextualization assumptions made by contemporary dialogue
management models. Moreover, rather than focusing only on the contextualization of explicit content, this contextualization model places equal emphasis on augmentation of implicit content through function application.

- **Dialogic Route Interpretation:** The second major contribution of this book
was an approach to the verbal route instruction interpretation problem. Verbal
route instructions exhibit many interesting features of situated language such as
the prevalence of spatial relationship and action description content. Moreover,
providing solutions to the problem of route interpretation is appealing as route
instructions are a specific instance of plan structures. However, existing route
interpretation models have focused on a monologic approach to the problem
without accounting for the role of dialogue in overcoming uncertainty and ambiguity. The contributed dialogic route interpretation model leveraged off the
agent-oriented dialogue architecture's contextualization process. Specific features of the contribution included:

- **The Navspace Spatial Representation Model:** Interpretation processes
require appropriate models against which interpretation can be made. In
the situated domain, and particularly in the robotics domain, the requirements placed on such a representation are diverse. To meet the requirements
placed on representation by the needs of exophoric reference resolution,
action interpretation, and the robotics-oriented requirements of localization
and navigation, a multi-tiered and ontologically diverse spatial representation was proposed.
- **Route Capability Inventory and Contextualization Functions:** To instantiate the functional contextualization process, an inventory of spatial
processes and related contextualization functions was also contributed.
These functions describe the constituent actions of route instructions, and

also the functions used in augmenting and enhancing explicit discourse content.

- **Computational Realizations:** The third major contribution of this book has been implementations of the agent-oriented dialogue processing architecture and route interpretation models. These implementations were developed not only as realizations of the models just described, but also to explore other theories of formal dialogue management and spatial language use. Two systems have been contributed and made publicly available:

  - **Daisie:** The *Diaspace Adaptive Information State Interaction Executive* provides a tightly coupled dialogue system development framework. The framework has been designed to be highly modular through the use of a plugin mechanism, and has been developed in a manner which allows it to act as a library of functionalities which can be embedded in standalone applications.
  - **Navspace:** The Navspace application builds on Daisie and captures a realization of the action-oriented route interpretation model. Moreover, Navspace is a true robot simulation environment which can be used as a general purpose tool for studying spatial language use in a controlled situated environment.

These models have pulled together themes of language processing, dialogue structure, intelligent agent theory, and spatial representation and reasoning into a coherent whole. As such, the contributions made have relevance to the fields of artificial intelligence and computational linguistics, and in particular to sub-fields including dialogue modelling and management, and cognitive robotics. That said, the contributions made here constitute only a basis for situated dialogue agents; this basis must be related to, and extended towards, existing theories in intelligent agency and language processing. We turn to such relations in the next section.

9.2 Related Frameworks

As indicated, the contributions of this book cover topics including dialogue management theory, spatial language processing, and intelligent agent architectures. The relationships between this book's contributions and comparable works in these domains have already been established in earlier chapters. Rather than reiterating such specific comparisons here, in this section we will relate the models of situated dialogue intelligence pursued here to theories which sit at the boundaries of this work. The motivation for setting out such relationships is twofold. First, it illustrates the difference in objectives and stance taken by these theories in comparison to the present work. And second, it also shows the directions in which the current work can be extended to move us towards ever more comprehensive architectures of situated conversational intelligence. We begin by considering the theory which shows most similarity and motivation to the objectives pursued in this book.

- **Discourse Semantics:** Discourse Semantics theories (see Section 2.5.1 for an introduction) are frameworks of language representation and interpretation that focus on the coherence-driven processing of textual phenomena such as anaphora and ellipsis within a semantics stratum. Arguably the most comprehensive such theory, Asher and Lascarides's [2003] SDRT, provides a highly modular theory that makes use of flattened semantics and a pragmatics interface in the resolution of eventuality relationships. SDRT's goals of ellipsis and anaphora resolution have much in common with the contextualization goals of this book. However, while this book has placed emphasis on the resolution of linguistic devices within a pragmatics stratum, Discourse Semantics such as SDRT focus on the textual resolution of such phenomena. That said, rather than being competing solutions, Discourse Semantics and the models pursued here are largely complementary. A natural extension to the models presented in this book would see a tractable Discourse Semantics layer integrated between the Language Interface and the Agent-Oriented Dialogue Management layers.
- **Frame-Based Dialogue Management:** As described in Section 3.3.2, frame-based dialogue management models are extensively applied in the conduction of dialogues where complex form structures must be filled as a prelude to a database query. Although the dialogue management model contributed in this book started out from the perspective of frame-based and the related information-state based approaches to dialogue management, the requirements of fine-grained situated contextualization moved the contributed model away from a coarse-grained form-filling methodology. However, coarse-grained form-filling models continue to play an important role in practical dialogue applications. Fortunately, form-based and the agent-oriented model presented here are to the most part congruous. Within the agent-oriented architecture, a frame-structure can be interpreted as a domain-specific structure that is operated on at a level above the AODM layer. Specifically, in such an interpretation, the AODM model retains responsibility for basic conversational management and contextualization processes, while a frame-filling process operates at a meta-interaction level. This process has responsibility for frame-filling – but does so in terms of contextualised dialogue moves rather than dialogue acts or messages.
- **Plan Recognition Theory:** Plan and Intention Recognition Theory (see Carberry [1988] for an overview) was a widely studied field of artificial intelligence and language processing in the 1980s. Plan Recognition Theory (PRT) looked at phenomena including indirect speech act recognition and the licensing of opportunistic cooperative behaviour by artificial interlocutors in conversation. Although PRT fell out of favour due to the computationally expensive inference techniques applied in early models, as well as the redirection of logic-oriented dialogue research towards Discourse Semantics and Information State based models in the 1990s, PRT has seen a resurgence of interest in recent years. This resurgence, seen almost entirely outside of the dialogue community, is due to more tractable grammar-based and probabilistic frameworks having being applied to the plan recognition problem. Re-evaluation of the relationship be-

tween PRT and agent-oriented dialogue models is thus likely overdue. In light of developments in conversational management seen in the past fifteen years, we would argue that PRT should however no longer be considered a type of dialogue management in the strict sense (as for example it is characterized by McTear [2002]). Instead, PRT is a higher-order form of conversational intelligence that sits above basic dialogue management mechanism as provided by the AODM model.

- **Language Generation:** Although this book has attempted to provide a relatively comprehensive framework for situated dialogue management, there has been an emphasis on language interpretation rather than generation. True language realization systems were made use of, but while a model of situated language contextualization was contributed, little was said concerning the processes that select distinguishing features in composing the semantics to be fed to the language realizer. The field of computational language generation (see Reiter and Dale [2000] for an overview of a classical language generation stack) has a considerable amount to offer dialogue processing frameworks. Although dialogue management models must remain responsible for determining if a dialogue move is to be made, language planning processes, and in particular Referring Expression Generation mechanisms [Dale, 2007], offer important solutions to the problems of selecting distinguishing characteristics for referenced entities that can be factored into dialogue act expressions. Bringing into tighter alignment the topics of dialogue planning and text planning for a situated domain will provide a significant improvement to the extensibility and versatility of current situated dialogue systems.

9.3 Future Directions

This book has contributed from both a theoretical and applied perspective to the topics of situated dialogue modelling and management. Since the management of situated dialogue draws on a range of models, resources, and processes, it would be a lengthy and perhaps thankless task to enumerate all potential refinements to the models contributed or relied upon by this book. Thus, instead, in the following we suggest a future research topic for each of the four major contributing chapters in this book (i.e., Chapters 5 through 8):

- **Systemic Functional Semantics:** As a constituent of the Language Interface, a linguistic semantics model was developed to act as a uniform syntactic/semantic interface. Although this linguistic semantics built on a well-motivated theory of ideational meaning organization, the linguistic semantics framework integrated only a fragmentary account of interpersonal meaning and made no account for the description of textual meaning. Such limitations diminish the expressive power of the semantics framework, and thus limit the level of control which can be achieved in moving towards more naturalistic uses of language. A useful research question will be to determine if, using the same linguistically motivated principles as were applied for ideation meaning representation, we can develop a more comprehensive *Systemic Functional Semantics* as a common interface to a range of grammars including not only systemic grammars themselves, but also lexical grammars such Combinatory Categorial Grammars. The extension of this semantics interface up towards a true dynamic semantics account would be a valuable contribution to the field of practical dialogue systems.

- **Dialogue Act Aggregation:** The AODM model's dialogue planning process may result in the adoption of multiple planned dialogue acts in response to the integration of user dialogue acts. In the current model, a simple sequencing solution was applied to handling the realization of multiple acts. However, this simple strategy can lead to unnatural dialogue contributions consisting of a sequence of disjoint acts. Such unnatural constructions will become all the more pronounced where dialogue planning is to be coupled to a language interface capable of supporting the realization of additional communicative functions such as turn management signalling. Rather than relying on the sequencing of acts, a linguistically motivated, and potentially systemic network driven, aggregation process that complexes individual planned dialogue acts in coherent and natural constructions would be extremely beneficial contribution to the quality of dialogue planning results.

- **Generalizable Contextualization Functions:** The situated contextualization process developed here relies on inventories of augmentation and resolution functions. While a number of functions have been proposed here for the interpretation of spatial constructs, these functions were hand-crafted for the domain of route interpretation. The investigation of a more complete and generalizable framework of spatial interpretation functions is an important research challenge following from the work presented here. The development of such a framework will not only require the generalization and extension of the class of projective and topological constructs considered earlier, but must also couple the quantitative spatial reasoning model pursued here with qualitative spatial reasoning. Such coupling is necessary in order to account for relational properties such as transitivity, but also to facilitate interpretation where complete metric spatial models may not be available. Indeed, it should be noted that the contextualization process pursued here is not in any way hard-coded to a quantitative reasoning assumption; thus, the wholesale substitution of the quantitative model pursued here in favour of a qualitative account might prove an interesting research goal.

- **Declarative Framework Development:** In order to allow the creation of a prototype which best met the particular requirements set out by both architectural and project requirements, the Daisie framework was developed independently of existing agency theories and logic programming systems. Although necessary, this methodology has two significant drawbacks. First, the Java language used to implement Daisie is far from a useful programming language for knowledge-oriented system development. Second, Daisie necessarily made use of a relatively elementary theory of intentionality and intention management in comparison to the state of the art available to cognitive robotics and the intelligent agent community. A necessary, if not rather practical future research objective, will be the recasting of the Daisie framework in terms of a more suitable development environment. Minimally, such recasting should leverage off a logic programming language, but may benefit most from integration with a more dedicated agent framework such as Levesque et al.'s [1997] Golog.

9.4 Closing Comments

In this book, I presented an implemented architecture for situated dialogue processing, and instantiated that architecture for the case of verbal route instruction interpretation. Novel features of this work included the modular integration of an intentionality model alongside an exchange-structure motivated organization of discourse, plus the use of a functional contextualization process that operates over both implicit and explicit content in user contributions.

The practical motivation for this work was to overcome perceived limitations in contemporary dialogue architecture and management models that rely on simplifying assumptions such as non-agentive relativistic domain contexts. But motivation also comes from a more philosophical perspective. Spatial language and physical action descriptions permeate our daily linguistic and cognitive processes, and it would thus seem that to truly understand dialogue processes, or at least the simulation of dialogue processes, it is necessary to pull together diverse theories such as spatial reasoning and representation, rational agency, and dialogue processes.

That said, the objective of pulling these different models together into a coherent whole necessarily led to breadth rather than depth in the contributions made in this book. That is, there is room for extension in many of the models presented here. For example, the dialogue management model proposed did not attempt to handle what are well understood dialogue processing phenomena such as accommodation, or turn taking. However, my goal was not to focus on one aspect of dialogue competence and develop it in isolation to the point where it could not be applied in connection with existing methods. The point, as with a theory such as SDRT, was to pull a number of different models together into a coherent whole, and contribute, I hope, a number of important pieces to the puzzle of situated dialogue system development.

References

Jan Alexandersson and Tilman Becker. Overlay as the basic operation for discourse processing in a multimodal dialogue system. In *Proceedings of the IJCAI Workshop : Knowledge and Reasoning in Practical Dialogue Systems*, Seattle (USA), 2001.

Jan Alexandersson and Tilman Becker. The formal foundations underlying overlay. In *Proceedings of IWCS-5, Fifth International Workshop on Computational Semantics*, 2003.

Jan Alexandersson, Bianka Buschbeck-Wolf, Tsutomu Fujinami, Michael Kipp, Stephan Koch, Elisabeth Maier, Norbert Reithinger, Birte Schmitz, and Melanie Siegel. Dialogue Acts in VERBMOBIL-2 (Second edition). Verbmobil Report 226, University of the Saarland, Saarbrücken, Germany, 1998.

James Allen and Mark Core. Draft of DAMSL: Dialog act markup in several layers. Unpublished manuscript, 1997.

James Allen, George Ferguson, and Amanda Stent. An architecture for more realistic conversational systems. In *Proceedings of the 2001 International Conference on Intelligent User Interfaces*, pages 1–8, 2001a.

James Allen et al. An architecture for a generic dialogue shell. *Natural Language Engineering*, 6(3):1–16, 2000.

James F. Allen and C. Raymond Perrault. Analyzing intention in utterances. In Karen Sparck-Jones, Barbara J. Grosz, and Bonnie Lynn Webber, editors, *Readings in Natural Language Processing*, pages 441–458. Morgan Kaufmann, Los Altos, 1980.

James F. Allen, Bradford W. Miller, Eric K. Ringger, and Teresa Sikorski. A robust system for natural spoken dialogue. In *Proceedings of the 34th Annual Meeting of the ACL*, 1996.

James F. Allen, Donna K. Byron, Myroslava Dzikovska, George Ferguson, Lucian Galescu, and Amanda Stent. Towards conversational human-computer interaction. *AI Magazine*, 22(4):27–38, 2001b.

Jens Allwood. An activity-based approach to pragmatics. In David Traum, editor, *Working Notes: AAAI Fall Symposium on Communicative Action in Humans and Machines*, pages 6–11, Menlo Park, California, 1997. AAAI, American Association for Artificial Intelligence.

Hiyan Alshawi, editor. *The Core Language Engine*. MIT Press, Cambridge, Massachusetts, 1992.

Hiyan Alshawi, David Carter, Björn Gambäck, and Manny Rayner. Swedish-English QLF translation. In Hiyan Alshawi, editor, *The Core Language Engine*, pages 277–319. MIT Press, 1992.

A. H. Anderson, M. Bader, E. G. Bard, E. H. Boyle, G. M. Doherty, S. C. Garrod, S. D. Isard, J. C. Kowtko, J. M. McAllister, J. Miller, C. F. Sotillo, H. S. Thompson, and R. Weinert. The HCRC Map Task corpus. *Language and Speech*, 34(4): 351–366, 1992.

Douglas E. Appelt. *Planning English Sentences*. Cambridge University Press, Cambridge, England, 1985.

Nicholas Asher and Alex Lascarides. *Logics of conversation*. Cambridge University Press, Cambridge, 2003.

J. L. Austin. *How to do things with words*. Harvard University Press, 1962.

F. Baader and T. Nipkow. *Term Rewriting and all that*. Cambridge University Press, Cambridge, 1998. ISBN 0521455200.

Franz Baader. Description logic terminology. In *Description Logic Handbook*, pages 485–495, 2003.

Kent Bach. Pragmatics and the philosophy of language. In Lawrence R. Horn and Gegory Ward, editors, *The Handbook of Pragmatics*, pages 463–487. Blackwell Publishing, 2004.

Kent Bach and Robert M. Harnish. *Linguistic Communication and Speech Acts*. The MIT Press, Cambridge, Massachusetts, 1979.

Colin F. Baker, Charles J. Fillmore, and John B. Lowe. The Berkeley FrameNet Project. In *Proceedings of the ACL/COLING-98*, pages 86–90, Montreal, Quebec, 1998.

Jason Baldridge and Geert-Jan Kruijff. Coupling CCG and Hybrid Logic Dependency Semantics. In *Proceedings of 40th Annual Meeting of the Association for Computational Linguistics*, pages 319–326, Philadelphia, Pennsylvania, 2002.

Thomas T. Ballmer and Waltraud Brennenstuhl. *Speech Act Classification: A Study in the Lexical Analysis of English Speech Activity Verbs*. Springer-Verlag, Berlin, 1981.

J. Bateman, J. Hois, R. Ross, T. Tenbrink, and S. Farrar. The Generalized Upper Model 3.0: Documentation. SFB/TR8 internal report, Collaborative Research Center for Spatial Cognition, University of Bremen, Germany, 2008.

John Bateman, Stefano Borgo, Klaus Luttich, Claudio Masolo, and Till Mossakowski. Ontological modularity and spatial diversity. *Spatial Cognition & Computation*, 7:1, 2007.

John A. Bateman. Enabling technology for multilingual natural language generation: the KPML development environment. *Natural Language Engineering*, 3(1): 15–55, 1997.

John A. Bateman and Scott Farrar. Modelling models of robot navigation using formal spatial ontology. In Christian Freksa, Markus Knauff, Bernd Krieg-Brückner, Bernhard Nebel, and Thomas Barkowsky, editors, *Spatial Cognition IV: Reasoning, Action, Interaction. International Conference Spatial Cognition 2004, Frauenchiemsee, Germany, October 2004, Proceedings*, pages 366–389, Berlin, Heidelberg, 2005. Springer.

Nicholas J. Belkin, Colleen Cool, Adelheit Stein, and Ulrich Thiel. Cases, Scripts, and Information Seeking Strategies: On the Design of Interactive Information Retrieval Systems. *Expert Systems and Applications*, 9(3):379–395, 1995. (Also available: "Arbeitspapiere der GMD", No. 875, Sankt Augustin, Germany 1994).

Margaret Berry. Systemic linguistics and discourse analysis: a multi-layered approach to exchange structure. In Malcolm Coulthard and Michael Montgomery, editors, *Studies in Discourse Analysis*, pages 120–145. Routledge and Kegan Paul, London, 1981.

Thomas Bittner and Barry Smith. A Theory of Granular Partitions. In Matthew Duckham, Michael F. Goodchild, and Michael F. Worboys, editors, *Foundations of Geographic Information Science*, pages 117–151. Taylor and Francis, London, 2003. URL http://wings.buffalo.edu/philosophy/faculty/smith/articles/partitions.pdf.

Patrick Blackburn and Johan Bos. *Representation and Inference for Natural Language: A First Course in Computational Semantics*. CSLI, 2005.

Peter Bohlin, Robin Cooper, Elizabet Engdahl, and Staffan Larsson. Information states and dialogue move engines. In Jan Alexandersson, editor, *Proceedings of the IJCAI-99 Workshop on Knowledge and Reasoning in Practical Dialogue Systems*, pages 25–31, Murray Hill, New Jersey, 1999. IJCAI.

Johan Bos and Tetsushi Oka. An inference-based approach to dialogue system design. In *COLING 2002*, 2002.

Johan Bos and Tetsushi Oka. Meaningful conversation with mobile robots. *Advanced Robotics*, 21(1):209–232, 2007. URL http://dx.doi.org/10.1163/156855307779293661.

Johan Bos, Ewan Klein, Oliver Lemon, and Tetsushi Oka. DIPPER: Description and Formalisation of an Information-State Update Dialogue System Architecture. In *4th SIGdial Workshop on Discourse and Dialogue*, 2003.

M.E. Bratman. *Intentions, Plans, and Practical Reason*. Harvard University Press, Cambridge, MA, USA, 1987.

Gillian Brown and Geoff Yule. *Discourse Analysis*. Cambridge University Press, Cambridge, 1983.

Kerstin Buecher, Yves Forkl, Gnther Grz, Martin Klarner, and Bernd Ludwig. Discourse and application modeling for dialogue systems. In *Proceedings of the 2001 Workshop on Applications of Description Logics*, 2001.

Dirk Buehler, Jochen Hssler, Sven Krger, and Wolfgang Minker. Flexible multimodal human-machine interaction in mobile environments. In *Proceedings of the ECAI 2002 Workshop on Artificial Intelligence in Mobile System (AIMS)*, Lyon, France, 2002.

Guido Bugmann, Ewen Klein, Stanislao Lauria, and Theocharis Kyriacou. Corpus-Based Robotics: A Route Instruction Example. In *Proceedings of IAS-8*, 2004.

Trung H. Bui, Mannes Poel, Anton Nijholt, and Job Zwiers. A tractable ddn-pomdp approach to affective dialogue modeling for general probabilistic frame-based dialogue systems. In *Proceedings of the 5th Workshop on Knowledge and Reasoning in Practical Dialogue Systems*, 2007.

Harry Bunt. Dimensions in dialogue act annotation. In *Proceedings of the Fifth International Conference on Language Resources and Evaluation (LREC 2006)*, Genova, Italy, 2006.

Harry C. Bunt. Context and dialogue control. *THINK Quarterly*, 3:19–31, 1994.

Christopher S. Butler. Discourse systems and structures and their place within an overall systemic model. In James D. Benson and William S. Greaves, editors, *Systemic perspectives on discourse: selected theoretical papers from the 9th International Systemic Workshop*. Ablex, Norwood, NJ, 1985.

Sandra Carberry. Plan Recognition and User Modelling. *Computational Linguistics*, 14(3), September 1988.

J Carletta, A. Isard, S.Isard, J. Kowtko, and G. Doherty-Sneddon. HCRC Dialogue structure Coding Manual. Technical report, Human Coomunication Research Centre, Scotland, 1996.

Jennifer Chu-Carroll and Sandra Carberry. Collaborative response generation in planning dialogues. *Computational Linguistics*, 24(3):355–400, 1998.

H. Clark. *Using Language*. Cambridge University Press, Cambridge, UK, 1996.

P. R. Cohen and C. R. Perrault. Elements of a Plan-Based Theory of Speech Acts. *Cognitive Science*, 3:177–212, 1979.

Philip R. Cohen and Hector J. Levesque. Confirmations and joint action. In *Proceedings of the International Joint Conference on Artificial Intelligence*, pages 951–959, 1991.

P.R. Cohen and H.J. Levesque. Intention is choice with commitment. *Artificial Intelligence*, 42:213–261, 1990.

A.G. Cohn and S.M. Hazarika. Qualitative spatial representation and reasoning: an overview. *Fundamenta Informaticae*, 43:2–32, 2001.

R. W. Collier and G. M. P. O'Hare. Modeling and programming with commitment rules in agent factory. In Giurca, Gasevic, and Taveter, editors, *Handbook of Research on Emerging Rule-Based Languages and Technologies: Open Solutions and Approaches*. IGI Publishing, 2009.

Rem W. Collier. *Agent Factory: A Framework for the Engineering of Agent Oriented Applications*. PhD thesis, University College Dublin, 2001.

Ann Copesteak, Dan Flickinger, Carl Pollard, and Ivan A. Sag. Minimal recursion semantics: An introduction. *Research on Language and Computation*, 3:281332, 2005.

Mark G. Core and James F. Allen. Coding dialogues with the DAMSL annotation scheme. In David Traum, editor, *Working Notes: AAAI Fall Symposium on Communicative Action in Humans and Machines*, pages 28–35, Menlo Park, California, 1997. AAAI.

R.M. Coulthard and D. Brazil. *Exchange Structure*. Number 5 in Discourse Analysis Monographs. University of Birmingham English Language Research, Birmingham, 1979.

Cowper. *A Concise Introduction to Syntactic Theory: The Government-binding Approach*. Chicago University Press, 1992.

Matteo Cristani and Anthony G. Cohn. SpaceML: A mark-up language for spatial knowledge. *J. Vis. Lang. Comput*, 13(1):97–116, 2002. URL http://www.idealibrary.com/links/doi/10.1006/jvlc.2001.0228.

Cycorp. OpenCyc 0.7.0. Technical report, Cyc Corp., 2004. URL http://www.opencyc.org.

R. Dale. The generation of referring expressions: Where we've been, how we got here, and where we're going. *Natural Language Processing and Knowledge Engineering, 2007. NLP-KE 2007. International Conference on*, pages 5–5, 30 2007-Sept. 1 2007. doi: 10.1109/NLPKE.2007.4368000.

Robert Dale, Sabine Geldof, and Jean-Philippe Prost. Using natural language generation in automatic route description. *Journal of Research and Practice in Information Technology*, 37(1):89–105, 2005. URL http://www.jrpit.acs.org.au/jrpit/JRPITVolumes/JRPIT37/JRPIT37.1.89.pdf.

M.P. Daniel and M. Denis. Spatial descriptions as navigational aids: a cognitive analysis of route directions. *Kognitionswissenschaft*, 7:45–52, 1998.

T. Daradoumis. Using rhetorical relations in building a coherent conversational teaching session. In R. Beun, M. Baker, and Reiner M., editors, *Dialogue and instruction*, pages 56–71. Springer, Heidelberg, 1995.

M. Denis. The Description of Routes: A Cognitive Approach to the Production of Spatial Discourse. *Cahiers de Psychologie Cognitive*, 16:409–458, 1997.

Daniel C. Dennett. *The Intentional Stance*. The MIT Press, Massachusetts, 1987.

Stephan Dilley, John A. Bateman, Ulrich Thiel, and Anne Tissen. Integrating Natural Language Components into Graphical Discourse. In *Proceedings of the Third Conference on Applied Natural Language Processing*, pages 72–79, Trento, Italy, 1992. Association for Computational Linguistics. 31 March - 3 April.

Myroslava O. Dizikovska, James F. Allen, and Mary D. Swift. Linking semantic and knowledge represenations in a multi-domain dialogue system. *Journal of Logic and Computation*, 2007.

Patrick Doherty, Gösta Granlund, Krzystof Kuchcinski, Erik Sandewall, Klas Nordberg, Erik Skarman, and Johan Wiklund. The WITAS Unmanned Aerial Vehicle Project. In W. Horn, editor, *ECAI 2000. Proceedings of the 14th European Conference on Artificial Intelligence*, pages 747–755, Berlin, 2000.

John Dowding, Jean Mark Gawron, Doug Appelt, John Bear, Lynn Cherny, Robert Moore, and Douglas Moran. GEMINI: a natural language understanding system for spoken-language understanding. In *Proceedings of the 31st Annual Meeting of the Association for Computational Linguistics*. Association for Computational Linguistics, 1993.

Markus Egg and Gisela Redeker. Underspecified discourse representation. In Anton Benz and Peter Khnlein, editors, *Constraints in Discourse*, 2007.

A. Egges, A. Nijholt, and R. op den Akker. Dialogs with BDP agents in virtual environments. In *Knowledge and Reasoning in Practical Dialogue Systems. Working Notes, IJCAI-2001*, 2001.

Suzanne Eggins. *An Introduction to Systemic Functional Linguistics*. Continuum, 2nd edition, 2004.

Ralph Engel. Spin: Language understanding for spoken dialogue systems using a production system approach. In *Proceedings of the 7th International Conference on Spoken Language Processing*, Denver, Colorado, 2002.

Carola Eschenbach. Geometric structures of frames of reference and natural language semantics. *Spatial Cognition and Computation*, 1(4):329–348, 1999.

S. Farrar and J. Bateman. General ontology baseline. SFB/TR8 internal report I1-[OntoSpace]: D1, Collaborative Research Center for Spatial Cognition, University of Bremen, Germany, 2004.

Robin P. Fawcett and Bethan L. Davies. Monologue as a turn in dialogue: towards an integration of exchange structure theory and rhetorical structure theory. In

Robert Dale, Eduard H. Hovy, Dietmar Rösner, and Olivero Stock, editors, *Aspects of automated natural language generation*, pages 151–166. Springer, 1992. (Proceedings of the 6th International Workshop on Natural Language Generation, Trento, Italy, April 1992).

Robin P. Fawcett and David Young, editors. *New developments in systemic linguistics: theory and application*, volume 2 of *Open Linguistics Series*. Pinter, London, 1988.

Robin P. Fawcett, Anita van der Mije, and Carla van Wissen. Towards a systemic flowchart model for discourse structure. In Robin P. Fawcett and David Young, editors, *New Developments in Systemic Linguistics: Volume 2*, pages 116–143. Pinter Publishers, London, 1988.

Christiane Fellbaum and George Miller, editors. *WordNet: An electronic lexical database*. The MIT Press, 1998.

J. Ferber. *Multi-Agents Systems - An Introduction to Distributed Artificial Intelligence*. Addison-Wesley, 1999.

George Ferguson and James F. Allen. TRIPS: An integrated intelligent problem-solving assistant. In *Proceedings of the 15th National Conference on Artificial Intelligence (AAAI-98) and of the 10th Conference on Innovative Applications of Artificial Intelligence (IAAI-98)*, pages 567–573, Menlo Park, July 26–30 1998. AAAI Press. ISBN 0-262-51098-7.

Raquel Fernndez and Matthew Purver. Information state update: Semantics or pragmatics? In *Proccedings of the 8th Workshop on the Semantics and Pragmatics of Dialogue*, 2004.

Richard E. Fikes and Nils J. Nilsson. STRIPS: A new approach to the application of theorem proving to problem solving. In D. C. Cooper, editor, *Proceedings of the 2nd International Joint Conference on Artificial Intelligence (IJCAI'71)*, pages 608–620, London, UK, September 1971. William Kaufmann. ISBN 0-934613-34-6.

C. Freksa. Using orientation information for qualitative spatial reasoning. In A. U. Frank, I. Campari, and U. Formentini, editors, *Theories and methods of spatio-temporal reasoning in geographic space*, volume 639 of *LNCS*, pages 162–178. Springer, Berlin, 1992.

Udo Frese. A discussion of simultaneous localization and mapping. *Auton. Robots*, 20(1):25–42, 2006. URL http://dx.doi.org/10.1007/s10514-006-5735-x.

Agne Frisch and Martin Stenberg. Navigo an in-vehicle navigation dialogue system. Master's thesis, University of Gothenburg, 2008.

M.P. Georgeff and A.L. Lansky. Reactive reasoning & planning. In *Proceedings of the Sixth Intenational Conference on Artificial Intelligence (AAAI-87)*, pages 677–682, Seatle, WA, USA, 1987.

J.J. Gibson. The theory of affordances. In R.Shaw and J.Brandsford, editors, *Perceiving, Acting, and Knowing: Toward and Ecological Psychology*, pages 62–82. Erlbaum, Hillsdale, NJ, 1977.

Jonathan Ginzburg. Dynamics and the semantics of dialogue. In J. Seligman, editor, *Language, logic and computation*, volume 1, pages 221–237. CSLI Lecture Notes, CSLI Stanford, 1996.

Peter Gorniak and Deb Roy. Situated language understanding as filtering perceived affordances. *Cognitive Science*, 31(2):197–231, 2007.

Juliana Goschler, Elena Andonova, and Robert J. Ross. Perspective use and perspective shift in spatial dialogue. In *Spatial Cognition VI: Learning, Reasoning and Talking about Space*, number 5241 in Lecture notes in Artifiicial Intelligence, pages 250–265. Springer, 2008.

Mark Green. A survey of three dialogue models. *ACM Transactions on Graphics*, 5(3):244–275, 1986.

H. Paul Grice. Logic and conversation. In Peter Cole and Jerry L. Morgan, editors, *Speech Acts*, volume 3 of *Syntax and Semantics*, pages 43–58. Academic Press, New York, 1975.

Barbara J. Grosz and Candace L. Sidner. Attention, Intentions and the Structure of Discourse. *Computational Linguistics*, 12(3):175–204, July-September 1986.

Barbara J. Grosz and Candace L. Sidner. Plans for discourse. In P.R. Cohen, J. L. Morgan, and M. E. Pollack, editors, *Intentions and Communication*, pages 417–444. MIT Press, Cambridge, MA, USA, 1990.

B.J. Grosz, A.K. Joshi, and S. Weinstein. Centering: a framework for modelling the local coherence of discourse. *Computational Linguistics*, 21(2):203–226, 1995.

Alexander Gruenstein. Conversational Interfaces: A Domain-Independent Architecture for Task-Oriented Dialogues. Master's thesis, Stanford University, 2002.

Iryna Gurevych, Robert Porzel, Elena Slinko, Norbert Pfleger, Jan Alexandersson, and Stefan Merten. Less is more: using a single knowledge representation in dialogue systems. In *Proceedings of the NAACL 2003 Workshop on Text Meaning*. Association for Computational Linguistics, 2003. URL http://acl.ldc.upenn.edu/W/W03/W03-0903.pdf.

Afsaneh Haddadi. *Communciation and Cooperation in Agent Systems: A Pragmatic Theory*. Number 1056 in Lecture Notes in Computer Science. Springer-Verlag: Heidelberg, Germany, 1996. ISBN 3-540-61044-8.

Michael A. K. Halliday. Language as code and language as behaviour: a systemic-functional interpretation of the nature and ontogenesis of dialogue. In Michael A. K. Halliday, Robin P. Fawcett, S. Lamb, and A. Makkai, editors, *The semiotics of language and cultbure*, volume 1, pages 3–35. Frances Pinter, London, 1984. (Open Linguistics Series).

Michael A. K. Halliday. *An Introduction to Functional Grammar*. Edward Arnold, London, 1985. (2nd. edition 1994).

Michael A. K. Halliday and Ruqaiya Hasan. *Cohesion in English*. Longman, London, 1976.

Michael A. K. Halliday and Christian M. I. M. Matthiessen. *Construing experience through meaning: a language-based approach to cognition*. Cassell, London, 1999.

Michael A. K. Halliday and Christian M. I. M. Matthiessen. *An Introduction to Functional Grammar*. Edward Arnold, London, 3rd edition, 2004.

Charles Hamblim. Questions in Montague English. *Foundations of Language*, 10 (1):41–53, 1973.

Zellig Harris. Discourse Analysis: a Sample Text. *Language*, 28:474–494, 1952.

C. A. R. Hoare. *Communicating Sequential Processes*. Prentice-Hall, Englewood Cliffs, New Jersey, 1985.

Jerry R. Hobbs, Mark E. Stickel, Douglas E. Appelt, and Paul Martin. Interpretation as abduction. *Artificial Intelligence*, 63:69–142, 1993. (ARTINT 1059).

J. Hois, Thora Tenbrink, R. Ross, and J. Bateman. The Generalized Upper Model spatial extension: a linguistically-motivated ontology for the semantics of spatial language. SFB/TR8 internal report, Collaborative Research Center for Spatial Cognition, University of Bremen, Germany, 2008. Version 3.0.

Joana Hois and Oliver Kutz. Counterparts in Language and Space - Similarity and 8-Connection. In Carola Eschenbach and Michael Grüninger, editors, *Proceedings of the International Conference on Formal Ontology in Information Systems (FOIS)*, pages 266–279, Amsterdam, 2008. IOS Press.

Laurence R. Horn. Implicature. In Lawrence R. Horn and Gegory Ward, editors, *The Handbook of Pragmatics*, pages 3–28. Blackwell Publishing, 2004.

Lawrence R. Horn and Gegory Ward, editors. *The Handbook of Pragmatics*. Blackwell Publishing, 2004.

Matthew Horridge and Peter F. Patel-Schneider. Manchester Syntax for OWL 1.1. *OWLED 2008, 4th international workshop OWL: Experiences and Directions*, 2008.

Eduard H. Hovy. Parsimonious and Profligate Approaches to the Question of Discourse Structure Relations. In *5th. International Workshop on Natural Language Generation, 3-6 June 1990*, Pittsburgh, PA., 1990. Organized by Kathleen R. McKeown (Columbia University), Johanna D. Moore (University of Pittsburgh) and Sergei Nirenburg (Carnegie Mellon University).

Yan Huang. *Pragmatics*. Oxford University Press, 2006.

U. Hustadt and R. A. Schmidt. MSPASS: Modal reasoning by translation and first-order resolution. In R. Dyckhoff, editor, *Automated Reasoning with Analytic Tableaux and Related Methods, International Conference (TABLEAUX 2000)*, volume 1847 of *Lecture Notes in Artificial Intelligence*, pages 67–71. Springer, 2000. ISBN 3-540-67697-X. URL http://www.cs.man.ac.uk/~schmidt/publications/HustadtSchmidt00b.html.

Arne Jönsson. Dialogue actions for natural language interfaces. In *Proceedings of the International Joint Conference on Artificial Intelligence (IJCAI)*, pages 1405–1413, 1995.

Aravind K. Joshi. An introduction to Tree Adjoining Grammar. In Alexis Manaster-Ramer, editor, *Mathematics of Language*, pages 87–114. John Benjamins Company, 1987.

Daniel Jurafsky. Pragmatics and computational linguistics. In Lawrence R. Horn and Gegory Ward, editors, *The Handbook of Pragmatics*, pages 578–604. Blackwell Publishing, 2004.

Daniel Jurafsky and James H. Martin. *Speech and Language Processing: An Introduction to Natural Language Processing, Computational Linguistics, and*

Speech Recognition. Prentice Hall, Englewood Cliffs, New Jersey, 2000. ISBN 0130950696.

Hans Kamp. A Theory of Truth and Semantic representation. In Jeroen A.G. Groenendijk, T.M.V. Janssen, and Martin B.J. Stokhof, editors, *Formal methods in the study of language*, number 136, pages 277–322. Mathematical Centre Tracts, Amsterdam, 1981.

Hans Kamp and Uwe Reyle. *From discourse to logic: introduction to modeltheoretic semantics of natural language, formal logic and discourse representation theory*. Kluwer Academic Publishers, London, Boston, Dordrecht, 1993. Studies in Linguistics and Philosophy, Volume 42.

Robert Kasper and Richard Whitney. SPL: A Sentence Plan Language for text generation. Technical Report forthcoming, Information Sciences Institute, 1989. 4676 Admiralty Way, Marina del Rey, California 90292-6695.

Michael Kay. XSL transformations (XSLT) version 2.0. W3C recommendation, W3C, January 2007. http://www.w3.org/TR/2007/REC-xslt20-20070123/.

S. Keizer. *Reasoning under Uncertainty in Natural Language Dialogue using Bayesian Networks*. Phd thesis, University of Twente, Enschede, September 2003. publisher: Twente University Press, publisherlocation: Enschede, ISSN: 1381-3617; No 03-55, ISBN: 90-365-1951-9, Numberofpages: 166.

Simon Keizer, Rieks op den Akker, and Anton Nijholt. Dialogue act recognition with bayesian networks for dutch dialogues. In *Proceedings of the 3rd SIGdial workshop on Discourse and dialogue*, pages 88–94, Morristown, NJ, USA, 2002. Association for Computational Linguistics.

J.D. Kelleher. Attention driven reference resolution in multimodal contexts. *Artificial Intelligence Review*, 25:21–35, 2007.

John Kelleher, Geert-Jan Kruijff, and Fintan Costello. Proximity in context: an empirically grounded computational model of proximity for processing topological spatial expression. In *Proceedings of Coling-ACL '06, Sydney Australia*, 2006.

Roberta L. Klatzky. Allocentric and egocentric spatial representations: Definitions, distinctions, and interconnections. *Lecture Notes in Computer Science*, 1404, 1998. ISSN 0302-9743.

Alexander Klippel, Carsten Dewey, Markus Knauff, Kai-Florian Richter, Dan R. Montello, Christian Freksa, and Esther-Anna Loeliger. Direction Concepts in Wayfinding Assistance Systems. In Jörg Baus, Christian Kray, and Robert Porzel, editors, *Workshop on Artificial Intelligence in Mobile Systems (AIMS'04), Proceedings*, number Memo 84 in SFB 378, pages 1–8, Saarbrücken, 2004.

M. Knees. Designing an anaphora resolution algorithm for route instructions. Master's thesis, The University of Edinburgh, 2002. URL http://www.inf.ed.ac.uk/publications/thesis/online/IM020002.pdf.

H. Knublauch, M. A. Musen, and A. L. Rector. Editing description logic ontologies with the protege owl plugin. In *International Workshop on Description Logics*, Whistler, Canada, 2004.

Kurt Konolige. COLBERT: A language for reactive control in Sapphira. In Gerhard Brewka, Christopher Habel, and Bernhard Nebel, editors, *Proceedings of the 21st Annual German Conference on Artificial Intelligence (KI-97): Advances in Arti-*

ficial Intelligence, volume 1303 of *LNAI*, pages 31–52, Berlin, September 9–12 1997. Springer. ISBN 3-540-63493-2.

Christopher Kray, J. Baus, H. Zimmer, H. Speiser, and A. Krüger. Two Path Prepositions: Along and Past. In D.R. Montello, editor, *Spatial Information Theory: Foundations of Geographic Information Science (International Conference Proceedings COSIT 2001 Morro Bay, CA, USA, September 19-23, 2001)*, Berlin and Heidelberg, 2001. Springer.

Joern Kreutel and Colin Matheson. Context-dependent interpretation and implicit dialogue acts. In *Perspectives on Dialogue in the New Millenium*. John Benjamins, 2003.

B. Krieg-Brückner, T. Röfer, H.-O. Carmesin, and R. Müller. A Taxonomy of Spatial Knowledge for Navigation and its Application to the Bremen Autonomous Wheelchair. In C. Freksa, C. Habel, and K.F. Wender, editors, *Spatial Cognition I - An interdisciplinary approach to representing and processing spatial knowledge*, pages 373–397. Springer, Berlin, 1998. URL http://link.springer-ny.com/link/service/series/0558/tocs/t1404.htm.

Bernd Krieg-Brückner and Hui Shi. Orientation Calculi and Route Graphs: Towards Semantic Representations for Route Descriptions. In M. Raubal, H. Miller, A. Frank, and M. Goodchild, editors, *Geographic Information Science - Fourth International Conference, GIScience 2006*, number 4197 in Lecture Notes in Computer Science, pages 234–250. Springer, Berlin, 2006.

Bernd Krieg-Brückner, Udo Frese, Klaus Lüttich, Christian Mandel, Till Mossakowski, and Robert Ross. Specification of an Ontology for Route Graphs. In Christian Freksa, Markus Knauff, Bernd Krieg-Brückner, Bernhard Nebel, and Thomas Barkowsky, editors, *Spatial Cognition IV: Reasoning, Action, Interaction. International Conference Spatial Cognition 2004, Frauenchiemsee, Germany, October 2004, Proceedings*, pages 390–412, Berlin, Heidelberg, 2005. Springer.

Ivana Kruijff-Korbayova and Mark Steedman. Discourse on information structure. *Journal of Logic, Language and Information: Special Issue on Discourse and Information Structure*, 12(3):249–259, 2003.

Ivana Kruijff-Korbayova, Stina Ericsson, Kepa Joseba Rodriguez, and Elena Karagjosova. Producing contextually appropriate intonation is an information-state based dialogue system. In *Proceedings of the 10th Conference of the European Chapter of the Association for Computational Linguistics (EACL)*, pages 227–234, ACL, Budapest, Hungary, 2003.

Benjamin Kuipers. The spatial semantic hierarchy. *Artificial Intelligence*, 19:191–233, 2000.

O. Kutz, C. Lutz, F. Wolter, and M. Zakharyaschev. ε-Connections of Abstract Description Systems. *Artificial Intelligence*, 156(1):1–73, 2004.

Staffan Larsson. Coding schemas for dialogue moves. Technical report, Göteborg University, 1998.

Staffan Larsson. *Issue-Based Dialogue Management*. Ph.d. dissertation, Department of Linguistics, Göteborg University, Göteborg, 2002.

Staffan Larsson and David Traum. Information state and dialogue management in the TRINDI Dialogue Move Engine Toolkit. *Natural Language Engineering*, pages 323–340, 2000. Special Issue on Best Practice in Spoken Language Dialogue Systems Engineering.

S. Lauria, T. Kyriacou, G. Bugmann, J. Bos, and E. Klein. Converting natural language route instructions into robot executable procedures. In *Proceedings of the 2002 IEEE Int. Workshop on Robot and Human Interactive Communication*, pages 223–228, Berlin, Germany, 2002.

Shaun Lawson, Emile van der Zee, and Laura Daley. Spatial language in computer mediated communication. In Phil Turner and Susan Turner, editors, *Exploration of Space,Technology, and Spatiality: Interdisciplinary Perspectives*. Hershey, Pennsylvania: Information Science Reference., 2008.

Oliver Lemon and Alexander Gruenstein. Multithreaded Context for Robust Conversational Interfaces: Context-sensitive speech recognition and interpretation of corrective fragments. *ACM Transactions on Computer-Human Interaction*, 11 (3):241–267, September 2004. ISSN 1073-0516.

Oliver Lemon, Alexander Gruenstein, and Stanley Peters. Collaborative Activities and Multi-tasking in Dialogue Systems. *Traitement Automatique des Langues (TAL)*, 43(2):131–154, 2002a. Special issue on dialogue.

Oliver Lemon, Prashant Parikh, and Stanley Peters. Probabilistic dialogue modelling. In *Proceedings of the 3rd SIGdial Workshop on Discourse and Dialogue*, pages 125– 128, 2002b.

Willem J.M. Levelt. The Speaker's Linearisation Problem. *Philosophical Transactions of the Royal Society, London*, B295:305–315, 1981.

H. Levesque, R. Reiter, Y. Lespérance, F. Lin, and R. Scherl. GOLOG: A logic programming language for dynamic domains. *Journal of Logic Programming*, 31:59–84, 1997.

H. J. Levesque, P. R. Cohen, and J. H. T. Nunez. On acting together. In Thomas Dietterich and William Swartout, editors, *Proceedings of the 8th National Conference on Artificial Intelligence (AAAI-90)*, pages 94–99, Boston, MA, USA, 1990. AAAI Press.

S. C. Levinson. *Pragmatics*. Cambridge University Press, Cambridge, England, 1983.

S. C. Levinson. Spatial language. In L. Nadel, editor, *Encyclopedia of Cognitive Science*, volume 3, pages 131–137. Nature Publishing Group., 2003a.

S.C. Levinson. Language and space. *Annual Review of Anthropology*, 25:353–382, 1996.

Stephen C. Levinson. *Space in language and cognition: explorations in cognitive diversity*. Cambridge University Press, Cambridge, 2003b.

M. Levit and D. Roy. Interpretation of spatial language in a map navigation task. *IEEE Trans. Systems, Man and Cybernetics*, 37(3):667–679, June 2007. URL http://dx.doi.org/10.1109/TSMCB.2006.889809.

Ian Lewin. A Formal Model of Conversational Game Theory. In *Fourth Workshop on the Semantics & Pragmantics of Dialogue*, 2000.

D. Lewis. Score keeping in a language game. *Journal of Philosophical Logic*, 8: 339–359, 1979.

Karen E. Lochbaum. A collaborative planning model of intentional structure. *Computational Linguistics*, 24(4):525–572, 1998.

Markus Lockelt, Tilman Becker, Norbert Pfeger, and Jan Alexandersson. Making sense of partial. In *Proceedings of the sixth workshop on the semantics and pragmatics of dialogue (EDILOG 2002)*, 2002.

Andy Luecking, Hannes Rieser, and Marc Staudacher. Sdrt and multi-modal situated communication. In *Proceedings of The 10th Workshop on the Semantics and Pragmatics of Dialogue (BRANDIAL)*, 2006.

K. Lüttich, B. Krieg-Brückner, and T. Mossakowski. Tramway Networks as Route Graphs. In E. Schnieder and G. Tarnai, editors, *FORMS/FORMAT 2004 – Formal Methods for Automation and Safety in Railway and Automotive Systems*, pages 109–119, 2004. ISBN 3-9803363-8-7.

Matthew Tierney MacMahon. *Following Natural Language Route Instructions*. PhD thesis, The University of Texas at Austin, 2007.

Brian MacWhinney et al. Cue validity and sentence interpretation in english, german and italian. *Journal of Verbal Learning and Verbal Behavior*, 23:127–150, 1984.

Evgeni Magid, Daniel Keren, Ehud Rivlin, and Irad Yavneh. Spline-based robot navigation. In *Proc. Of the IEEE/RSJ International Conf. on Intelligent Robots and Systems*, October 2006.

C. Mandel, K. Huebner, and T. Vierhuff. Towards an autonomous wheelchair: Cognitive aspects in service robotics. In *Proceedings of Towards Autonomous Robotic Systems (TAROS 2005)*, pages 165–172, 2005. URL http://www.taros.org.uk/.

C. Mandel, U. Frese, and T. Röfer. Robot navigation based on the mapping of coarse qualitative route descriptions to route graphs. In *Proceedings of the IEEE/RSJ International Conference on Intelligent Robots and Systems (IROS 2006)*, 2006.

William Mann and Jrn Kreutel. Speech acts and recognition of insincerity. In *The 8th Workshop on the Semantics and Pragmatics of Dialogue*, 2004.

William C. Mann. Dialogue analysis for diverse situations. In Johan Bos, Mary Ellen Foster, and Colin Mathesin, editors, *EDILOG 2002: Proceedings of the Sixth Workshop on the Semantics and Pragmatics of Dialogue*, pages 109–116. Cognitive Science Centre, University of Edinburgh, Edinburgh, 2002.

William C. Mann and Sandra A. Thompson. Rhetorical Structure Theory: Toward a Functional Theory of Text Organization. *Text*, 8(3):243–281, 1988.

D. Martin, A. Cheyer, and D. Moran. The Open Agent Architecture: a framework for building distributed software systems. *Applied Artificial Intelligence*, 13(1/2): 91–128, 1999.

James R. Martin. *English text: systems and structure*. Benjamins, Amsterdam, 1992.

Claudio Masolo, Stefano Borgo, Aldo Gangemi, Nicola Guarino, and Alessandro Oltramari. Ontologies library (final). WonderWeb Deliverable D18, ISTC-CNR, Padova, Italy, December 2003. URL http://wonderweb.semanticweb.org/deliverables/documents/D18.pdf.

Colin Matheson, Massimo Poesio, and David R. Traum. Modelling grounding and discourse obligations using update rules. In *ANLP*, pages 1–8, 2000. URL `http://acl.ldc.upenn.edu/A/A00/A00-2001.pdf`.

Nicolas Maudet, Philippe Mullerr, and Laurent Prvot. Conversational gameboard and discourse structure. In *Proceedings of the International Workshop on the Semantics and Pragmatics of Dialogue*, 2004.

J. McCarthy. Ascribing mental qualities to machines. Technical Report STAN-CS-79-725, Stanford University, 1979.

John McCarthy. Situations, actions and causal laws. In M. Minsky, editor, *Semantic Information Processing*, pages 410–417. MIT Press, 1968.

Michael F. McTear. Spoken dialogue technology: Enabling the conversational user interface. *ACM Computing Surveys (CSUR)*, 34(1):90 – 169, 2002.

Marie W. Meteer. *The Generation Gap: the problem of expressability in text planning*. PhD thesis, Computer and Information Sciences Department, University of Massachusetts, Amherst, Amherst, Massachusetts, February 1990.

Marvin Minsky. *The Society of Mind*. Simon and Schuster, New York, 1985.

T. Misu and T. Kawahara. Speech-based interactive information guidance system using question-answering technique. *Acoustics, Speech and Signal Processing, 2007. ICASSP 2007. IEEE International Conference on*, 4:IV–145–IV–148, April 2007. ISSN 1520-6149. doi: 10.1109/ICASSP.2007.367184.

Roser Morante, Simon Keizer, and Harry Bunt. A dialogue act based model for context updating. In *Proceedings of the 2007 Workshop on the Semantics and Pragmatics of Dialogue (DECALOG)*,, 2007.

R. Moratz, K. Fischer, and T. Tenbrink. Cognitive Modelling of Spatial Reference for Human-Robot Interaction. *International Journal On Artificial Intelligence Tools*, 10(4), 2001.

Reinhard Moratz, Jochen Renz, and Dietrich Wolter. Qualitative Spatial Reasoning about Line Segments. In W. Horn, editor, *Proceedings of the 14th European Conference on Artificial Intelligence (ECAI00)*, Amsterdam, 2000. IOS Press.

Boris Motik. Kaon2. http://kaon2.semanticweb.org/, 2007.

Philippe Muller, Laurent Prvot, and Nicolas Maudet. Social constraints on rhetorical relations in dialogue. In *Proceedings of the 2nd Workshop on Constraints in Discourse*, Ireland, 2006.

Robin R. Murphy. *Introduction to A.I. Robotics*. The MIT Press, 2000.

Daniele Nardi and Ronald J. Brachman. An introduction to description logics. In *Description Logic Handbook*, pages 1–40, 2003.

I. Niles and A. Pease. Toward a Standard Upper Ontology. In Chris Welty and Barry Smith, editors, *Formal Ontology in Information Systems (FOIS)*, Ogunquit, Maine, 2001. Association for Computing Machinery. URL `http://projects.teknowledge.com/HPKB/Publications/FOIS.pdf`.

Michael O'Donnell. A dynamic model of exchange. *Word*, 41(3):293–328, December 1990.

Michael O'Donnell. Dynamic representation of exchange structure. Technical report, Departments of Linguistics and Electrical Engineering, University of Sydney, 1992.

Michael O'Donnell. *Sentence analysis and generation: a systemic perspective*. PhD thesis, University of Sydney, Department of Linguistics, Sydney, Australia, 1994. URL http://www.wagsoft.com/Papers/Thesis/index.html.

Michael J. O'Donnell. Context in dynamic modelling. In M. Ghadessy, editor, *Text and Context in Functional Linguistic*, (CILT Series IV), pages 63–99. Benjamins, Amsterdam, 1999.

J. Kevin O'Regan. How to build consciousness into a robot: The sensorimotor approach. In Max Lungarella, Fumiya Iida, Josh C. Bongard, and Rolf Pfeifer, editors, *50 Years of Artificial Intelligence*, volume 4850 of *Lecture Notes in Computer Science*, pages 332–346. Springer, 2006.

C. Raymond Perrault and James F. Allen. A plan-based analysis of indirect speech acts. *American Journal of Computational Linguistics*, 6(3–4):167–182, 1980.

Norbert Pfleger, Jan Alexandersson, and Tilman Becker. Scoring functions for overlay and their application in discourse processing. In *Proceedings of KONVENS*, Saarbrcken (Germany), 2002.

Norbert Pfleger, Jan Alexandersson, and Tilman Becker. A robust and generic discourse model for multimodal dialogue. In *IJCAI-03 Workshop on Knowledge and Reasoning in Practical Dialogue Systems*, 2003.

Martin J. Pickering and Simon Garrod. Towards a mechanistic psychology of dialogue. *Behavioural and Brain Sciences*, 27(2):169–190, 2004.

Massimo Poesio and David Traum. Towards an axiomatization of dialogue acts. In *Processings of TWENDIAL, the Twente Workshop on the Formal Semantics and Pragmatics of Dialogues*, 1998.

Massimo Poesio and David R. Traum. Conversational actions and discourse situations. *Computational Intelligence*, 13(3):309–347, 1997.

A. Pokahr, L. Braubach, and W. Lamersdorf. Jadex: Implementing a bdi-infrastructure for jade agents. *in: EXP - In Search of Innovation (Special Issue on JADE)*, 3(2):76–85, September 2003.

Carl Pollard and Ivan A. Sag. *Head-Driven Phrase Structure Grammar*. University of Chicago Press and CSLI Publications, Chicago, Illinois, 1994.

Heather Pon-Barry, Brady Clark, Elizabeth Owen Bratt, Karl Schultz, and Stanley Peters. Evaluating the effectiveness of scot: a spoken conversational tutor. In *Proceedings of ITS 2004 Workshop on Dialogue-based Intelligent Tutoring Systems*, Maceio, Brazil., 2004a.

Heather Pon-Barry, Brady Clark, Karl Schultz, Elizabeth Owen Bratt, and Stanley Peters. Advantages of spoken language in dialogue-based tutoring systems. In *Proceedings of ITS 2004, 7th International Conference on Intelligent Tutoring Systems*, 2004b.

Heather Pon-Barry, Brady Clark, Karl Schultz, Elizabeth Owen Bratt, and Stanley Peters. Contextualizing learning in a reflective conversational tutor. In *Proceedings of The 4th IEEE International Conference on Advanced Learning Technologies*, Joensuu, Finland, 2004c.

R. Porzel, N. Pfleger, S. Merten, M. Loeckelt, I. Gurevych, Engel R., and J. Alexandersson. More on less: Further applications of ontologies in multi-modal dialogue systems. In *Proceedings of the IJCAI Workshop on Knowledge and Reasoning in Practical Dialogue Systems*, Acapulco, Mexico, 2003.

Robert Porzel and Iryan Gurevych. Towards context adaptive utterance interpretation. In *Proceedings of the 3rd SIGDial Workshop on Discourse and Dialogue. Association for Computational Linguistics, Philadelphia*, 2002.

Robert Porzel, Berenike Loos, and Vanessa Micelli. Making relative sense: From word-graphs to semantic frames. In *Proceedings of the HLT-NAACL 2004 Workshop on Scalable Natural Language Understanding*, 2004.

R Power. The organisation of purposeful dialogues. *Linguistics*, 17:107–151, 1979.

Laurent Prévot. Topic structure in route explanation dialogues. In *Proceedings of the workshop "Information structure, Discourse structure and discourse semantics" of the 13th European Summer School in Logic, Language and Information*, 2001.

Eric Prud'Hommeaux and Andy Seaborne. SPARQL query language for RDF. World Wide Web Consortium, Recommendation REC-rdf-sparql-query-20080115, January 2008.

Stephen G Pulman. Conversational games, belief revision and bayesian networks. In *Proceedings of the 7th Computational Linguistics in the Netherlands*, 1996.

Matthew Purver, Ronnie Cann, and Ruth Kempson. Grammars as parsers: Meeting the dialogue challenge. *Research on Language and Computation*, 4(2-3):289–326, October 2006.

L. R. Rabiner. A tutorial on hidden Markov models and selected applications in speech recognition. *Proceedings of the IEEE*, 77(2):257–285, February 1989.

A. Ramsay. *Formal Methods in Artificial Intelligence*. Cambridge University Press, 1988.

D.A. Randell, Z. Cui, and A.G. Cohn. A spatial logic based on regions and connection. In *Proceedings of the 3rd. International Conference on Knowledge Representation and Reasoning*, pages 165–176, San Mateo, 1992. Morgan Kaufmann.

Ehud Reiter and Robert Dale. *Building Natural Language Generation Systems*. Cambridge University Press, Cambridge, U.K., 2000.

Charles Rich, Candace L. Sidner, and Neal Lesh. Collagen: Applying collaborative discourse theory to human-computer interaction. *AI Magazine*, 2001.

Kai-Florian Richter. *Context-Specific Route Directions - Generation of Cognitively Motivated Wayfinding Instructions*, volume DisKi 314 / SFB/TR 8 Monographs Volume 3. IOS Press; Amsterdam; http://www.iospress.nl, 2008. ISBN 978-1-58603-852-6. URL http://www.iospress.nl/html/9781586038526.php.

Kai-Florian Richter and Alexander Klippel. A model for context-specific route directions. In *Proceedings of Spatial Cognition 04*, 2004.

Thomas Röfer and Axel Lankenau. Route-based robot navigation. *Künstliche Intelligenz*, 2002. Themenheft Spatial Cognition.

Massimo Romanell, Tilman Becker, and Jan Alexandersson. On plurals and overlay. In *Proceedings of DIALOR*, Nancy, France, 2005.

Robert Ross. MARC - applying multiagent systems to service robot control. Master's thesis, University College Dublin, 2004.

Robert Ross, Rem Collier, and G.M.P. O'Hare. Demonstrating social error recovery with agentfactory. In *Proceeedings of The Third International Joint Conference on Autonomous Agents and Multi Agent Systems*, 2004a.

Robert Ross, Rem Collier, and G.M.P. O'Hare. AF-APL Bridging Princples & Practices in Agent Oriented Languages. In *PROMAS 2004*, volume 3346 of *Lecture Notes in Artificial Intelligence*. Springer-Verlag, 2004b.

Robert J. Ross and John Bateman. Agency & Information State in Situated Dialogues: Analysis & Computational Modelling. In *Proceedings of the 13th International Workshop on the Semantics & Pragmatics of Dialogue (DiaHolmia)*, Stockholm, Sweden, 2009.

Robert J. Ross, Rem Collier, and G.M.P. O Hare. ALPHA A Language for Programming Hybrid Agents. Presented at Second European Workshop on Multi-Agent Systems (EUMAS 04), Dec 2004c.

Robert J. Ross, Hui Shi, Tillman Vierhuf, Bernd Krieg-Bruckner, and John Bateman. Towards Dialogue Based Shared Control of Navigating Robots. In *Proceedings of Spatial Cognition 04*, Germany, 2004d. Springer.

Stuart J. Russell and Peter Norvig. *Artificial Intelligence: a modern approach*. Prentice Hall, Englewood Cliffs, NJ, 1995.

Stuart J. Russell and Peter Norvig. *Artificial Intelligence: a modern approach*. Prentice Hall, Upper Saddle River, NJ, 2nd international edition, 2003.

H. Sacks, E. Schegloff, and G. Jefferson. A simplest systematics for the organisation of turn-taking for conversation. *Language*, 50:696–735, 1974.

M. David Sadek, Philippe Bretier, and E. Panaget. ARTIMIS: Natural dialogue meets rational agency. In *IJCAI (2)*, pages 1030–1035, 1997.

Jerrold Sadock. Speech acts. In Lawrence R. Horn and Gegory Ward, editors, *The Handbook of Pragmatics*, pages 53–73. Blackwell Publishing, 2004.

John Saeed. *Semantics*. Blackwell Publishers, Oxford, 1996.

D. Schlangen. *A Coherence-Based Approach to the Interpretation of Non-Sentential Utterances in Dialogue*. PhD thesis, School of Informatics, University of Edinburgh, 2003. URL http://www.inf.ed.ac.uk/publications/thesis/online/IP030019.pdf.

Marc Schröder and Jürgen Trouvain. The German text-to-speech synthesis system MARY: A tool for research, development and teaching. In *Proceedings of the 4th ISCA Tutorial and Research Workshop on Speech Synthesis, August 29 - September 1*, pages 131–136, Perthshire, Scotland, 2001. URL http://www.ssw4.org/papers/112.pdf.

Karl Schultz, Elizabeth Owen Bratt, Brady Clark, Stanley Peters, Heather Pon-Barry, and Pucktada Treeratpituk. A scalable, reusable spoken conversational tutor: Scot. In Vincent Aleven, Ulrich Hoppe, Judy Kay, Riichiro Mizoguchi, Helen Pain, Felisa Verdejo, and Kalina Yacef, editors, *AIED 2003 Supplementary Proceedings*, pages 367–377, Sydney, Australia, 2003.

J. Searle. *Speech Acts: An Essay in the Philosophy of Language*. Cambridge University Press, Cambridge, England, 1969.

J. R. Searle. A taxonomy of illocutionary acts. In K. Gunderson, editor, *Language, Mind and Knowledge*, pages 344–369. University of Minnesota Press, Minneapolis, 1975.

Hui Shi and Thora Tenbrink. Telling Rolland where to go: HRI dialogues on route navigation. In *WoSLaD Workshop on Spatial Language and Dialogue, October 23-25, 2005*, 2005.

Hui Shi, Robert Ross, and John Bateman. Formalising Control in Robust Spoken Dialogue Systems. In Bernhard K. Aichernig and Bernhard Beckert, editors, *Proceedings of Software Engineering and Formal Methods 2005*, IEEE, pages 332–341. IEEE Computer Society, 2005. ISBN 0-7695-2435-4.

Yoav Shoham. Agent Oriented Programming. *Artificial Intelligence*, 60:51–92, 1993.

Candice L. Sidner. An artificial discourse language for collaborative negotiation. In Barbara Hayes-Roth and Richard Korf, editors, *Proceedings of the Twelfth National Conference on Artificial Intelligence*, pages 814–819, Menlo Park, California, 1994. American Association for Artificial Intelligence, AAAI Press.

John Sinclair and R. Malcolm Coulthard. *Towards an Analysis of Discourse: the English used by teachers and pupils*. Oxford University Press, London, 1975.

Evren Sirin, Bijan Parsia, Bernardo Cuenca Grau, Aditya Kalyanpur, and Yarden Katz. Pellet: A practical OWL-DL reasoner. *J. Web Sem*, 5(2):51–53, 2007. URL http://dx.doi.org/10.1016/j.websem.2007.03.004.

Stefan Sitter and Adelheit Stein. Modeling Information-Seeking Dialogues: The *C*onversational *R*oles Model. *Review of Information Science*, 1(1), 1996. URL http://www.inf-wiss.uni-konstanz.de/RIS. (On-line journal; date of verification: 20.1.1998).

Michael K. Smith, Raphael Volz, Deborah McGuiness, and Christopher Welty. Web Ontology Language (OWL) Guide Version 1.0. Technical report, W3C World Wide Web Concortium, 2002. URL http://www.w3.org/TR/2002/WD-owl-guide-20021104/.

Mikhail Soutchanski. An on-line Decision-Theoretic golog interpreter. In Bernhard Nebel, editor, *Proceedings of the seventeenth International Conference on Artificial Intelligence (IJCAI-01)*, pages 19–26, San Francisco, CA, August 4–10 2001. Morgan Kaufmann Publishers, Inc.

R. Sproat, A. Hunt, M. Ostendorf, P. Taylor, A. Black, K. Lenzo, M. Edgington, and Sun Microsystems. Sable: A standard for tts markup. In *Third ESCA/COCOSDA Workshop on Speech Synthesis*, 1998. URL http://citeseer.ist.psu.edu/263544.html;http://www.cs.cmu.edu/~awb/papers/ICSLP98-sable.ps.

Robert C. Stalnaker. Assertion. In P. Cole, editor, *Pragmatics*, volume 9 of *Syntax and Semantics*, pages 315–332. Academic Press, New York, 1978.

Mark J. Steedman. *The syntactic process*. MIT Press, Cambridge, Massachusetts, 2000.

A. Stein, J. A. Gulla, A. Müller, and U. Thiel. Conversational interaction for semantic access to multimedia information. In M. T. Maybury, editor, *Intelligent*

Multimedia Information Retrieval, pages 399–421. AAAI/The MIT Press, Menlo Park, CA, 1997.

Adelheit Stein, Jon Atle Gulla, and Ulrich Thiel. User-tailored planning of mixed-initiative information-seeking dialogues. *User Modeling and User-Adapted Interaction*, 9(1-2):133–166, 1999.

Thomas A. Stocky. Conveying routes: Multimodal generation and spatial intelligence in embodied conversational agents. Master's thesis, Massachusetts Institute of Technology, 2002.

Maite Taboada and William C. Mann. Rhetorical Structure Theory: Looking Back and Moving Ahead. *Discourse Studies*, 8(3), 2006.

Leonard Talmy. How language structures space. In H.L. Pick and L.P. Acredolo, editors, *Spatial Orientation: Theory, Research, and Application*, pages 225–282. Plenum Press, New York, 1983.

Leonard Talmy. *Towards a cognitive semantics*. A Bradford Book, MIT Press, Cambridge, MA, 2000.

Holly A. Taylor and Barbara Tversky. Perspective in spatial descriptions. *Journal of Memory and Language*, 35:371–391, 1996.

Paul Taylor, Alan W Black, and Richard Caley. The architecture of the festival speech synthesis system. In *The Third ESCA Workshop in Speech Synthesis*, pages 147–151, 1998.

Elke Teich, Eli Hagen, Brigitte Grote, and John A. Bateman. From communicative context to speech: integrating dialogue processing, speech production and natural language generation. *Speech Communication*, 21(1–2):73–99, February 1997.

Stefanie Tellex and Deb Roy. Spatial routines for a simulated speech-controlled vehicle. In Michael A. Goodrich, Alan C. Schultz, and David J. Bruemmer, editors, *HRI*, pages 156–163. ACM, 2006. ISBN 1-59593-294-1. URL http://doi.acm.org/10.1145/1121241.1121269.

Stefanie Tellex and Deb Roy. Grounding language in spatial routines. In *AAAI Spring Symposia on Control Mechanisms for Spatial Knowledge Processing in Cognitive / Intelligent Systems.*, 2007.

Thora Tenbrink. *Localising objects and events: Discoursal applicability conditions for spatiotemporal expressions in English and German. Dissertation.* PhD thesis, University of Bremen, FB10 Linguistics and Literature, Bremen, 2005.

Geoff Thompson. *Introducing Functional Grammar*. Edward Arnold, London, 1996.

D. Traum and E. Hinkelman. Conversation Acts in Task-Oriented Spoken Dialogue. *Computational Intelligence*, 8(3):575–599, 1992.

David Traum and Staffan Larsson. The Information State Approach to Dialogue Management. In Ronnie Smith and Jan van Kuppevelt, editors, *Current and New Directions in Discourse and Dialogue*, pages 325–353. Kluwer Academic Publishers, Dordrecht, 2003.

David R. Traum. *A Computational Theory of Grounding in Natural Language Conversation*. Ph.D. dissertation, Department of Computer Science, University of Rochester, Rochester, New York, 1994.

David R. Traum and James F. Allen. Discourse obligations in dialogue processing. In *32nd. Annual Meeting of the Association for Computational Linguistics*, pages 1–8, New Mexico State University, Las Cruces, New Mexico, 1994.

David R. Traum and Carl F. Andersen. Representation of dialogue state for domain and task independent meta-dialogue. In *Proceedings of the IJCAI'99 Workshop on Knowledge And Reasoning In Practical Dialogue Systems*, pages 113–120, Stockholm, 1999. URL http://www.cs.umd.edu/users/traum/Papers/ijcai99-dial.ps.

Barbara Tversky and Paul U. Lee. How space structures language. *Lecture Notes in Computer Science*, 1404:157–176, 1998. ISSN 0302-9743.

Barbara Tversky and Paul U. Lee. Pictorial and verbal tools for conveying routes. *Lecture Notes in Computer Science*, 1661, 1999. ISSN 0302-9743.

Barbara Tversky, Paul Lee, and Scott Mainwaring. Why do speakers mix perspectives? *Spatial cognition and computation*, 1:399–412, 1999.

D. Vanderveken and S Kubo. *Essays in Speech Act Theory*. John Benjamins Publishing Company, 2002.

Eija Ventola. *The Structure of Social Interaction: A Systemic Approach to the Semiotics of Service Encounters*. Frances Pinter, London, 1987.

Eija Ventola. The logical relations in exchanges. In James D. Benson and William S. Greaves, editors, *Systemic Functional Approaches to Discourse: Selected Papers from the Twelfth International Systemic Workshop*, pages 51–72. Ablex, Norwood, New Jersey, 1988.

Philip Wadler. A formal semantics of patterns in XSLT. In *Markup Technologies*, Philadelphia, December 1999.

Gerd Wagner, Grigoris Antoniou, Said Tabet, and Harold Boley. The abstract syntax of ruleML - towards a general web rule language framework. In *Web Intelligence*, pages 628–631. IEEE Computer Society, 2004. ISBN 0-7695-2100-2. URL http://doi.ieeecomputersociety.org/10.1109/WI.2004.134.

Wolfgang Wahlster. Smartkom: Symmetric multimodality in an adaptive and reusable dialogue shell. In *Proceedings of the Human Computer Interaction Status Conference*, 2003.

Jan Oliver Wallgrün. Hierarchical Voronoi-based Route Graph representations for planning, spatial reasoning and communication. In *Proceedings of the Fourth International Cognitive Robotics Workshop*, 2004.

Wayne Ward and B. Pellom. The cu communicator system. In *IEEE Workshop on Automatic Speech Recognition and Understanding*, 1999.

S. Werner, B. Krieg-Brückner, and T. Herrmann. Modelling Navigational Knowledge by Route Graphs. In C. Freksa, W. Brauer, C. Habel, and K.F. Wender, editors, *Spatial Cognition II - Integrating Abstract Theories, Empirical Studies, Formal Methods, and Practical Applications*, pages 295–316. Springer, Berlin, 2000.

M. Wessel and R. Möller. A high performance semantic web query answering engine. In I. Horrocks, U. Sattler, and F. Wolter, editors, *Proc. International Workshop on Description Logics*, 2005.

Jason D. Williams and Steve Young. Partially observable Markov decision processes for spoken dialog systems. *Computer Speech & Language*, 21(2):393–422, 2007.

Carsten Winkelholz and Christopher M. Schlick. Modeling human spatial memory within a symbolic architecture of cognition. In Thomas Barkowsky, Markus Knauff, Gérard Ligozat, and Daniel R. Montello, editors, *Spatial Cognition*, volume 4387 of *Lecture Notes in Computer Science*, pages 229–248. Springer, 2006.

William A. Woods. Transition Network Grammars for Natural Language Analysis. *Communications of the Assocation for Computing Machinery*, 13, 1970.

M. Wooldridge. *Reasoning about Rational Agents*. Intelligent Robots and Autonomous Agents. The MIT Press, Cambridge, Massachusetts, 2000.

X3D.   X3d   architecture   and   base   components   edition   2. http://www.web3d.org/x3d/specifications/, Dec 2007.

George Yule and H. G. Widdowson. *Pragmatics*. Oxford Introductions to Language Study. Oxford University Press, 1996.

www.ingramcontent.com/pod-product-compliance
Lightning Source LLC
Chambersburg PA
CBHW071405050326
40689CB00010B/1763